Kathy Davis

POWER UNDER THE MICROSCOPE

1988
FORIS PUBLICATIONS
Dordrecht - Holland/Providence RI - U.S.A.

Published by:
Foris Publications Holland
P.O. Box 509
3300 AM Dordrecht, The Netherlands

Sole distributor for the U.S.A. and Canada:
Foris Publications USA, Inc.
P.O. Box 5904
Providence RI 02903
U.S.A.

CIP-DATA

The Editors are grateful to the Henny Verhagen Fonds, a Dutch Foundation for Women and Health Care and Vrouwen VU Hulp of the Free University in Amsterdam. The Publication of this book has been facilitated by their generous grants.

ISBN 90 6765 381 0

Printed in The Netherlands by ICG Printing, Dordrecht.

POWER UNDER THE MICROSCOPE

Toward a grounded theory of gender relations
in medical encounters

POWER UNDER THE MICROSCOPE

CONTEXT:
Sociological Studies on Text, Discourse and Conversation

The CONTEXT Series is designed to bring into focus sociological studies on text, conversation and discourse. Each volume in the series will explore why and how various approaches to text, discourse or conversation analysis can contribute empirically or theoretically to the understanding of contemporary issues in the social sciences.

Editors
Joke Haafkens
Tony Hak
Gerhard Nijhof
Brian Torode

FOREWORD

I began the research for this book in the context of the first women's studies project ever to be set up at a Medical Faculty in The Netherlands. The project 'Woman/patient and man/physician' (Vrouw/patient en man/huisarts), which was launched at the end of 1982, was a collaborative effort of myself and my colleague, Thea Dukkers van Emden, aided and abetted by our respective Department of Behavioral Sciences and the Institute for General Practice at the Free University in Amsterdam. It was not initially my plan to wrap up my end of the project with a dissertation, mainly because I had every intention of returning to my original vocation, psychotherapy, as soon as it was over. However, somewhere in the early stages of the project, I made the fateful discovery that female 'doctors' are something of a scarcity in The Netherlands. In that particular year (1983/1984), only 9.2% of the dissertations were written by women. This fact 'got my back up,' as we say in my homeland, and I became determined to do my bit towards altering this regrettable state of affairs. Had I known what I now know, it is possible that I would not have been quite so enterprising....

Having reached the end of my struggle - a struggle which was in some senses just as dialectical and full of small acts of heroism as the struggles of the patients in this book, there are quite a few people who have to be thanked - persons without whose assistance, this book could not have been written.

For their help in the initial phases of the inquiry, when I was trying to shape up the research problem and work out a plan for investigating it, special thanks go to: Mary Fahrenfort, Hanne van Herwerden, Thea Dukkers van Emden, Marianne Derks, Wien Meeuwesen, Peter Bal, Zsuzsa Baross, Livia Polanyi, Doro Franck, en Paul ten Have.

Jan Thomassen of the Institute for General Practice was kind enough to allow me the use of his collection of taped consultations and provided kind assistance throughout the inquiry.

For the gruelling work of sorting through the tapes for instances of storytellings as well as collecting literature on general practice, my thanks go to Yen Ni Yap and Netteke Schouten-van Meeteren.

A special note of thanks belongs to Vrouwen VU-hulp, who went to bat for me when it looked like my research appointment would not be extended long enough for me to complete the dissertation. And, of course, my thanks to the medical faculty and particularly to Professor Louwrens Menges for offering me the use of the facilities of the Department of Behavioral Sciences until I was finished and keeping me well supplied with examples of medical sexism, both from the popular press as well as the more scientific variants.

I am indebted to my colleagues in the Department of Women's Studies in Leiden for countless stimulating discussions on gender and power. The final chapter of this book could not have been written without them.

For reading and commenting on earlier drafts of various chapters, I am especially grateful to Nelly Steffens, Aafke Komter, Marianne Grunell, Jantine Oldersma, Tony Hak, Thea Dukkers van Emden, and Jan Thomassen.

For welcome support in the final stages with proof-reading, my thanks go to Mary Lommerse, Jantine Oldersma, Antoinette Geelhoed, Aafke Komter, Lena Inowlocki, and Annemiek Pels.

I would especially like to thank Karen Haasdijk, who put hours and hours into typing the manuscript and, later, word-processing the camera-ready version. Not only did she valiantly do battle with the new computer and various rebellious printers, but she polished up her English so that she could correct my spelling. If the final version is correctly spelled, it is to her credit rather than mine.

I am grateful to Tony Hak, Gerben van der Heide, and Harrie Jansen of the Institute for Preventive and Social Psychiatry at the Erasmus University in Rotterdam for invaluable help in printing the final manuscript.

And, last but not least, my most heart-felt thanks go to three people without whom this book very probably could not have been written.

Professor Jenny de Jong-Gierveld was unparallelled as supervisor and my most rigorous and exacting critic throughout the inquiry. She was indefatigable in discovering methodological weak spots or inconsistancies in my argumentation as well as figuring out how to fix them. I would especially like to thank her for helping me stay on the track and encouraging me to make the most that I could make of this inquiry. Any deficiencies in the final product are entirely of my own making.

Many, many thanks go to Dr. Sue Fisher for bringing her own expertise and creativity in the field of gender and medical interaction to this inquiry and providing invaluable help with the actual analysis, encouraging me to do conversation analysis innovatively, and convincing me not to wait until the very end with the punchline. She was a stimulating and supportive conversational partner and a joy to work with.

And, finally, I would like to thank Willem de Haan who, many years ago, introduced me to conversation analysis when I, as clinical psychologist, thought research could only be done with an inventory; who, several years later, as I was floundering about in social theories on power, suggested that I might like to try Giddens; and who has, in the past 14 years, helped me untangle more knots, research- and otherwise, than bear thinking about.

Kathy Davis
May 1988

I

THE POWER PROBLEM:

SOME METHODOLOGICAL AND THEORETICAL CONSIDERATIONS

1

INTRODUCING THE RESEARCH PROBLEM

GENDER AND MEDICAL ENCOUNTERS

The present inquiry is about women as patients and men as physicians and what happens when they find themselves facing one another across a desk, ready to embark on a general practice consultation. In the past two decades, it has been a frequently-heard claim that women as patients occupy a specific position in the health care system (Dreifus 1977; Ruzek 1978; Ehrenreich and English 1979; Scully 1980; Roberts 1981b,c; Fee 1983; Stromberg 1984; Lewin and Olesen 1985; Fisher 1986). Women not only have their own health care problems and needs, but they are apparently being treated in specific and, more importantly, negative ways within the various health care services. It would seem that encounters between female patients and male physicians involve asymmetrical relations of power, which are based, at least in part, on gender inequality.<1> Most of the available literature on the subject of gender and health care has emerged as a result of the feminist movement, but, more recently, we can find (social) scientists addressing the issue as well.<2> In order to understand how the assumption could be made that gender and power are routinely involved in medical encounters, it makes sense to take a closer look at what both these sources have had to say on the subject.

The feminist critique

Since the advent of the feminist movement in the late '60's, feminist scholars have been concerned with issues surrounding women's health and the care they receive in the medical system. Not only have they had a great deal to say on the subject, but, as Fee (1983) notes, most of it has been of a critical bent. Women, it would seem, do not fare well at the hands of the (male) medical establishment.

The feminist critique of the health care services falls roughly into three categories:

First, women's bodies and reproductive functions are subject to rampant *medicalization*. Normal conditions or processes like menstruation, pregnancy, and menopause are defined in terms of health and illness, thereby making them matters requiring almost constant medical surveillance. From the moment she enters puberty, a woman becomes, by virtue of her female-ness, sick (Arms 1975; Culpepper 1978; Ehrenreich and English 1979; Oakley 1979; Standing 1980; Roberts 1981c; Laws 1983; Riessman 1983). Women's control over their bodies and fertility has been usurped by the medical profession.

Second, whereas women's normal bodily processes are subject to too much medical attention, many of their most *common health problems* scarcely seem to merit any attention at all. Problems like menstrual cramps and irregularities, cystitis, or just the business of having to find a method of contraception, which is not only effective, but also safe are *not taken seriously*. This is often simply a matter of omission. Neither individual practitioners nor medical research as a whole have bothered to find solutions for these common problems (Homans 1985; McPherson 1988).<3> There is, however, also a notable tendency within medical circles to regard women's complaints as caused, or, at the very least, aggravated by psychological factors. Thus, dysmenorrhea can be attributed to the patient's 'faulty outlook,' infantile colic to the mother's 'insecurity' or 'anxiety' (Lennane and Lennane 1973) and chronic vaginal discharge to 'neurotic self-observation' (Kloosterman 1983). Accompanying the notion that 'it's all in her head,' is the negative stereotype of the female patient taking up the busy doctor's time with a host of vague symptoms and endless complaining (Scully and Bart 1973; Lorber 1976; Cooperstock 1978; Fidell 1980; Lipset 1982). Taken together, the impression is inescapable that women's complaints are not taken seriously and may, therefore, not be receiving adequate medical attention.

Third, the medical profession is - and has been historically - an instance of the *social control* over women. It is the vehicle for eliminating or, at least, controlling problematic experiences by defining them as deviant. Physicians have frequently and with considerable success managed to channel women's dissatisfaction with their living and working situations away from active or collective protest to the safer paths of endurance and acceptance (Cloward and Piven 1979; Ehrenreich and English 1979; Donzelot 1979). By means of benevolent advice, women are helped to view their difficulties as instances of personal pathology, essentially internal to the individual. Boredom or anxiety may be smoothed away with a chat, a prescription for Valium, or the promise of another appointment (Prather and Fidell 1975; Barrett and Roberts 1978; Cooperstock 1978; Standing 1980). Using their authority, physicians can lull women into accepting or adapting to problems stemming from the oppressive contingencies of their lives.

In conclusion, this very cursory look at the feminist critique of the health care system indicates that asymmetrical gender relations do not come to a halt in front of the doctor's office. Just as gender inequality is an integral feature of most existing social relations, it seems to be implicated in medical encounters as well. This initial foray into the feminist literature on women and health care indicates, then,that medical encounters are a good starting point for investigating asymmetrical relations between the sexes.

Empirical research on women as patients

Feminists have not been the only ones, who have busied themselves with the lot of the female patient. In view of the extensive critique of the medical system in terms of how women fare as patients, it is not surprising that empirical investigations on the subject began to appear. Social scientists, ever alert to signs of commotion, have made it their business to find out 'what's going on out there' and to supply the empirical evidence for making claims about it. Consequently, several years after the emergence of the feminist movement, the first investigations on women (but also on men) as patients made their entrance.

Whereas this research might have taken a variety of forms, it turned out to be largely of the same variety; namely, sex differences research.<4> Male and female patients were compared in terms of their respective complaints (type, numbers, manner of presentation), their illness behavior, and their utilization of the various health care services. Several prominent examples of this type of research in the field of health care are Nathanson (1975, 1977), Gove and Hughes (1979); Verbrugge (1976, 1979), Mechanic (1976, 1979, 1980) and, more recently, Hibbard and Pope (1986). After more than a decade of research,<5> there are two sex differences which have been discovered with any consistency or regularity. Women apparently distinguish themselves from their male counterparts on two counts:

1. *Women use health services more frequently than men do.*
   In most surveys, women as a group appear to pay more visits to physicians than men. It is estimated to be about one and a half

times more (Nathanson 1977).<6> This higher rate of utilization holds true, even when visits in conjunction with contraception, pregnancy, or sick children are taken into account.

Explanations for why women go to doctors more often than men vary. Some suggest that they are, in fact, 'sicker' than men; for example, due to their higher incidence of genito-urinary infections or stress stemming from the demands placed on them as wives and mothers: the 'nurturant role hypothesis' put forward by Gove and Hughes (1979).

Others attribute the frequency to women's 'role behavior' (Nathanson 1975, 1977; Verbrugge 1976, 1979; Mechanic 1976, 1979). In this view, women are not 'really' sicker than men; they are just more likely to define themselves or behave 'as though' they were sick. The account provided is that the 'sick role' is compatible with women's role as housewife and the absence of 'fixed role obligations' or that traditional female roles make it acceptable to be dependent, ask for help, or generally be preoccupied with one's body and bodily changes (female narcissism).

Of course, physician behavior may also be a factor, influencing the frequency with which women utilise the health care services. For example, physicians display a marked tendency to prescribe more drugs to women, particularly psychotropic ones (Cooperstock 1978). Gynecological surgery like hysterectomies may be performed, whether necessary or not (Scully 1980; Fisher 1986). Physicians tend to make more follow-up appointments with their female patients (Fidell 1980).

2. *Women are more prone to 'vague illness' than men.*<7>
The second sex difference is even less straightforward than the first. Although the term 'vague complaints' may be encountered regularly in literature on female patients, authors are often less than explicit about what they mean by it.<8> A look at the differences in symptoms manifested in men and women does not help. For example, typical female complaints include rheumatic disorders or genito-urinary complaints, neither of which seems more 'vague' than the male respiratory disorders and heart disease (Centraal Bureau voor de Statistiek 1985:54-55). Bremer (1980a), who has specifically investigated 'vague, unclear complaints' also discovered sex differences in complaint patterns. Women have more headaches and cold-related symptoms. They also display vague complaints at a younger age <9> than men - a fact which the author attributes to stress-related symptoms upon entering marriage (!) and the tendency of women to bring up

> vague complaints when they consult a physician for contraceptives (Bremer 1980a:56). Men, on the other hand, express vague complaints later in life and these complaints are related to alcohol consumption and the tendency to 'get angry' (Bremer 1980a:57).
>
> Although these sex-related variations in complaints are interesting in and of themselves, it is not immediately clear why women's complaints should be considered more 'vague' than men's. The appellation 'vague' seems rather to indicate that physicians are having difficulties pinpointing and diagnosing the cause of women's distress in strictly medical terms.
>
> The term 'vague' may not refer to the type of complaint, however, but to the fact that women are considered to exhibit more 'psychiatric disorder' than men. Authors sometimes suggest that women 'have' more psychological problems or that they tend to talk more about emotional distress, thus opening themselves up to psychiatric labelling (Phillips and Segal 1969; Dohrenwend and Dohrenwend 1969; Gove 1978; Verbrugge 1984). Of course, physicians may also take the view from the outset that women are more likely to be neurotic, which would account for their readiness to provide psychiatric diagnoses for women's complaints (Broverman *et al* 1970).
>
> And, finally, 'vague complaints' may be connected with differences in the way men and women talk about their complaints. Not only do women present more complaints at a time, but they apparently introduce 'extraneous' information as well (Cooperstock 1978; Bremer 1980a; Tielens *et al* 1986). Physicians apparently also prefer to talk about psychosocial matters with their female patients, giving them not only more time to engage in this kind of talk, but actively eliciting it as well. (Smits *et al* 1983).<10>

It goes beyond the scope of the present inquiry to deal with the this kind of research in any detail. However, a few words are in order concerning how a consideration of its limitations informed the way the present inquiry was conceived.

Beginning with the most conclusive finding of sex differences research, the claim could certainly be made that, indeed, women will visit the doctor more often than men. This has been demonstrated so often that it could, I would suggest, be accepted as, for all intents and purposes, 'proven.' The next question, however, which unavoidably follows this insight, is: so what? Women may also be visiting their hairdressers more often than men, but this event has, up until now, not been regarded as something meriting extensive investigation nor, more importantly, do we find it relevant for understanding gender inequality. The explanations for the difference might have provided a clue for its relevance, but, as we have seen, these proved sadly inconclusive. After decades of research, we are no closer than we ever were to knowing why women seek the services of medical professionals more often than men do. And, more to the point, we do not know much about what happens once they do.

The second difference raises even more questions. The term 'vague complaints' is itself notoriously vague, hovering as it does between functional illness, psychological disorder, and differences in speech style. It is unclear how much of the 'vagueness' is a part of the woman's complaint or may be attributed to the practitioner doing the labelling, who, after all, may even be actively soliciting that sort of complaint from his female patients. Or vagueness might simply be an artefact of the medical model, which demands that a specific etiology be identified or identifiable in order for a complaint to qualify as a disease meriting medical attention. In any case, the term itself is invested with derogatory connotations, making it of questionable and problematic value as a description of possible sex differences in the kinds of complaints patients have.

In conclusion, the current investigations on women (and men) as patients have produced very little empirical backing for some of the issues that had been raised by feminist scholars. Assuming that their original intention was in some way emancipatory,<11> the results of the research have been disappointing. In order to discover how gender inequality works in medical encounters, it clearly requires being looked for explicitly. This is, however, not what most investigators have done. Thus, the feminist critique remains, more or less, where it began - without the grounding, which empirical research might have provided.

Applied to the present investigation, the following conclusions may be drawn:

1. If the comparison between male and female patients has not yielded relevant information about how gender inequality works in medical encounters, then it is time to try another approach. The question should no longer be whether or not women and men are different, but rather how differences in power - which was, after all, the original issue - are manifested and sustained in actual medical encounters between male physicians and female patients.
2. The issue of 'vague' complaints suggests that the process of diagnosing and, ultimately, treating women's problems does not occur *strictly* according to medical criteria. Whereas the research in the field of sex differences has not been able to explain how

and why this appellation comes to be attached to women and their complaints, it does provide a rationale for taking it as a topic meriting further investigation. This is precisely what I propose to do.

Such a proposal entails taking the view that a diagnosis is a social construction. This is, by no means, an original notion; nor is it limited to the diagnosing of women's complaints. I shall now direct my attention to what other social scientists have had to say on the subject.

DIAGNOSES AS SOCIAL CONSTRUCTIONS

Traditionally, medicine has treated diagnostic categories as though they referred to biophysical states or the non-functioning of specific organs or tissues in the human body. More recently, 'social factors' were added to the list of possible illness etiologies. In both cases, the general idea was that discrete and neutral categories could be found for illness. The process of diagnosis essentially involved the matching of these categories by means of scientific methods and procedures to individuals manifesting the appropriate symptoms. Whereas the procedures for discovering and naming disease might be imperfectly applied by medical practitioners, medical knowledge itself as well as the legitimacy of the methods were taken for granted. They were treated as self-evident and, consequently, unproblematic.

In the late sixties and early seventies, social scientists began to take issue with this stance. In particular, Berger and Luckmann (1966) formulated a position called 'social constructionism,' which provided the basis for a new perspective on the definition of illness as well as the possibility for a critique of the medical monopoly of this defining.

Social constructionism rejected the idea of a neutral reality to which scientists might have access by means of objective methods. On the contrary, reality, including medical reality, could not be considered in isolation from the person or the context in which it was being generated or elaborated.

As Conrad and Schneider note:

> ... illness is a social construction based on human judgments of some condition in the world. In some fashion, illness, like beauty (and like deviance), is in the eye of the beholder. Although it is based partly on current cultural conceptions of what disease is, and more often than not in Western society

> grounded in biophysiological phenomena, this social evaluative process is central rather than peripheral to the concept of illness and disease. It follows logically that both diagnoses (as systematized classifications) and treatments are founded on these social judgments; they cannot be separated (1980:31-32).

Taking a social constructionist stance has several immediate consequences for how the process of defining and diagnosing complaints in medical encounters can be viewed: <12>

First, it treats *medical knowledge as problematic*. Medical categories are no longer regarded as self-evident or of interest only in terms of whether or not they are adequately applied. Biomedical reality becomes a topic in its own right. The question becomes not which people are subject to what illness, but why and how certain categories emerge at a particular historical juncture for describing such bodily processes or experiences in the first place. <13>

If medical knowledge is always constructed in a social context, then it makes sense to take a closer look at how this is accomplished. This means, among other things, that the process of defining and diagnosing women's complaints is not something to be taken for granted, but becomes itself a suitable and relevant topic for further investigation.

Second, the diagnostic *categories* themselves *do not correspond with natural phenomena, nor are they 'discovered' by means of neutral or rational methods*. The very act of discovering and defining a person's experience is a social event, taking place in a specific social context. In a world of 'multiple realities,' defining behavior is a political process; i.e. definitions have to be negotiated. The boundaries between what counts as medical knowledge and what does not are neither stable nor given. The more powerful are more likely to implement a version of reality in accordance with their interests. To suggest that this is otherwise, is to disguise the ongoing struggles and the consequences attached to the construction of legitimate knowledge. In this sense, the process of defining people's problems becomes a political issue (Conrad and Schneider 1980:22-28). <14>

Third, medical knowledge is not only a social product, but it *organizes social relations* as well. Medical knowledge and practice do not stand outside social relations, but rather mediate them. To call a particular experience 'medical,' places it in a specific relationship to other areas of social life. Medical reality interacts with prevailing ideologies, taking on particular contours and meanings depending on the social group for which it is being used.

As long as we assume that diagnoses do nothing more than assign a name to what the patient actually has, then investigating how women's complaints are labelled can be of epidemiological value only. A picture can be developed of the *kinds* of complaints women tend to have, which is, in fact, what current literature on women as patients has, up until now, provided. By taking a social constructionist stance, it becomes possible to regard diagnoses as constructions rather than simple reflections of some independent reality. If they are socially produced, then they will also be inextricably tied up with other social structures and relations as well. The relevant question becomes not *what* a woman has, but rather *how* she came to be defined as having it. Of further concern is, of course, how that particular definition, as well as the way in which it came into being, is connected to asymmetrical gender relations and women's subordination.

In conclusion, the social constructionist perspective provides a welcome solution to some of the difficulties encountered in current empirical research on women as patients. It does this primarily by opening up a new field for further investigation; i.e. how women's complaints are being defined and diagnosed in medical encounters. Moreover, it provides a rationale for taking a closer and, more importantly, critical look at how this process may be contributing to the gender inequality which has so often been cited in the feminist critique of how women fare at the hands of male physicians.

Before indicating how these background considerations were instrumental in setting up the present inquiry, a brief look will be taken at a specific medical context; namely, general practice, as a good starting point for the investigation of asymmetrical gender relations in medical encounters.

GENERAL PRACTICE AS CONTEXT FOR DEFINING WOMEN'S PROBLEMS

General practice in The Netherlands

Since World War II, General Practice has occupied an increasingly difficult position within the Dutch health care system. This can be attributed primarily to two changes.

The first concerns the rapid rise in the number of specialists as opposed to general practitioners. Whereas in 1963 their relative numbers were equal, there has since then been a veritable boom in specialists, whereby there are at present nearly 1,5 times as many specialists as GPs (Knottnerus 1982:67). This goes hand-in-hand with an increasing trend toward hospital-based medicine as well as massive developments in the field of medical technology, which led to a catapulting of the GPs into a minority position- both numerically and in terms of prestige - within the health care system. GPs not only do not enjoy the same material and status benefits as their specialist colleagues in the hospitals, but they are frequently referred to derogatorily as 'talk doctors,' implying that they might not even really belong to the ranks of the *medical* profession (Spreeuwenberg 1983).

A second change involves the conditions under which general practitioners currently must perform their work. Before the war, patients did not have medical insurance, which meant that they were dependent for medical treatment on whether they could afford it or how sympathetic the family doctor was to their predicament. In 1941, compulsory health insurance was instituted, making health care available to large groups of patients, who had formerly been left to their own devices. Unsurprisingly, the number of patients consulting GPs increased three-fold between 1938 and 1954 (Knottnerus 1982:69).

Not only did GPs find themselves confronted with heavy patient loads, but they were no longer equipped to deal with the problems their patients were coming in with. Patterns of disease and suffering have changed markedly in Western countries within this century. Gone are the days when infectious diseases exacerbated by poor nutrition, housing, and sanitation were the main health problem. This is the age of the chronic illness, psychosomatic complaints, and, of course, that elusive but ubiquitous phenomenon 'stress.' Such illnesses lack clear-cut somatic etiologies. Moreover, they are singularly unresponsive to medical-technical solutions.<15> For many illnesses, the only option available to medical practitioners is to monitor them as they make their intermittent, but inexorable, progression from bad to worse. Care has joined cure as the medical order of the day.

None of this has served to lighten the GP's load, of course. Moreover, this is accompanied by the unfortunate fact that GPs are poorly-equipped to deal with the increasing demands being placed on them. Medical training remains primarily oriented toward a specialist and hospital-based medicine with its emphasis on somatic etiology and technical solutions.

In view of the above, it is hardly surprising that general practitioners became increasingly dissatisfied with their position in the health care system. A process of self-reflection was launched, beginning with Buma's (1950) attempt to define the field as well as specific tasks of the family doctor. This was followed by the founding of the first professional organization for general practitioners (Nederlands Huisartsen Genootschap) in 1956. After that, there was a general renaissance of interest in the field as illustrated by conferences, a professional journal for General Practitioners, the founding of a research institute for general practice in 1964 and, finally, in 1966 the installment of the first professor for General Practice at the University of Utrecht. General Practice was well on its way toward becoming a full-fledged medical specialism. Programs were established at eight Dutch universities and in 1973 it became obligatory for every GP to have at least one year of vocational training as a general practice specialist (Stephen 1979).

With the development of specialized vocational training programs, there was a marked shift in how GPs thought about diagnosing and treating their patients' problems. Whereas training programs differed in their emphasis and orientation, they had several features in common:

First, the medical model was no longer deemed appropriate for diagnosing and treating patients problems. It was abandoned in favor of more consultative or therapeutic approaches which emphasized problem clarification and solving. The GP was no longer a medical technician, but had joined the ranks of the professional helpers (Weterman-Hezemans 1982).<16>

Second, theoretical frameworks were developed for understanding illness and its etiology in a more comprehensive way (Menges 1986). Integral and holistic approaches to medicine made their entrance. The psychosomatic as well as the psychosocial problem took a place of honor in the GP's diagnostic arsenal. Consequently, GPs could expand their expertise from mere bodily ills to psychological difficulties and even problems of living.

Third, along with more comprehensive theories for coming to terms with illness, practical skills needed to be acquired. Whether in Balint groups <17> or training in 'Methodisch Werken,' as developed by Van Aalderen and cohorts, <18> the modern GP was expected to be able to talk to patients. Training programs set out to improve GPs' listening skills, help them develop an empathetic attitude toward their patients and provide reflections in lieu of 'just' giving advice. The gap between the medical professional and the psychotherapist was rapidly narrowing.

Fourth, more attention was paid to the pitfalls inherent in the business of helping and of which the GP might well expect a fair share. Discussions abounded concerning the problematic area of 'responsibility' and GPs were warned not to take on too much of it. Ultimately, it was the patient who was responsible for her own health (Kocken and Til 1982). GPs were warned not to be overhasty in diagnosing problems. Problems required 'clarification' first in 'cooperation' with the patient (Hollenbeek Brouwer 1982). A step in the right direction was developing the right kinds of 'attitudes' or attributes, which would make the path that the professional helper had to take just a bit less thorny (Sigling 1982). Whereas the consultation continued to be the responsibility of the GP, the patient was awarded a more equal position in it than before, as 'conversational partner,' a 'member of the team,' and as someone, who, at least in theory, should have some say in matters concerning her illness and how it was to be treated.

In conclusion, General Practice has been and continues to be in a state of ongoing upheaval and change. Or, as Knottnerus (1982:67) notes: crisis might be considered the hallmark of the discipline. With the emergence of specialist training programs for General Practice, a marked shift has occurred in medical thinking on the nature and causes of their patients' problems as well as the GP's expertise in treating them.

This has not been without effect on the way women's complaints are dealt with in general practice.

Women and General Practice

General practice occupies the important position of gatekeeper in the Dutch health care system. It is the first place a patient will go if she needs medical help. In the interest of keeping costs at a minimum, the GP will attempt to handle as many complaints as possible himself, referring patients to specialists or for examinations only when absolutely necessary. GPs obviously play a central role in defining patients' complaints. Confronted with the initial complaint presentation, it is their job to shape it up into something which can be treated medically (or to reject it as unsuitable for medical attention). Thus, they will always have first shot at defining and diagnosing the patient's problems.

Women make up the majority of the GP's patients. Not only do women visit the GP more frequently than men, but GPs have more consultations, on the average, with women than with men. This fact alone makes general practice a good setting for investigating gender and medical encounters. However, women not only make up the majority of the patients in general practice, but the GPs are, in an overwhelming majority, *men*. Despite the steady increase of female GPs in recent years, 91% of the GPs are, in fact, still men.<19> This means that the *typical* general practice consultation will involve a male GP and a female patient. If this is the most routine format for general practice, it is an obvious choice for investigating how gender relations are 'routinely' constructed in medical contexts.

General practice plays a pivotal role in the social control of women. In their study on GPs and the consulting patterns of middle-aged women, Barrett and Roberts discovered that:

> In consultation after consultation the GP smooths away the surface anxiety and adjusts the woman to the limitations of a life located totally in a home from which the children have moved away. In this respect the institution of medicine legitimates and endorses the status quo in relation to the position of women, and in so doing it fulfills an ideological function as an agency of disguised social control. We found that frequently doctors would use the authority of their medico-moral language to offer not neutral, clinical advice but a set of prescriptions based on the conventional wisdom of their social milieu (1978:42).

This suggests that, when faced with women's problems, GPs will make use of unspoken assumptions about the 'nature' of women (and men) and these will constrain how they define her problems and the kind of treatment they suggest.

In conclusion, general practice is not only a good context for examining how patients' complaints are diagnosed, but it is a particularly suitable setting for investigating how women's complaints are being handled, since women make up the majority of the patients. Moreover, in view of the fact that most GPs are men, it is a useful starting point for the study of how asymmetrical gender relations can be constructed and maintained in medical encounters.

A PRELIMINARY FORMULATION OF THE RESEARCH PROBLEM

An attempt has been made to explicate some of the sources, which provided the backdrop for the present inquiry. Beginning with the feminist critique of the health care system, the assumption was made that gender and, more specifically, women's subordinate social position would be part and parcel of any medical encounter, particularly when the physician is a man. Whereas this critique has pinpointed several areas of general concern and produced myriad illustrations of medical malpractice and derogatory treatment of women as patients, little empirical backing has been provided concerning precisely how gender and power are involved in doctor-patient relationships. Survey research, by far the most common type of inquiry in this field, has persistently limited itself to mapping out the differences between male and female patients. As we have seen, this has done little to deepen our understanding of how power works in medical encounters between male physicians and female patients.

Discussions around the supposed 'vagueness' of women's illnesses, as well as social constructionist insights into the social and political nature of the process of defining problems in professional encounters indicated that taking a closer look at how women's complaints were being defined, diagnosed and treated in interaction with male physicians could be a useful starting point for an investigation of gender and power in medical contexts. And, finally, as medical context for investigating this process, general practice consultations seemed a particularly good choice. Not only does general practice occupy a key position in terms of how patients' problems are initially defined, but the most routine general practice consultation will involve an asymmetrical gender relation; namely, a male physician and a female patient.

These considerations led to the formulation of the research problem for the present inquiry as follows:

> *How are women's complaints defined, diagnosed, and treated in consultations with male general practitioners and how does this process reflect, and, in fact, construct asymmetrical power relations?*

With this background and preliminary formulation of a research problem, I am now ready to embark on the present investigation. In the remaining chapters of *Part I*, I shall be showing how I selected my material and a suitable research method (*Chapter 2*), went about the preliminary analysis of the material (*Chapter 3*), and how I ended up, somewhat unexpectedly, scouting around in theories on power in order to find a theoretical framework, which could help me come to terms with some of the difficulties encountered in the preliminary analysis (*Chapter 4*).

In *Part II*, attention will be paid to the problem of operationalizing power at the level of medical interaction (*Chapter 5*) - also a tricky undertaking - and to the selection of suitable analytic categories for organizing the consultations for further analysis (*Chapter 6*). This having been done, a structural description will be provided of those consultations, which, ultimately, were chosen for the final analysis *(Chapter 7)*.

Part III marks the beginning of the analysis proper. After explaining how I propose to analyze power at the level of interaction *(Chapter 8)*, I will proceed to show how asymmetrical relations of power are ongoingly produced, reproduced, but also undermined in the course of consultations between male GPs and female patients (*Chapter 9-12*).

In *Part IV*, some empirical (*Chapter 13*) and theoretical (*Chapter 14*) conclusions are drawn on the basis of the present inquiry concerning how women's problems are defined, diagnosed, and treated in general practice consultations and how this process reflects, sustains, and transforms asymmetrical power relations between male GPs and female patients. Returning to the original formulation of the research problem, some necessary (but painful) revisions will be proposed concerning how gender and power should be conceptualized, both in research on medical interaction, as well as within feminist theory in general.

One final note is in order concerning my modus operandi in the present inquiry. Throughout the book, an attempt has been made not simply to find an answer to the research problem, but to delineate in step-by-step fashion how I went about finding it. As will be shown, this was an arduous process in which many of my initial ideas had to be abandoned or substantially revised before I could proceed. Finding and locating power proved a difficult task, requiring the development of an instrument, a kind of microscope, in order to accomplish it. It was only after I had developed this 'microscope' that I was able to find, delineate, and, of course, explain the working of power at the level of the face-to-face interaction between male GPs and female patients. The result is the beginning of what might be called a 'grounded' theory of gender relations in medical encounters.

NOTES

1. 'Sex' is a biological term, whereas 'gender' refers to the psychological and cultural meanings attached to being a woman or a man in a specific social and historical context. It was one of the initial contributions of feminist scholarship to make this distinction, paving the way for an explanation of 'sex differences' which was not based on biology alone. See Oakley (1972); Chodorow (1971).

2. In the present inquiry, I have concerned myself primarily with what social scientists have had to say about women as patients. This is partly because I am a social

scientist myself and partly because social science has been more inclined to assume a critical stance toward women and health care issues. This does not mean, however, that the medical profession has not been concerned with the female patient. Historically, in fact, there has been no end to the medical theorizing and subsequent treatment of the female patient. See, for example, Barker-Benfield (1972); Smith-Rosenberg (1972; 1973); Smith-Rosenberg and Rosenberg (1973); Ehrenreich and English (1973); Wood (1974); Verbruggen (1976); Donnison (1977); Summey and Hurst (1986).

3. The obvious case in point is contraception. Despite its importance for women, we have yet to see the contraceptive which is safe, effective, and convenient to use. For a good critical account of some of the issues involved in contraceptive research, the reader is referred to Gordon (1974), Seaman and Seaman (1977), Petchesky (1986), Arditti *et al* (1984), and Homans (1985).

4. Sex differences research has a long tradition within feminist scholarship. As Bernard (1975) has pointed out, it began with the explicitly emancipatory intention of showing that if women and men are different, this was not due to any innate biological difference, but rather to differences in their respective social positions. The first investigations dealt with intelligence and provided empirical support for the conclusion that women were, in fact, just as intelligent and competent as men.

5. A thorough discussion of this tradition would go beyond the scope of the present inquiry. In an earlier inquiry, I have discussed it in more detail, Davis (1984b). Even in the period between that inquiry and the writing of this chapter, more investigations of the same ilk have emerged. See, for example, Hibbard and Pope (1986), titled "Another Look at Sex Differences in the Use of Medical Care: Illness Orientation and the Types of Morbidities for Which Services Are Used. They discovered the same differences in 1986 as were found ten years ago by Nathanson (1977), indicating that 'another look' yields the same information.

6. This figure is used in international publications as well. In The Netherlands it holds true for the frequency with which women and men visit general practitioners, where 76% of the women as compared to 65,8% of the men consult their general practitioner (Centraal Bureau voor de Statistiek). Of this percentage, women have, on the average, 4,2 consultations per year as compared to 2,8 of the men. Each woman has slightly more contacts per GP than her male counterpart. Interestingly, in The Netherlands, male patients visit specialists more often and are more likely to be admitted to hospitals than women.

7. For an excellent discussion of some of the difficulties associated with the notion of 'vague complaints,' the reader is referred to Derks and Baart (1981).

8. The term 'vague complaints' as well as the more elegant designations like 'hyperaesthetic-emotional syndrome,' 'neurasthenia,' 'surmenage,' or 'functional,' 'psycho-social' or 'stress-induced' disorders refer to complaints which cannot be attributed to a specific medical etiology or illness. Such complaints are a familiar, if not altogether welcome, occurrence in contemporary general practice. Possibly due to their intractibi-

lity as far as diagnosis and standard methods of treatment go, they have been the subject of considerable discussion (Brouwer and Touw-Otten 1974; Bremer 1980a; van Es 1980; Jacobs 1980; de Haan 1982). Interestingly, patients also refer to their complaints as 'vague.' This sometimes indicates that they, like physicians, do not know what is causing their distress. It can, however, also mean that the patients, unlike physicians, do not find the complaints particularly significant or bothersome (de Haan 1982).

9. In Bremer's (1980a) study, the average age of male patient's with vague complaints was 39 as compared to the average age of 29 for female patients. Although this is a small study (N=50), it runs counter to the popular stereotype of the 'chronic complainer' as a menopausal woman in the throes of the empty nest syndrome (Barrett and Roberts 1978).

10. This tendency was also exhibited in the collection of general practice consultations from which my own material was taken. Thomassen (1984) discovered that GPs not only spent more time talking to women than to men (more than one minute more, on the average) and that more 'social conversation' occurred with women than with men. It seems unlikely that this can be attributed solely to the patient. Thomassen suggests that GPs seem to enjoy talking to female patients more (personal communication).

11. It is my contention (Davis 1984a) that this is what underlies sex differences research in the first place. Demonstrating that women and men are different is the first step towards establishing that their positions are unequal. Unfortunately, the second step cannot be 'proven' by means of the first for the simple reason that it is tied up with all sorts of political, ethical and evaluative considerations. If we want to demonstrate how gender inequality works, we will probably have to give up trying to convince the non-believers of its existence and just get down to the business of investigating it. This recommends treating it as a background assumption rather than something to be demonstrated empirically.

12. For a good view of this perspective, particularly as it is employed in the field of medical sociology, see Bury (1986). He attempts to unravel the basic themes inherent in the position, whether of the original Berger and Luckman variety or, more recently, more Foucaultian-inspired writings. The present discussion draws heavily from his article.

13. Take, for example, the very cornerstone of the medical model that disease can be located in organs of the body. This notion is of comparatively recent origin and had, prior to the late 17th century, no meaning at all. Our own view of 'psychosomatic' complaints is of even more recent date. Armstrong (1983) provides an interesting constructionist treatment of this development, including its relevance to and mediating influence on the emergence of general practice as a full-fledged medical speciality.

14. A good example of this is childbirth, where women and medical professionals often display competing definitions of the situation. Whether or not pregnancy and childbirth are treated as 'medical' or 'normal' has far reaching consequences for how

individual women will experience their pregnancy as well as how childbirth is socially organized (Oakley 1979; Rothman 1982).

15. Quoting Querido, Van Aalderen (1982:207) notes that among hospital patients, somatic complaints are the least important of their problems. Analyses unearthed the surprising finding that 70% of diagnoses can be made on the sole basis of a chat with the physician, 20% of the diagnosis required a simple clinical examination, and only 10% necessitated laboratory, X-ray or other technical diagnostic procedures.

16. There is no English equivalent for the Dutch word 'hulpverlening,' used to designate this new model for professional help in medical settings. It refers to any kind of service: medical, juridical, or psychiatric or, as is the case here, more (psycho-)therapeutic or problem-solving approaches to medical consultations.

17. Balint, a Hungarian psychiatrist working in England, developed psychoanalytically-oriented training groups for physicians to help them come to terms with 'problem patients.' Considerable emphasis is given to the physician's functioning. These groups have been particularly well received in The Netherlands and it is estimated that 10% of all Dutch GPs have taken part in Balint groups. See Balint (1970).

18. 'Methodisch Werken' is a method developed by Van Aalderen to enable GPs to clarify patients' questions and problems by proceeding in a systematic, step-by-step way towards a solution. It is more than a program for better provision of help, however. It also rests on a series of assumptions about the nature of peoples' problems, their fundamental right to make decisions affecting their health and well-being, and the potentially emancipatory role of GPs, both for themselves and their patients. See Van Aalderen en Hollenbeek Brouwer (1982); Hollenbeek Brouwer (1982).

19. As of December 1985, female physicians comprised 23% of the total population of physicians in The Netherlands. Among GPs, however, only 9% were female. Although this percentage appears to be increasing, it still is safe to say that the number of female GPs is so small as to be almost negligible (information from the Ministry of Welfare and Public Health, Department of Epidemiology).

2

SELECTING MATERIAL AND A METHOD FOR ANALYSIS

The research problem could have been tackled in various ways. Before describing how I went about it; i.e. how I selected my data as well as a suitable method for analyzing it, a few words are in order of a preliminary nature.

To begin with, the phenomenon being investigated placed some restrictions on the choices to be made. The aim of the inquiry was to show how women's initial presentation of their complaints are diagnosed and treated in consultations with male GPs and how this process reflects and constructs asymmetrical power relations. Such a transformation is an *interactional* event, occurring in and through medical talk. For that reason, it made sense to undertake the investigation of this event where it was actually taking place rather than, for example, asking physicians - after the fact - how they had arrived at their diagnosis or interviewing patients about what they thought had been happening.

Second, whereas current literature in the field of general practice does concern itself extensively with the process of diagnosing and treating patients' complaints, it tends to shy away from using real *consultations as data.*<1> Authors describe how physicians can best elicit relevant information from patients, the steps required for medical history-taking, and, finally, what labels should be attached to which symptoms (Royal College of General Practitioners 1972; Bremer *et al* 1974; Wulff 1976; Van Es 1980). The focus is on what practitioners *should* be doing rather than on what they, in actual fact, *are* doing. Whereas this may be helpful for the beginning GP, it tells us very little about how diagnoses come into being in concrete consultations and, more to the point, what happens when the process does not proceed as it, ideally speaking, should. In an investigation like the present one, it seemed safer to start with observable instances of the process in question rather than relying on the prescriptive accounts or idealized examples of interventions so often found in medical textbooks.

Third, it is inevitably a *complicated endeavor to investigate power.* Power cannot be neatly extracted and examined on its own, but is invariably 'confounded' by other variables - in this case, most notably, gender and institutional authority. This makes it difficult, if not impossible, to sort out in advance what might be happening 'because' the GP is a man (and the patient a woman) and what is simply an artefact of any encounter between a professional expert and a layperson.

In a similar vein, the feminist critique of the health care system suggests *that* power will be routinely implicated in encounters between male physicians and female patients. There have been, with some exceptions,<2> relatively few attempts, however, to show explicitly *how* power is involved. For example, what would 'medicalization'<3> look like in an actual encounter between a GP and his patient? How might a female patient's 'control over her body' be 'usurped' in the course of a consultation? Questions about the 'how's' of power and gender in medical encounters are complicated and do not lend themselves to over-hasty operationalizations to be tested by familiar quantitative research approaches. The present inquiry required a more qualitative approach,<4> which would enable an in-depth examination of processes occurring at the level of face-to-face interaction.

Based on these considerations, two conclusions could be drawn. In order to investigate how power works in the process of defining and treating women's complaints in general practice, an approach was required which

- used actual (taped) consultations between GPs and female patients as data
- employed in-depth, qualitative methods for analyzing interaction.

These choices are, of course, inter-dependent. I shall now examine each one in more detail. After indicating how I obtained data for the present inquiry and why it was suited to the task at hand, I shall describe the method selected for investigating it; i.e. conversation analysis. Since this is part of a more general research orientation, some of its basic tenets and procedures will be discussed, briefly.

ASSEMBLING THE MATERIAL

Normally, it would have been necessary at the outset of this inquiry to collect suitable general practice consultations as data. This would have entailed contacting a group of GPs and arranging for them to tape consultations in their practices, particularly with female patients. I was spared this extremely time-consuming procedure by the presence of a large collection of audio-taped general practice consultations, which had been systematically assembled in 1980 in the context of the TRIPT-project (Thomassen and Woltring 1982). The project was set up at the HIVU (Institute for General Practice at the Free University) in Amsterdam with the aim of investigating the quality of the communication between GPs and patients in order to improve the general practice training programs. The collection was to be made available for other research projects as well.

I shall now describe the TRIPT-collection in more detail. This will be followed by a discussion of its representativity for general practice consultations as well as its suitability as data for the present inquiry.

Description of the material from the TRIPT-collection

The TRIPT-project was not concerned with the general population of Dutch GPs: in 1980, 5,468 (Centraal Bureau voor de Statistiek 1985). Since the express aim was to investigate the quality of communication in consultations, the members of the project were primarily interested in GPs, who had some vocational training. For this reason, the population was limited to GPs who had completed their medical exams after 1969 and had been practicing for at least 1½ years at the outset of the inquiry (Thomassen and Woltring 1982). In the interests of avoiding large differences in the practices themselves, the population was restricted to GPs practicing in urban settings; i.e. Amsterdam, Rotterdam, Groningen, Arnhem, Eindhoven. All of the GPs were male.<5>

These restrictions having been taken into account, a potential population of 262 GPs was contacted and asked if they would be willing to cooperate with the project. More than half responded to this request and 30% agreed to participate. Fourteen had to be rejected for practical reasons (not being available at the time of the inquiry, etc.), leaving a sample of 70 GPs (25%), who took part in the project (Thomassen and Woltring 1982.)

The GPs made audio-tapes during their office hours for a period of one to several weeks. Since the observation period extended over a fairly long time span, it seemed unlikely that the tape-recorder had a disturbing effect on the GP's behavior (Thomassen 1986:83). For each patient, administrative data, dates, and times were recorded on a separate form. In nearly all cases, the patient was directly asked for permission to have the consultation taped.

This procedure yielded a total of 2,023 audio-taped consultations. 60% (1,232) were selected on the basis of the following criteria: that the patient speak Dutch, be between 17 and 65 years of age, and introduce at least one new complaint <6> in the course of the consultation. Following this initial selection of tapes, it was discovered that only 40% were of sufficient quality to be transcribed (complete, audible, etc.).<7>

Since it was necessary for the project to have at least 5 consultations per GP in order to discover regularities in his communicative behavior, 5 to 10 consultations were chosen at random for each GP. This resulted in a final selection of 315 tapes from 52 GPs. Transcriptions were made and a catalogue containing general information concerning the type of practice, the GP, and the patient was assembled (Thomassen and Woltring 1982).

It is from this collection that I selected material for my own inquiry. Before I describe how I went about this, however, a few words are in order concerning the suitability of the collection itself.

Representativity of the sample and its suitability for the present inquiry

Obviously, the TRIPT-collection is not representative of the larger population of Dutch GPs. However, it does provide a good representative sample of one specific segment of this population; namely, the young, urban, male GP, who has received recent and up-to-date specialist training. Whereas it is impossible to determine whether this training has actually effected the kind of health care he provides his patients, it seems safe to assume that he will at least have been exposed to training in good listening skills, more integral approaches to diagnosis, and some critical reflection about problems which can arise in professional service encounters (Chapter 1). In Thomassen's view (1986: 83), the only thing that could be considered 'a-typical' is that the GPs were especially willing to take a critical look at their own behavior in consultations with patients. This could lead to the conclusion that the GPs in the TRIPT-sample were not only younger and better trained than the run-of-the-mill GP, but that they were also more idealistic and actively concerned about providing optimal health care for their patients.

Whereas a non-representative sample can be a thorn in the side of many a research endeavor, it is my contention that it does not have to be a disadvantage for the present inquiry. In fact, it may even prove to be an advantage. There are two reasons for this:

The first concerns my underlying assumption about the nature of gender inequality. As previously mentioned, feminist scholars have already devoted considerable effort towards uncovering extreme cases of medical malpractice towards women as patients. Whereas I do not doubt that similar cases could also be found in Dutch general practice, I do not want to limit my investigation to them. On the contrary, if asymmetrical gender relations are a 'social fact' and not simply the product of the evil intentions of the individual male physician, then they should be observable in even the most ordinary and unproblematic encounter. In that case, instances of medical domination ought to be available in consultations with female patients even when the individual GP is bursting with good listening skills and empathy. Thus, selecting a sample of GPs, which is - if anything - positively slanted, corresponds with my theoretical assumption that gender inequality is an everyday occurrence, part and parcel of any encounter between the sexes.

A second, related consideration is of a more methodological nature. Qualitative research methods are generally geared toward uncovering patterns or relationships between phenomena rather than measuring or counting the frequency with which a specific event or constellation of events occurs. To this end, an open-ended and comprehensive approach to the setting or activity under investigation is required. In order to come to terms with the complexity of the phenomenon in question, it is necessary to have a small sample of 'cases' which can then be subjected to in-depth analysis.<8> The criteria, then, for selecting a sample is not that it be representative of some larger population, but that it be relevant or interesting to developing theory (Glaser and Strauss 1967) or that it serve as a good 'take-off point' for 'finding one's way into the phenomenon' (Schwartz and Jacobs 1979:293).<9>

One difficulty, inherent in qualitative methods, concerns the issue of the reliability and generalizability of the research findings. In other words, can they offer anything of a more general relevance for some larger population? Whereas this kind of criticism is the unavoidable concomitant of the kind of research I am doing, it is my contention that the use of extreme cases of medical sexism would leave me even more open for this kind of attack than I would normally be. It would be a relatively easy matter for readers to distance themselves from my findings, under the motto that the GPs being investigated were probably exceptional, the a-typical 'bad apple' in the otherwise exemplary bushel. However, by selecting exceptionally well-trained GPs as a starting point for investigating power and gender in medical encounters, I can make a stronger case for the existence of gender inequality and relations of domination and subordination as a routine and general feature of consultations between male practitioners and female patients.

Seen in this light, the possible lack of representativity of the TRIPT-collection for the general population of GPs did not seem particularly problematic for the present inquiry. On the contrary, the collection seemed to provide a more than adequate source of general practice consultations for my investigation.

Selecting material for the preliminary analysis

Having satisfied myself that the collection as a whole was suitable for my research problem, my next task was to select a corpus of tapes for further analysis. Obviously, it was not feasible to use all the tapes of the TRIPT-collection in my preliminary analysis. At this point, the only criteria guiding the selection were that the patient be female <10> and that she spend some time talking during the consultation. I rejected very brief consultations (1-2 minutes) where the patient's contribution was extremely limited or where the GP's main activity consisted of writing out a prescription or providing a referral card.<11> Since the research problem concerned the process of negotiating definitions and diagnoses of patient's complaints, I wanted to make sure there would be enough interaction available to delineate it. I proceeded to make an initial selection from the 315 tapes, which had been audible enough for transcription. This was done by thumbing through the transcripts and selecting 30 at random. Later, after listening to the tapes, several had to be thrown out because they were still not audible enough for the detailed kind of transcription, which my own analysis would entail.<12> Finally, I was ready to embark on my preliminary analysis with 20 audio-taped consultations between male GPs and female patients.

The results of this preliminary analysis will be discussed in the next chapter. I shall now describe the research method selected for the analysis.

SELECTING A RESEARCH METHOD

As research method for the present inquiry, conversation analysis was selected. Conversation analysis is a methodology for the empirical investigation of social interaction, using naturally-occurring conversations as data. An off-shoot of ethnomethodology and, more generally, the tradition of 'interpretative sociology,' <13> its aim is

> the analysis of the *competences* which underlie ordinary social activities. Specifically it is directed at describing and explicating the competences which ordinary speakers use and rely on when they engage in intelligible, conversational interaction. At its most basic, the objective is to describe the procedures and expectations in terms of which speakers produce their own behaviour and interpret the behaviour of others (Heritage 1984:241).

Or, as the late Harvey Sacks, the undisputed founder of the discipline more simply put it, 'to describe the methods persons use in doing social life' (Sacks 1984a:21).

From its inception in the early 1970's, conversation analysis has had a distinct impact on debates within sociology about methodology as well as stimulated a wealth of empirical studies in various disciplines (linguistics, medical sociology, social psychology). A wide variety of substantive concerns have been investigated using conversation analytic procedures. To date, this has yielded an impressive arsenal of analytic concepts and procedures, available for anyone interested in analyzing conversational interaction in any social setting.<14>

Basically, conversation analysis picked up one of several themes <15> inherent in ethnomethodology; namely, that members <16> of society employ ordinary methods and routine procedures for finding their way about in their daily lives. These activities should be treated as valuable in their own right and subjected to serious and careful investigation on the part of social scientists. There was a call, then, for abandoning *a priori* categorizations of what participants 'typically' did (in a particular setting or as a result of their 'role') in favor of straightforward, naturalistic descriptions of what they, in actual fact, were doing. These descriptions could be provided by attending to actual instances of conversational interaction.

The term 'conversation analysis' might mistakenly be construed as a unified research methodology. Unfortunately, this is not the case. In actual practice, there are a variety of ways to go about it and, unsurprisingly, conflicting opinions about which way is best. In the next chapter, I shall be dealing with some of these variations and, in fact, proposing some modifications of my own.

At this point, however, an attempt will be made to familiarize the reader with some of the basic assumptions underlying conversation analysis as it was originally developed as well as describing the basic concepts and procedures, which have been developed for the analysis of conversational interaction. In order to do this, I have chosen to rely primarily on the work of Heritage (Heritage 1984, 1985; Heritage and Atkinson 1984) for the following reasons:

It is exceedingly difficult to find a more general methodological discussion of how (or how not) to go about it. Research in the field of conversation analysis has always tended toward the empirical description of talk. For the beginning analyst, this can mean laboriously wading through various empirical studies and attempting to figure out how it was done. The basic tenets were originally formulated by Harvey Sacks in his lecture notes and these were widely-distributed in stencil form to investigators in the field. With some exceptions (Sacks 1973; 1974a,b; Schegloff and Sacks 1974; Sacks *et al* 1974), they were not published before his death in 1975.<17>

Meanwhile, however, Heritage, has emerged as something of a modern-day exegist of conversation analysis and ethnomethodology. Although he began doing the usual empirical description of conversational structure, he has since taken upon himself the task of making the methodological and theoretical underpinnings of the field available to a wider audience than had hitherto been the case. After providing a comprehensive and eminently accessible account of Garfinkel's ethnomethodology (Heritage 1984), he set about describing the most important contributions and recent developments in the field of conversation analysis as well (Atkinson and Heritage 1984; Heritage 1985). Although his perspective is a selective one, it does provide a systematic and thorough account of the basic assumptions underlying conversation analysis as methodology, true to the spirit of its original formulation.

I shall, therefore, begin by describing some of these basic assumptions. This will be followed by a brief treatment of the basic concepts and findings in the field of conversation analysis as well as some of the more recent developments, especially those which are relevant to my own research problem. In conclusion, the advantages of conversation analysis as research method for the present inquiry will be discussed.

Whereas this chapter will be almost exclusively devoted to the advantages of conversation analysis, I shall not be leaving it at that. The reader is asked to persevere until the next chapter when, on the basis of the preliminary analysis of the data, I shall be making some reservations known. This will lead to a critical re-evaluation of conversation analysis as method for the present inquiry and some conclusions about how it needs to be modified in order to make it applicable for a research endeavor like my own.

Some background assumptions of conversation analysis

1. *Language is a medium for social activity.*
The most basic tenet underlying conversation analysis is that language, and, more specifically, the activity of conversing, is a good starting point for the study of social life. This may seem self-evident, but, in fact, it represents a rather different orientation to both the way language should be studied as well as how social life can best be approached.

Linguists and sociolinguists had been studying language long before conversation analysis made its entrance. Language, however, was treated as a structured entity, which could be described by means of idealized models. Chomsky's (1965) transformational grammar is a well-known example. The person's innate knowledge of the structures and rules of language, her transformative 'competence' for generating and understanding speech were what needed to be delineated. The 'performance' end of things, i.e. the actual activity of talking, was a lesser concern. This was probably due to the unfortunate fact that talk tends to deviate messily from the models in real life, ever a 'degenerate realization of linguistic competence' (Atkinson and Heritage 1984:20).

Social scientists have also traditionally displayed an awareness of the importance of language: as symbolic system reflecting and determining our perception of the world, as the basic ingredient for our human-ness, or as social skill effecting and constraining our life chances. Finding language *relevant* for social life is one thing; making it a *topic* for social research is, however, quite another. Like linguists, social scientists also found actual instances of talk far too 'random' or 'idiosyncratic' to provide a focus for analyzing institutions or social organizations or any of the other topics deemed important for understanding how society works.

Conversation analysis marks a radical departure from both sociological and linguistic understandings of language. Unlike other sociological research orientations, it gives the study of talk primacy in understanding how social life works. Unlike linguistic approaches, it does not concern itself with the structure of language *per se*, but rather with the structure of talk as a social activity. In short, its focus is on the various practical social activities, which can be accomplished using language as a medium.

2. *Interaction is systematically structured.*
A second assumption is that social action and interaction are

organized in such a way that they exhibit stable, recurrent patterns or structures.<18> The fact that social activities are accomplished by means of accountable sets of methods and procedures is what gives interaction its orderly, regularized character.

As Schegloff and Sacks put it:

> We have proceeded under the assumption (an assumption born out by our research) that in so far as the materials we worked with exhibited orderliness, they did so not only to us, indeed not in the first place for us, but for the co-participants who had produced them. If the materials (records of natural conversation) were orderly, they were so because they had been methodically produced by members of society for one another, and it was a feature of the conversations we treated as data that they were produced so as to allow the display by the co-participants to each other their analysis, appreciation and use of that orderliness (1974:290).

This attitude has as immediate effect that it provides the analyst with a basis for investigating social interaction. She does not have to resort to speculations about why people behave as they do. Nor does she need to 'clean up' the data in advance in order to investigate it. On the contrary, if members are ongoingly and regularly engaged in the production of structured and organized activities, all the analyst has to do is observe how they go about it.

3. *Social interaction cannot be understood without reference to the context.*

A third assumption draws upon ethnomethodological notions concerning the contextual nature of social interaction (Garfinkel 1967; Garfinkel and Sacks 1970). The meaning of any utterance is always dependent upon the context of its use; that is, the person who produces the utterance, particulars taken from her biography, where and when the utterance occurs, etc. Taken on their own, most utterances tend to be hopelessly vague, equivocal, and, in many cases, quite unintelligible. Members need to draw upon the situation in which the talk takes place in order to create and sustain orderly and meaningful interaction.

This contextuality is a double one (Heritage 1985:242). Social interaction is *context-shaped* because no utterance can be understood without reference to sequences of talk immediately preceding it. Members construct their own talk with reference to prior talk. Social interaction is also *context-renewing*; i.e. any 'current action' will automatically inform the 'next action' in the sequence, thereby providing the framework within which that action can be understood. In this way, a prevailing sense of context or 'what the situation is all about' can be ongoingly produced, sustained, or altered.

Taking the view that conversation is an ongoingly situated practice has consequences, of course, for how it may be analyzed. Conversation analysis requires the immediate context in which the talk occurs as locus for the analysis. In other words, utterances cannot be separated from the sequences in which they occur within a concrete conversation.<19>

4. *Data should be collected from naturally occurring occasions of everyday social interaction.*

A fourth assumption which emerges from the previous three is that naturally occurring instances of interaction must provide the data for analysis. At the time, this was a rather revolutionary departure from traditional methods of data collection within the social sciences. Prior to conversation analysis, actual talk was not used as data. There was always some degree of intervention or manipulation by the investigator in the process of selecting material, whether through interview techniques, experimental methodologies, pre-coded observational schemes, or field notes.

In the interests of a truly empirical approach to the study of interaction, conversation analysts insist on using actual instances of naturally-occurring talk as data. The influence of the investigator at the stage of data collection should be minimized as much as possible. To this end, everyday conversations are recorded, either audio-taped or video-taped, and transcribed. Transcriptions are made according to extremely time-consuming and painstaking methods whereby details like simultaneous talk, overlapping utterances, intervals, characteristics of speech delivery, and even transcriptionist doubt are all recorded (Heritage and Atkinson 1984:ix-xvi). The idea is to provide the next best thing to a 'listener's transcript.'<20>

The use of transcripts is a step in the direction of eliminating the investigator's as well as the subject's pre-conceived ideas about what is going on in favor of a more empirical view. Moreover, they enable a repeated examination of the interaction, making a detailed and in-depth analysis possible. And, last but not least, since transcripts are directly available to public scrutiny, any claims put forth by the analyst can be immediately checked out and verified by others.

In short, by using careful transcriptions of naturally-occurring conversations as data, sociology as a 'natural observational science', as Sacks (1984a:21) put it, can be developed.

5. *Social analysis of interaction needs to concentrate on the details of talk, putting the 'big issues' aside until later.*
A final assumption is that in order to come up with a truly empirical description of social interaction, it is essential for the investigator to be open to empirical phenomena, no matter how small and seemingly trivial at first glance. This requires that the investigator get rid of any abstract or theoretical concepts or assertions and assumptions about what is 'typical' for a particular situation before embarking on the analysis. These can only get in the way of an open and detailed examination of what is actually happening. This kind of open-mindedness is the prerequisite to understanding how social interaction is being constructed.

> We will be using observation as a basis for theorizing. Thus we can start with things that are not currently imaginable, by showing that they happened. We can then come to see that a base for using close looking at the world for theorizing about it is that from close looking at the world we can find things that we could not, by imagination, assert were there. We would not know that they were 'typical.' Indeed, we might not have noticed that they happen (Sacks 1984a:25).

With this stance, Sacks broke with the tendency within sociology to limit itself to what he called the 'big issues.' This had led to the unfortunate inattention to the details of talk. However, precisely these details have an intrinsic significance for how social life is constituted and organized. Thus, in a sense, Sacks calls for a 'time out' and suggests that investigators might profitably subject the small, so-called trivial phenomena of social life to some 'close looking.' Such close looking can eventually provide invaluable insights into sociological concerns like institutional realities or social problems.

I shall be returning to this point in more detail in the next chapter, particularly in terms of how Sacks' original injunction has been used (and abused) by his latter-day followers.

In summary, conversation analysis takes naturally-occurring instances of conversational interaction as the object for an analysis of social life. An attempt is made to develop a method, which is rigorously 'data-driven.' Conversational activity or the 'details of talk' are treated as the central resource for analysis. Any *a priori* speculation concerning members' motives and orientations as well as the investigator's presuppositions about what might be going on are strongly rejected in favor of a strongly empirical approach to the study of social interaction.

Having set out the basic assumptions underlying conversation analysis, I shall now provide a brief treatment of some of its central research findings. The aim is to familiarize the reader with the research tradition as well as delineate some of the basic conceptual tools, which can be used for analyzing the structure of conversational interaction in any setting.

Conversational analysis: research findings

Since the most fundamental aim of conversation analysis has been to delineate the socially organized features of talk in context (Atkinson and Heritage 1984), it is not surprising that 'ordinary conversations' were selected as a place to start. The original studies demonstrate a marked preference for everyday chats between friends, haphazard snatches of talk picked up by the analyst as s(he) wandered about her daily life activities, or for the, by now infamous, telephone call (Schegloff 1972; 1979; Schegloff and Sacks 1974). They discovered that even the most simple interactional ritual, like the greeting sequence in a telephone call, is subject to highly intricate organizational work on the part of the participants. By uncovering these activities, analysts assume that the methods can be delineated, which will also be employed for more complicated interactional tasks. This is a necessary first step towards studying talk in institutional encounters or, for that matter, how it is implicated in any of those other issues of social life, which had hitherto interested sociologists. Thus, after several years of painstaking pioneering work on the part of people like Sacks, Schegloff and Jefferson, a kind of 'basic conceptual scaffolding' (Heritage 1985:3) could be developed which is, theoretically at least, applicable to the analysis of any instance of conversational interaction.

I shall now direct my attention to the most basic elements of this 'scaffolding'. This will be followed by a look at some of the more recent findings, particularly those which are of concern for the present inquiry.

Some basic features of conversational organization

One of the first things discovered was that talk is *sequentially organized*. Utterances do not stand alone, but are always related in some way to prior talk. Each utterance projects a relevant next activity or range of activities to be accomplished by the next speaker. This means that utterances can be analyzed in terms of how they indicate the speaker's preparedness or willingness to forward, acquiesce in, or resist the course of action proposed by an earlier speaker (Atkinson and Heritage 1984:6). For obvious reasons, sequential organization will be tied up with the construction and regulation of social relationships. These can also be investigated through the study of how talk is sequentially ordered.

To account for the sequential organization of talk, Sacks *et al* (1974) proposed a *turn-taking system* as the most basic means for producing everyday talk. It was discovered, for instance, that members take turns at talking. Apparently, the conversational floor is something of a scarce resource and as such must be distributed among speakers. This is done by means of rules and procedures for allocating turns at talk. The turn-taking model is used to account for a variety of observable everyday phenomena like, for example, the fact that generally only one party will speak at a time; that the position of speaker has to be transmitted from one person to the next; that these transitions occur primarily without overlap or overly-long pauses; that the length of the speaker's turn can be regulated without having to specify this in advance, and more. Turn-taking procedures account for how participants enter (Schegloff 1972), exit from (Schegloff and Sacks 1974) and suspend (Sacks 1974b) their conversation.

The turn-taking model led to the specification of a series of rules,<21> collectively shared by participants and recursively applied, enabling them to transfer and regain the right to speak in the course of a conversation with precision and a surprising minimum of interactional muss and fuss.

One of the main ways this is done is through the *adjacency pair* (Schegloff and Sacks 1974). The adjacency pair is a fundamental and effective means for selecting a next speaker. It refers to a two-part sequence in which the first utterance constrains the selection of the next utterance by the second speaker to a same pair type. Examples are greeting-greeting, question-answer, formulation-decision (confirmation/disconfirmation). It is a simple, but elegant device for ensuring that the conversation proceeds in an orderly, step-by-step fashion.

If the adjacency pair can be used to account for much of the smooth-running quality of conversations, it is neither invariably nor unproblematically implemented. Within conversation analysis, there has always been a particular concern for the 'deviant case;' i.e. that moment when the greeting is not returned or the question is left dangling (Schegloff 1972).<22> It was discovered that such problems occur, in fact, on a rather regular basis. Fortunately, members have all sorts of routine procedures for dealing with problems in talk. For example, they display great proficiency at *repair*, preferably self-correction occurring within the very same utterance in which the blunder occurs (Schegloff et al 1977). Moreover, forestalling devices were uncovered, whereby potentially undesirable responses could be 'headed off at the pass.' For example, a *pre-sequence* like 'Are you busy tonight?' could allow participants just that extra bit of interactional leeway, enabling them to manoeuvre around the possible difficulties which the refusal of an invitation could entail (Sacks 1974a; Schegloff 1984; Drew 1984).

It appears that members display considerable *indirectness* in the course of their talk, constructing their requests, but also challenges and criticisms, in such a way that a minimum of interactional disruption would ensue (Brown and Levinson 1978; Labov and Fanshel 1977). Moreover, they have definite interactional *preferences* <23> regarding the 'next actions' they selected in the course of a conversation. There is, for example, a notable tendency to avoid disagreements, rejections, and refusals (Pomerantz 1975, 1984; Atkinson and Drew 1979), observable in both the empirical (in)frequency with which such actions occur as well as in the way they are handled, should they, after all, arise. All of this research empirically highlights issues developed earlier by Goffman (1955) and, later by Brown and Levinson (1978) and Labov and Fanshel (1977) about the lengths to which members will go in order to avoid threat to 'face' (their own and others') as well as to delay or, preferably, avoid altogether any outright conflict in the course of their interaction.

In summary, conversation analysis represents a general approach to the analysis of social interaction, which can be applied to a wide variety of topics and problems. Its central achievement has been to advance our understanding of how social interaction is organized and the ways that a variety of mundane social activities are performed. It demonstrates how members arrive at shared understandings, displaying them ongoingly within their talk. And, finally, it provides insight into how social relations as well as a sense of 'context' can be constituted in everyday instances of conversational interaction.

Some recent developments in conversation analysis

In his attempt to delineate recent developments in the field, Heritage (1985) suggests that conversation analysis has moved well beyond its original intention to supply some basic tools for understanding and analyzing everyday conversations. Three of these more recent developments are of particular interest in terms of the present inquiry; namely, topic organization, the analysis of 'long' stretches of talk, and interaction in institutional settings. I shall touch on each of them briefly here as they will be dealt with in more detail later on in the inquiry.

1. *Topic organization*
   Even in the early days, conversation analysts did not limit themselves to the utterance-by-utterance level of conversational organization. From the outset, topicality was an issue as well. They were concerned with how topics are initiated, maintained, shifted or closed off in the course of a conversation. Typically, the content of the topic *per se* is not addressed, but rather the procedural bases underlying the structuring activities of topic maintenance and change. Investigations show how complicated it is to ensure that one topic flow into another during a conversation.

   Topic organization can occur by means of a step-wise transition, sometimes in a boundaried fashion with one topic being closed down before the next is initiated (Jefferson 1984b; Button and Casey 1984). Topics apparently have to be initiated in specific and methodical ways. This can not be done abruptly or with a direct reference to the speaker's wish to talk about a particular topic, but rather requires a careful display of 'other-attentiveness' (Jefferson 1980; Jefferson and Lee 1981). Moreover, it is not always easy to get 'off' the topic, once launched (Jefferson 1978b; 1984b).

2. *'Long' stretches of talk*
   Whereas speakers tend to organize their talk on a turn-by-turn basis, there will always be exceptions; i.e. occasions where the turn-taking machinery is temporarily suspended. This occurs when one or more speakers engage in a protracted stretch of talk, encompassing many turns.

   From the beginning, there has been some interest in this type of talk on the part of conversation analysts. Sacks (1970; 1974a; 1978), for example, set the stage for analyzing 'long' stretches of talk in his work on storytelling and jokes. Later, this was expanded by Jefferson (1978a; 1980; 1984a,b), Jefferson and Lee (1981) in the investigation of troubles-telling, a specific type of 'story.'

Because this kind of talk extends over many turns, it presents specific organizational problems for the speakers. It must be announced, for example, so that the next speaker will be prepared to relinquish her rights, temporarily at least, to the conversational floor (Sacks 1970; Jefferson 1980). Moreover, both the teller and the recipient will be ongoingly concerned with eliciting or displaying affiliation and alignment. This will be one of the major interactional tasks required for the accomplishment of the activity of storytelling. Despite all attempts at achieving alignment, such talk often proves difficult to sustain for the length of a conversation as Jefferson and Lee's (1981) work on troubles-talk convincingly shows.

The analysis of long stretches of talk yields considerable insight into the complexities of sequential organization - more than the analysis of greeting rituals could ever hope to accomplish. Moreover, it marks an important departure from the previous tendency to avoid the content of talk in conversation analysis. Whether a story, a pun or a troubles-telling, the talk is organized around some substantive content. The beginning of a fusion between *what* is being talked about and *how* the talk is being ordered is represented in this type of work.

3. *Interaction in Institutional Settings*

Originally, conversation analysis dealt with 'mundane conversation' as 'the primary and prototypical form of language usage' (Heritage 1985:7). Whereas this continues to be the over-riding tendency, in recent years there has been an increased interest in conversations which are not just something the investigator happened to overhear somewhere. Analysts have actively sought out conversations occurring in specific; i.e. institutional, settings. Thus, a diversity of studies have emerged on interaction in classroom, courtroom, medical, and social work settings.<24>

Institutional talk is generally approached from two vantage points. An attempt is made to show how institutional interaction deviates from ordinary conversation. The assumption is, of course, that all of the insights gleaned from random, 'ordinary' instances of talk will provide a kind of 'standard baseline,' from which other; i.e. less normal, talk may then diverge. In accordance with deviant-case philosophy, uncovering the differences will, ultimately, prove the rule; i.e. establish the 'normalcy' of ordinary talk. <25>

The second dimension involves showing how institutional contexts as such are produced and sustained recursively in the course of the interaction (Heritage 1985:7). For example, constructing an interview as a medical one will require specific activities on the part of the participants. They will need to make use of 'categorization devices' and to 'typify' events as belonging to the specific interactional event in question (Sacks 1974b). Thus, the interactional conduct of the physician is categorized by both him as well as the patient as belonging to the class of 'what physicians do,' thereby facilitating, but also placing constraints on the interaction which follows.

These studies provide relevant information for understanding how institutions create and negotiate their organizational categories as well as label the individuals coming under their jurisdiction (Heritage 1985:8).

In summary, the last decade shows a tendency within the field of conversation analysis to branch out and away from the analysis of simple, ritual, and everyday instances of interaction. Investigators are now concerning themselves with applying the basic analytic tools derived from earlier analyses to more complicated interactional phenomena. More important, however, is that there is a move towards regarding conversation as something to be described only in terms of rules and procedures to a consideration of what the talk is about as well. Whereas this is by no means the central focus of conversation analysis, it opens up possibilities for approaching sociological issues which, up until now, fell under the category of something to be 'put aside until later.'

Advantages of conversation analysis for the present inquiry

I shall now discuss some of the reasons, which led to my initial selection of conversation analysis as a useful method for coming to terms with the present research problem. In other words, what advantages does it have for showing how women's complaints are defined, diagnosed and treated in general practice consultations and delineating the workings of power and gender in this process?

1. The first advantage concerns the kind of data to be used. Conversation analysis has always insisted on 'naturally-occurring conversations' as the only proper material for the analysis of social interaction. If there was one thing that I had a lot of at the outset of this inquiry, it was 'naturally-occurring conversati-

ons.' In fact, I had access to a veritable data-paradise, a conversation analyst's dream. I would be a hypocritical (and, in fact, stupid) investigator, if I were to suggest that this did not influence my choice of method. The kinds of materials available at the outset of any inquiry and the method selected to investigate them exert an irresistible and decidedly mutual influence upon one another.

2. The second advantage is connected to the research problem itself, which is about a process occurring at the level of talk. The patient's complaints are being defined and treated within the conversational interaction between her and the GP. There are, of course, various ways of analyzing discourse, conversation analysis being but one possibility. <26> It is, however, a good method for studying interactive processes; i.e. *how* social activities are accomplished in talk, rather than, for example, the content of the talk or the formal features of discursive structures. I was primarily interested in how physicians and patients travel from *her* initial presentation of what is wrong to *his* assessment of the same. It is not irrelevant, of course, whether this turns out to be 'a case of gall stones' or 'a nervous question.' In keeping with the considerations mentioned in Chapter 1, however, the process of constructing problems could use a bit of empirical investigation itself, independent of the end-product; i.e. the final categorization of the patient's complaint in the consultation. <27>

3. Conversation analysis provided a wealth of empirical studies, which were useful for further investigations. A collection of usable concepts and means for describing how talk in any situation is organized had already been developed. It was, however, particularly the more recent developments, which caught my eye. Both the studies on topicality as well as 'longer' stretches of talk seemed especially handy in terms of my own research problem. After all, the patients will be arriving with a problem or problems, which will be not be dealt with in one turn, but rather over long stretches of talk. Whether these problems are treated as 'topics' or as 'troubles' or stories,<28> it is useful to have the empirical support of prior studies in order to study them.

4. Since 1975, conversation analysis has branched out into the field of medical interaction. Already myriad investigations are available, showing how medical talk differs from 'ordinary talk' and constructs a consultation as specifically medical event. This

research can provide a useful backdrop for my own investigation into one particular aspect of medical interaction; i.e. diagnosing patient's problems.

5. With a few exceptions, power is not a topic dealt with by conversation analysts. Nevertheless, I did not see, at the outset of the inquiry, why it could not be a suitable object for analysis, should the investigator wish to make it one. Since I suspected that power would be involved in any consultation between a male GP and a female patient and I wanted to know how it worked at the level of interaction, I assumed that it would be possible to use the conceptual apparatus provided by conversation analysis for this purpose; i.e. to uncover how members go about constructing asymmetrical power relations in the course of their talk. Thus, taking the late Harvey Sacks at his word that the 'big issues be put aside until later,' I decided that 'later' had, at long last, arrived.

In conclusion, conversation analysis appears to be a suitable methodological starting point for the present inquiry. It enables me to use actual consultations between physicians and patients as data rather than having to rely on second-hand accounts from the participants in question. Moreover, it is a rigorously empirical approach, allowing for a detailed and in-depth analysis of social interaction. For this reason, it is particularly appropriate for delineating how women's complaints were being defined and treated in consultations. Over the years, conversation analysts have unearthed a host of useful concepts and procedures for describing how everyday talk is organized, even tackling medical talk and showing some of the specific conversational features involved in this particular setting. Although power relations are rarely an issue for conversation analysis, I saw no reason at the outset of the inquiry why its findings and analytic techniques could not be successfully applied to an in-depth analysis of how gender and power work at the level of interaction.

As it turned out, I was wrong about that. It is to this issue that I shall now direct my attention.

NOTES

1. Some exceptions are Brouwer and Dijkhuis (1967) and, more recently, Van Staveren (1984), where excerpts are used from general practice consultations. In *Praten met patiënten*, the authors inform the reader at the outset, however, that these examples have been 'fixed up' so that they could serve as 'good illustrations' of the points they wish to make (Brouwer and Dijkhuis 1967:11).

2. In Chapter 3, I shall be dealing in more detail with some analyses of medical interaction which explicitly deal with power from a feminist point of view.

3. The term medicalization is used for two interrelated processes. In the first sense, specific conditions or behaviors are given medical definitions; i.e. they are defined in terms of health and illness. In the second sense, medicalization refers to the ever-expanding radius of control of the medical profession. More and more experiences come to be defined as medical, thereby making them subject to medical surveillance and control (Riessman 1983:4). Social scientists, who have been concerned with medicalization are: Freidson, (1970); Zola, (1972); Ehrenreich and Ehrenreich, (1978); Conrad and Schneider, (1980) and, in The Netherlands, Fahrenfort and Klinkert (1978), Klinkert (1982).

4. For a general discussion of qualitative methods of data collection and analysis, the reader is referred to Lofland, (1971); Schatzman and Strauss, (1973); Bogdan and Taylor, (1975), Schwartz and Jacobs, (1979) and, more recently, Silverman (1985).

5. This is, as previously noted, the 'routine' case anyway. In The Netherlands, at the time the TRIPT collection was being assembled, 94% of the GPs were male. Interestingly, the only reference made to this fact was that it was necessary in order to have a homogeneous sample of practices (Thomassen 1986:63). The sample is apparently regarded as a representative sample of young urban GPs with no further explanation required concerning their sex.

6. 'Complaint' was defined as follows (my translation):

   > the present or experienced sensations or perceptions regarding one's body, which the patient regards as an indication of deviations in the structure of or disturbances in the functioning of his or her body (Thomassen and Woltring 1982:25).

   'New' meant that it was unfamiliar to the GP at the time of presentation.

7. Thomassen and Woltring (1982) contrast the high rate of response among the GPs approached by the TRIPT project and the disappointing quality of the material itself. The reasons for the poor quality of many of the tapes were varied: the recorders weren't set up properly; instructions weren't followed for operating them; recorders

weren't switched on until the consultation was already well underway; desk drawers housing the tape recorders weren't opened, making it impossible to understand what participants were saying; and, of course, there was often simply too much background noise from other appliances, children crying, assistants entering, etc.

8. As Schwartz and Jacobs note:

> What is often done is to use some preselected topic to locate a single case, which is from the outset treated as a single case of 'something' - storytelling, misunderstanding, rule violation, or some other topic. Then one produces a theory for why this event or action came about as it did ... For example, it becomes 'obvious' from a transcript that two people misunderstood each other. Now, why it is obvious, exactly how does one define a misunderstanding, and what enables a reader, a speaker, or a hearer to detect that such a thing happened in the world (1979:290)?

In this kind of inquiry, the data being collected consist of 'examples' or 'cases,' which provide a starting point for the analysis precisely because they are good illustrations of whatever the investigator wants to analyze.

9. For a good discussion of the differences between qualitative and quantitative research methods and their ramifications for data collection, see Komter (1985).

10. Of the 315 transcripts, 63% of the consultations were with female patients. This is in accordance with national (and international) figures, which estimate the percentage of female patients at roughly 2/3 of the entire patient group.

11. On the average, the consultations took about 7 minutes, whereby the talk concerned not only complaints being presented, but also what Thomassen (1986) calls 'conversation fragments;' i.e. the weather, vacations, leisure activities, etc.

12. See Chapter 7 for a discussion of the transcription system used as well as the considerations leading to the ultimate selection of material for the final analysis.

13. 'Interpretative sociology' refers to such diverse schools of thought as philosophical hermeneutics (Schutz), ethnomethodology (Garfinkel) and its offshoots, conversation analysis (Sacks, Schegloff, Jefferson), and, more radically, cognitive sociology (Cicourel). In spite of their differences, they share the following: a conception of interaction which depends on interpretation; a belief that the investigator draws upon the same resources as laypersons do in her attempt to make sense of human conduct; stocks of knowledge employed by members of society to create a meaningful social world are largely taken-for-granted and pragmatically oriented; and the concepts employed by the social scientist are linked to those used by laypersons in sustaining interaction. (Giddens 1976:52-53; see also Giddens 1976: 23-70.) For the original formulation of this methodological position in terms of interaction, see Wilson (1970).

14. The most comprehensive list of references from the field of conversation analysis may be found in Heritage (1985) with well over 200 titles. For a good picture of the standard kinds of analysis being done, the reader is referred especially to Schenkein (1978); Psathas (1979); and Atkinson and Heritage (1984).

15. The other theme of central importance within ethnomethodology was that no description of social life could be made independent of the person doing the describing. Interpretation was inevitably and invariably involved in social inquiry. This theme disappears completely, however, from conversation analysis with its concern for 'objective' descriptions of practical reasoning procedures. Thus, whereas they pay lip service to the fact that the people they are investigating are engaged in interpreting, they seem to exonerate themselves from that very same activity. I shall be returning to this issue in more detail in the next chapter.

16. 'Member' is ethnomethodological jargon for any person, competent enough to produce and interpret behaviors appropriate for a particular social group, world or setting. It is shorthand for 'member of society,' meaning any socialized adult.

17. Gail Jefferson, who was Sacks' collaborator up until his death, has since taken upon the task of editing his transcribed lectures for publication. Publications after 1975 are a part of this project. For a complete list of these and forthcoming papers, see Heritage (1985).

18. In order to avoid confusion, the term 'structure' is used differently in conversation analysis than has traditionally been the case in the social sciences. 'Social structure' refers to the sequential organization of social interaction, i.e. the way that utterances are fitted into prior talk, thus setting the stage for the talk to follow. Whereas this notion has been expanded by some, most notably, Cicourel (1972; 1980) - to include interpretative activities, social structure is regarded as a strictly interactive accomplishment. It has nothing to do with 'social structures,' existing independently of members' activities as stabilized institutionalized arrangements. This idiosyncratic usage of social structure within the 'interpretative sociologies' has become increasingly problematic - a point to which I shall be returning in later chapters.

19. Like 'structure,' 'context' has a specific meaning here. It refers to *local* context, i.e. the fact that utterances can only be understood with reference to one another. Whereas this local context reflects and helps to construct larger contexts, the emphasis in conversation analysis is decidedly on local rather than institutional or organizational contexts. I shall be discussing this conception critically in the next chapter.

20. Transcripts are, of course, never identical to the talk itself. There is always some intervention of the investigator in her data. For example, an audio-tape does not permit consideration of all the non-verbal factors, which are inevitably involved in social interaction, something my not being present at the taping precluded anyway. Moreover, no matter how painstakingly a conversation is transcribed, a transcription remains a selection and interpretation of the talk. Finally, the very act of listening to

tapes, re-winding them or stopping at intervals to analyze what is happening cannot seriously be equated with a 'natural' method. These activities and, in fact, even the taperecorder itself, exert an unmistakable influence on the data and what is construed by the investigator as a 'conversation' in the first place. I am indebted to Doro Franck for this observation.

21. The conception of 'rule' used by conversation analysts should not be confused with other sociological or linguistic notions of rule as something which 'guides' or 'causes' persons to act in systematic ways in a given situation. Garfinkel, most notably, proposed viewing rules as reasoning procedures or methods, which are drawn upon by members in complex, tacit, and inductively-based ways in order to create a sense of order in their interaction. They are never prescriptive or exhaustive, but based on myriad arrays of 'considerations' and 'awarenesses' (Heritage 1984:128). As 'on the spot' productions, they help organize the situation and should be analyzed as such. This is not to say, that rules and procedures cannot appear with great regularity in many social situations. The turn-taking machinery, as conversation analysts have convincingly demonstrated, is a case in point.

22. In the deviant case, a particular response becomes viewable as 'due' or officially absent, thereby necessitating a protracted series of negotiations on the part of the participants as in the following 'deviant case' to the adjacency pair structure, quoted in Heritage (1984:248):

    Child: Have to cut the:se Mummy.
    (1.3)
    Child: Won't we Mummy.
    (1.5)
    Child: Won't we
    Mummy: Yes.

    This 'exception to the rule' provides even stronger evidence for the normative character of the adjacency pair structure than a more unproblematic rendition ever could.

23. Preference in the sense it is used within conversation analysis has nothing to do with the personal, subjective or psychological desires or dispositions of the speakers. It is associated with the empirically observable event that speakers have alternative courses of action and that these alternatives are not equivalent in terms of their interactional consequences (Atkinson and Heritage 1984: 53).

24. See Heritage (1985) for a comprehensive bibliography from a strict conversation analytic perspective. Applied to medical interaction, the next chapter will provide additional references.

25. A good example is Turner's analysis of what makes therapy- talk different from other kinds of talk. This was accomplished by comparing the conversational interaction among patients before the session starts with what happens after therapy begins.

From the moment the session 'begins,' patients' utterances are to be treated as 'data,' and not honored as the activities they might be seen by patients as constructed to do ... Thus, to ask patients 'how they feel' about the meeting time *within* therapy is to entitle them, for example, to hear the question as inviting such things as an account of 'neurotic' problems with punctuality rather than as 'face-value' talk concerning scheduling sessions ... 'patients' as well as the 'therapist' are expected to attend therapy-talk as possessing distinctive properties (Turner 1972:396).

26. For a comprehensive look at other types of discourse analysis as well as a thoughtful discussion of the differences between conversation and discourse analysis, the reader is referred to Levinson (1983) and Thompson (1984).

27. There were other methodological reasons for concerning myself with the process rather than the content of diagnoses. Take, for instance, the 'vague complaint' discussed in Chapter 1. As starting point for an analysis, it would have been necessary to rely upon either the GP's categorization or my own that the patient's symptoms or her presentation of them were, indeed, 'vague'. It was, however, precisely the fact that the term was being attributed at all, which interested me. For this reason, it made more sense to start out with the process and forget about the content of the definitions except as part of the ongoing negotiations within the consultations.

28. This will be dealt with in more detail in Chapter 5.

3

GETTING STARTED: A PRELIMINARY ANALYSIS

INTRODUCTION

The aim of my investigation is to show how power works in medical talk and, in particular, how it is exercised in the definition, diagnosis, and treatment of women's complaints. Up until now, I have been primarily concerned with the background assumptions leading to the formulation of the research problem (Chapter 1) as well as the method I intended to use for investigating it (Chapter 2). I have, however, said very little about the issue of power, despite its obvious centrality to the present inquiry.

In this chapter, I shall be describing my preliminary analysis of how power works in the interaction between male GPs and female patients.

I began the undertaking with the assumption that I 'knew,' more or less, what power was. This assumption, which I shall call my 'common sense notion' of power,<1> was heavily influenced by what I had been reading about health care, particularly for women as patients. As the critics of modern medicine have pointed out, not only is the structure of medicine based on dependency and inequality (Zola 1972a,b; Illich 1976; Navarro 1976; Ehrenreich 1978), but the doctor-patient relationship is apparently a most suitable location for the exercise of social control over patients (Freidson 1970; Ehrenreich and Ehrenreich 1978). I took as my starting point that power would be a routine feature of any medical encounter. Moreover, I expected that the physician would be the one exercising it. He would be defining and treating the patient's complaints in accordance with *his* professional system of relevance. The patient, as layperson, would not have access to this professional knowledge and would, therefore, be relegated to a position of relative dependency and powerlessness. In view of the massive feminist critique of the health care system (Chapter 1) as well as the overwhelming empirical fact that physicians are men, I assumed that medical encounters would reflect asymmetrical gender relations as well. Thus, having familiarized myself with this literature, I expected that power would inevitably be implicated in the production of medical talk. All I needed to do, then, was to 'locate' it in my data.

Unfortunately, this proved by no means a straightforward affair. My initial confrontation with my data, in fact, irrevocably shattered any hopes I might have entertained of discovering clear-cut cases of medical domination with the GP indisputably cast in the role of the Omnipotent Representative of the Medical Establishment to the patient's Helpless Victim. Whatever else power was about, it was certainly not going to be that simple.

My preliminary analysis convinced me that, contrary to my original expectations, I could not make do with my (or others') common sense notions about power in medical settings. A more sophisticated theoretical framework was required to account for what I saw happening in my data. This entailed interrupting the empirical investigation of my consultations, already underway, and inserting an additional cycle into my research process. Obviously, this discovery had important methodological implications as well. It became clear that conversation analysis, as it has been described in the previous chapter, would not be adequate for coming to terms with my research problem.

In this chapter, I shall be showing how I reached this conclusion. After describing the preliminary analysis of the data, a brief look will be taken at how power is dealt with (or not dealt with) in other studies on medical interaction. This will be followed by a critical re-evaluation of conversation analysis as suitable research method for the present inquiry. Finally, conclusions will be drawn for how the research problem might better be approached.

INITIAL CONFRONTATION WITH MY DATA

I began my investigation with a preliminary analysis of 20 audio-taped consultations selected from the original TRIPT-collection, where the patient was a woman and the GP a male. In this initial phase, I engaged in what Glaser and Strauss (1967) call 'theoretical sampling.<2> The purpose was to familiarize myself in a general way with what was happening in the consultations. It was a way to start thinking about how my data could be organized for further analysis and the kinds of analytic categories, which were needed. At the same time, I wanted to keep my eyes open for some of the potential 'problems areas' as indicated by the feminist critique of the health care system (Chapter 1).

Although I intended to use the feminist critique of the health care services as a general perspective, I also wanted to approach my material in a flexible, creative way, open to the possibilities it had to offer. Thus, whereas I expected to discover some of the 'problem areas' I had been reading about, I did not want to categorize them in advance. As it turned out, what my material had to offer me was, to put it mildly, something of a shock.

My initial reaction after listening to the tapes several times was, first and foremost, this: the GPs seemed unabashedly and unswervingly *nice*! They were friendly, interested, empathetic, and quite obviously doing their best to provide their patients with quality care. I could easily detect traces of training in good listening skills as well as a familiarity with the tenets of some of the more progressive, comprehensive medical approaches. Many were noticeably reticent about prescribing psychotropic drugs.

My response was mixed. At first, I was inclined to reject these tapes as exceptional. They did not correspond with what I had been reading about women's experiences with physicians, so they must be in some way unusual. Upon returning to the larger collection of tapes and listening to them, however, I discovered that, by and large, they all shared this same quality of intimacy and niceness.

At this point, I might simply have concluded that the problem of medical sexism no longer exists. Or, the treatment of women patients in The Netherlands is, perhaps, exceptional. General practice training programs may, indeed, have succeeded in instilling non-sexist attitudes in their trainees.

I am a skeptic, however. I decided to have another look. After re-examining my tapes at length, I discovered that they were, in fact, full of instances of the kinds of phenomena the feminist critique of health care system had brought to light: GPs making moral judgments about women's roles as wives and mothers, psychologizing women's problems, not taking their complaints seriously, massively prescribing tranquillizers, usurping women's control over their reproduction - in short, it was all there, just waiting to be investigated.

Considering the - by now extensive - body of literature on the subject, this is not surprising. What *is* surprising, however, is that these instances, while available in my material, were not immediately so. In fact, they nearly escaped my attention altogether.

It was precisely the intimate, pleasant quality of the medical encounter itself that made issues like power and control continually seem like something else. For example, was the GP telling a joke in order to put the patient at ease or was he interrupting her troubles-talk, conveying that what she had to say was irrelevant to the business at hand?

Ultimately, it was this experience of sensing that *something* was wrong without being able to put my finger on the source of that feeling which caused me to stop dead in my tracks. A kind of conversational *déjà vu*. I was reminded of the countless, everyday experiences of being ignored, not taken seriously, the minor insults and jokes which are part of women's conversational lot in life. And, of course, I recognized the all-too-familiar occurrence of emerging from conversations feeling vaguely dissatisfied or even humiliated, *without being able to say why*. It is this last element which particularly interests me here.

The micro-insults, which are part of the fabric of talk between men and women, are bad enough in and of themselves. When they are also unidentifiable as such, however, the problem becomes infinitely more serious. Should one venture a protest, the other party may immediately retreat behind the other, also-present face of the interaction ('Only joking, lady!'), leaving one unconvinced, but without a leg to stand on ('Just another humorless female'). As a result of my preliminary analysis, I arrived at the following conclusions:

The Janus-quality, which permeated the general practice consultations of my data, played an important, if not essential, role in the way relations of domination and subordination could be constructed between male GPs and female patients. This was, by no means, going to be an easy matter to investigate. Moreover, my 'common sense' notions on power as well as what I had been reading about in the critical and feminist literature on health care would not help me much in coming to terms with what I saw happening in my data. At the very least, I would need a way of conceptualizing power which did not reduce it to a straightforward matter of the GP exercising control or constraint over the activities of the patient. Clearly, it would have to also take into account the subtle and even benevolent workings of power. Thus, a conceptualization of power was required which would fit my data and help me to make sense of the puzzling phenomena just described.

This led to my decision to take a more explicit and systematic look at how power could be conceptualized. As following step, I decided to examine how power is dealt with in the field of medical interaction. In other words, how do these studies deal with the problem of power and can they provide a possible working definition for my own investigation of power in general practice consultations between male physicians and female patients?

CURRENT LITERATURE ON MEDICAL INTERACTION

Studies on communication in medical settings

Doctor-patient communication has become a topic of increasing interest within medical sociology in recent years. West (1984) attributes this to the fact that the recognition has finally dawned among scholars and laypersons alike that 'something is wrong' with the way doctors and patients talk to one another. Attention needs to be paid to issues like patient compliance, patient satisfaction, and physician control of the communication (Davis 1968; Korsch and Negrete 1972; Cartwright *et al* 1974; McKinlay 1975; Stone 1979; Pendleton 1983; Waitzkin 1983).

In medical sociology, communication between doctors and patients has generally been approached from a social roles perspective.<3> Participants interact in accordance with standardized sets of behavioral expectations and obligations. Patients, for example, are expected to want to get well and seek the services of a professional. On the other hand, physicians, by virtue of their specialist training, are obligated to discover what is wrong with patients, supply a diagnosis, and, if possible, alleviate the problem. This role division automatically places the physician in a dominant position within the medical encounter; he or she has more control over what happens there. Although a social roles perspective does address the issue of asymmetry, it is not considered, in and of itself, as particularly problematic. In fact, the asymmetry between patients and practitioners can be seen as the *sine qua non* of medical therapy (Parsons 1951).

The literature on doctor-patient communication was not particularly helpful for the present inquiry. In addition to the much-debated drawbacks of using role-theory to explain unequal relations of power in *any* context (Lukes 1974; 1979a; Giddens 1976; 1979), these studies tend to treat communication in an abstract fashion, resorting to generalizations about how doctors and patients talk to one another rather than taking a look at the actual talk. Fortunately, in recent years investigators have begun to use medical interaction as a resource for investigation rather than just a topic. In other words, the focus has shifted from how various verbal and non-verbal activities could be characterized by investigators or the participants themselves to the study of actual conversational practices. Medical talk itself became the data for research.

I shall now turn to studies of medical interaction to see how they deal with the issue of power.

Conversation analysis of medical talk

Current research in the field of *medical interaction*, using actual consultations as data, can be divided roughly into 'mainstream' approaches, where quantitative research methods are employed, and 'alternative' approaches, which draw their research methodologies from sociolinguistics, ethnomethodology and conversation analysis or phenomenology.<4> In the former, general and fairly abstract categories are established in advance as being of relevance to describing medical interaction. The criterion for selection is that they will lend themselves readily to quantification and statistical testing. A well-known example of this type of research is Byrne and Long's (1976) study of general practice consultations in the UK. Not only does it distinguish itself from others of its ilk by virtue of having the largest data base hitherto available (1800 consultations), but it sports a remarkably elaborate and detailed coding system with no less than 60 categories for describing physician behavior in consultations. Whereas excerpts taken from the consultations themselves are sprinkled liberally through the book, they are of a strictly illustrative nature. In other words, they appeared as examples of the items in the code or, as Mishler (1984:26) rather unkindly put it, as a kind of 'instruction manual' for applying the coding system.

Alternative approaches to medical interaction, whether of a conversation analytic or pragmatic bent,<5> are likely to reject standard coding systems as useless for studying human activities and experience, particularly at the level of language. Communicative behavior in these studies is not categorized in terms of lexical units or separate speech acts, but rather the focus is on the structure of talk. These investigators examine talk within the framework of a particular model of discourse; for example, as 'turn-taking system' (Sacks *et al* 1974) or a system of 'rules' for producing and understanding discourse (Labov and Fanshel 1977). Actual instances of medical talk, carefully transcribed, are used not only as data, but as the starting point for the analysis itself.

As diverse as mainstream and alternative approaches to medical interaction are, they display a remarkable similarity as well. One of the first things, for example, which is discovered is that medical talk is organized along specific and recurrent lines. Typically, a consultation will go through the following stages: opening - history taking - examination - discussion of diagnosis - discussion of treatment or further examination or referral - termination (Byrne and Long 1976). The terms may vary, but the scheme remains essentially the same. It resembles, in fact, what any medical student might expect to find in a textbook with instructions on how to take a medical history (Fahrenfort 1984). Thus, in most analytic descriptions, the consultation emerges as a task for the physician, to be managed in accordance with his or her professional system of relevance (Davis 1978; Conte 1981; Heath 1981; Drass 1982).

In the 'alternative' approaches, lip service is invariably paid to the 'negotiated character' of the talk. However, it is inevitably the perspective and understandings of the professional, which are being negotiated. Consider, for example, the following excerpt taken from Heath's conversation analysis of how medical interviews are opened:

> The patient and the doctor collaborate in the production of a description which will provide for the categorization of the patient's illness. The patient's description is designed to give the doctor the resources through which an assessment can be accomplished (1981:79).

This is not to say that Heath and cohorts describe consultations as proceeding without an occasional snag. On the contrary, physicians are displayed as encountering myriad management problems in the course of medical interviews due to differences in 'belief systems' (Cicourel 1983), 'territorial disputes' (Lacoste 1981), or the 'limited competence' of the patient (Hughes 1982). They must cope with 'unwanted patient initiatives' (Bliesener and Siegrist 1981), unravel the 'routine complications' of the medical consultation (West 1984), and generally ensure that 'ordered sequences of topically relevant talk' are produced (Hughes 1982). Moreover, there are various 'relationship tasks,' which need to be performed like 'displaying sincerity' (Maseide 1981) or 'maintaining credibility' (Skopek 1979). Nevertheless, these difficulties are managed within the framework of the consultation as an essentially harmonious enterprise. The assumption remains that participants design their talk in accordance with 'shared knowledge' about what the encounter is all about and what they should be doing there (Heath 1981:81).

Returning to my initial concern, the question now becomes what these studies have to offer concerning how power works in medical interaction. At first glance, the answer would appear to be: very little. On the whole, there is an overriding tendency to emphasize the 'normal appearances' of the medical interview, whereby normalcy tends to be equated with the production of unproblematic discourse.<6> Moreover, despite differences in their activities and behavior, the patient and physician are essentially treated as peers in the interaction game, engaged in the mutual and cooperative production of an 'orderly' and 'accountable' social world (Garfinkel

1967). Power relations do not enter the picture. It would seem that they are neither relevant nor integral to the interaction.

This could lead to the conclusion that, however valuable these studies have been in delineating the way medical talk is organized, they simply do not have much to say when it comes to the workings of power in medical settings. That particular topic seems to have been bracketed out of the research. Using the well-worn argument of academic freedom when it comes to selecting one's research topic, we could reason that asymmetrical relations of power may be *one possible* feature of medical interaction, but certainly not the only one. In that case, it would be perfectly legitimate for investigators to concern themselves with matters other than power.

This conclusion would be warranted *if* investigators had, indeed, omitted power from their studies of medical interaction. Unfortunately, that was not the case. The main problem with the above-mentioned studies is not so much that they do not deal with the issue of power in medical consultations, but rather that they do deal with it, while pretending that they don't. Under the guise of providing 'neutral descriptions' of talk, they, in fact, reproduce dominant definitions of what the situation is about. It is not coincidental that the 'normal appearances' of the medical consultation which are 'objectively' described, time and again, resemble the institutional requirements of the encounter. Nor is it any wonder that the tasks to be accomplished by both participants, 'cooperatively' and 'unproblematically,' fit neatly into the physician's agenda.

It would appear that, unless the investigator entertains the possibility that members may have different and possibly divergent interpretations of the nature of the encounter as well as their respective positions, it will be difficult for her to discover struggles between participants *even when they occur.* The investigator runs the risk of being automatically, and perhaps unwittingly, 'aligned *with* the professional looking at the client' (Murcott 1981:132).<7> Thus, 'empirically derived' descriptions of the ongoing talk become little more than illustrations of dominant, professional world-views.

The conclusion I drew from these studies was that, whether the investigator wants to or not, she will encounter power in medical interaction. This is simply because it is implicated in any institutional encounter. Since power cannot be avoided, it would seem to make better (scientific) sense to be explicit about one's ideas and background assumptions at the outset of the inquiry.

Feminist analyses of medical interaction

Up until now, I have dealt with studies of medical talk which do not explicitly conceptualize power relations between doctors and patients. Whereas these constitute the majority of the research efforts in the field of medical interaction, there are some notable exceptions and it is to these I shall now direct my attention. In recent years, a number of investigators have addressed some of the problems, which asymmetrical relations of power entail in medical contexts (Fisher 1986; Fisher and Todd 1983, 1986; Fisher and Groce 1985; West 1984). These studies are, unsurprisingly, often feminist in orientation and attempt to uncover some of the subtle and not-so-subtle forms of interactional power being exercised over patients, especially female patients. Their objective is to show that

> language use in the medical setting is micropolitical in its asymmetry: It both reflects the larger structural context and helps to sustain the context (Fisher and Todd 1983:13).

It is demonstrated, for example, how physicians exercise almost complete control over the unfolding consultation. Not only do they ask most of the questions, but the patient is continually being interrupted and thereby prevented from supplying information which may be crucial for the provision of quality health care (West 1984). Decisions on contraception or abortion are largely shaped by the physician, often in accordance with his stereotypic views on women (Fisher 1983; Todd 1983; Fisher and Groce 1985; Fisher 1986). The patient appears to have very little effect on what happens in the medical interview. She emerges as the inevitable recipient of his not-so-benevolent ministrations. The only time she may exercise any control over the decisions being made in the consultation is when she is 'allowed to' do so by the physician. This is the case, for example, when she is upper-class and conforms with his cultural assumptions of what constitutes a 'good woman' (Fisher and Groce 1985).

Clearly, these studies are a step forward within conversation and discourse analysis in terms of the way medical interaction is being investigated. Not only are issues like institutional authority and gender inequality re-instated as part and parcel of most medical interaction, but valuable insights are provided into some of the difficulties faced by women in medical encounters. More important, however, these studies mark a pioneer endeavor to connect socially structured relations of power between practitioners and patients to medical talk as an interactional accomplishment.

Whereas I was relieved to see that power had finally become an acknowledged feature of medical interaction, I was left feeling somewhat dissatisfied about the *way* it had been done. It seemed that power had become a topic, but only at a price.

The first cost was the patient as competent and knowledgeable social actress. In their commitment toward uncovering power imbalances in medical interaction, feminist investigators tend to focus on the dominant interactional partner; i.e. the physician. His activity becomes the locus for delineating how asymmetrical power relations are ongoingly constructed in consultations with patients. The patient's activity becomes - almost automatically, it would seem - less visible and well-defined. Although this time the cause is attributed to her 'oppression' rather than her 'incompetence,' the end effect is, in one respect, the same. *Her* contribution to the encounter is obscured, making her a rather shadowy partner to the ongoing interaction.

The notion that social life is everywhere and always the *active* and knowledgeable accomplishment of its members is something of a shibboleth among interpretative sociologists. This is in accordance with Garfinkel's well-known claim that members are not 'judgmental dopes', blindly driven by forces beyond their control, but usually know pretty well how to get about in their daily activities. Consequently, studies within this tradition have been devoted to uncovering just how competent members are at 'doing social life,' regardless of the situation in which they happen to find themselves. Ironically, and perhaps unwittingly, as soon as investigators begin to tackle the thorny problem of power and gender head on, this ethnomethodological stance seems to recede into the background and we lose sight of the patient in the interaction.<8>

The second cost, which is not unrelated to the first, concerns the linkage of interactional power to domination, coercion or repression. In their attempt to come to terms with the potentially problematic dimensions of medical interaction, the focus of analyses from a feminist perspective has been on authoritarian control by (male) physicians over (female) patients. This leaves the patient's power, however situationally restricted, unexplored and tends toward a theoretical conception of power, which is top-down and repressive.

Traditionally, conversation analysis and ethnomethodology before it, have emphasized the cooperative nature of social interaction. Interaction is a co-production, the participants being inevitably 'in it together,' unburdened by such divisive accoutrements as gender, class or ethnic background. This stance was carried over into the analysis of institutional interaction, which was also treated as an essentially harmonious affair. Presumably based on their shared orientation to this 'best of all possible worlds' (Silverman and Torode 1980), patients and physicians construct the consultation together, unproblematically, as a medical event.

By rejecting the notion that medical events are entirely unproblematic, or even harmonious co-productions, feminist analysts seem to imply the asymmetrical relations involving power and gender preclude amicable, pleasant or even intimate instances of interaction. The exercise of power automatically becomes, at least by implication, a rather nasty business, in which the dominant party (the physician) subordinates, silences or otherwise prevents the weaker party (the patient) from acting as she might have done when left to her own devices.

As useful as this conception of power may be for coming to terms with much of the interaction in institutional settings and even some of the interaction between men and women, it is of little use for studying encounters which have as their most obvious and distinguishing feature displays of friendliness, sociability and empathetic concern. Such interaction - and such is the interaction between the GPs and patients in the present inquiry - clearly requires another way of thinking about power.

In conclusion, despite their important contribution to the field of medical interaction, the model of power employed - whether implicitly or explicitly - in many current feminist studies proved inadequate for the purposes of the present inquiry. In delineating how power relations were being ongoingly produced and reproduced in interaction between male GPs and female patients, I wanted to explore how my patients could actively and knowledgeably participate in the struggle for power. With all due respect to the inequities inherent in these encounters, I did not see how a top-down, repressive model of power relations between patients and physicians could help me come to terms with what I saw happening in my own data. Both my preliminary analysis as well as my initial foray into studies in the field of medical interaction convinced me that a 'time out' was in order. I would need to pay more serious attention to how power could be investigated at the level of interaction. Since this had consequences for my using conversation analysis as it has been described in Chapter 2, another look at the method of research for the present inquiry is required.

A CRITICAL APPRAISAL OF CONVERSATION ANALYSIS FOR INVESTIGATING POWER IN MEDICAL INTERACTION

Conversation analysis, as described in Chapter 2, cannot, in and of itself, provide an adequate methodological basis for delineating how power works in consultations between male GPs and female patients for the following reasons:

- it forestalls any attempts to address the issue of power in social interaction
- it neglects the investigator's own interpretative activities during the analysis
- it takes a negative stance toward the development of theoretical concepts prior to empirical analysis.

I shall address each of these criticisms separately. Some remedies will be proposed, which will, hopefully, enable me to use the valuable analytic tools provided by conversation analysis for investigating interaction while, at the same time, embarking on a study of power at the micro-level of talk.

The centrality of power in social interaction

Power as well as the related issues of gender inequality and institutional authority are not treated as an integral feature of social interaction within mainstream conversation analysis. It is not so much that such things do not 'exist' for conversation analysts, but rather that they propose to bracket them out of their empirical studies. This is in keeping with the notion that it is possible to investigate the details of talk, while putting the 'big issues' aside until later.

My initial confrontation with current literature in the field of medical interaction made me question this practice. I began to doubt whether it was, in fact, possible to describe how medical talk was organized without taking power into account. It is my contention that power can only be 'left aside until later' if the investigator makes certain assumptions about the nature of social interaction as well as how it comes to be organized. And these assumptions are faulty.

Conversation analysis takes as its starting point that social order is not something 'out there,' but rather is produced by and embedded in the conversational practices of members. Treating social life as a 'social accomplishment' rests on another presupposition; namely, that everyone brings the same interactional equipment into the interaction and that conversational resources are equally available to all participants. In accordance with a liberal-pluralistic world-view, (interactional) power becomes something everyone has, *in principle*. If power is not problematic, it does not require further investigation. These assumptions form the necessary backdrop and rationale for ignoring power and concentrating on the details of talk in order to find out how social life is constituted.

To this end, a universal, formal system of rules and procedures- most notably, the turn-taking system (Sacks *et al* 1974) - could be proposed and subsequently applied to any concrete conversation. The turn-taking system has as distinguishing feature that it is both 'context free' (invariantly present in any conversation) and 'context sensitive' (flexibly attuned to the particulars of the specific situation). It is this conception of context, defined as either universal or highly local, which justifies the claim that *any* stretch of talk can provide a good starting point for uncovering the workings of conversational organization.

Whereas early studies were, in fact, concerned with 'any kind of talk' (hence the telephone conversation), the same stance was maintained as investigators ventured out into the world of institutional interaction. They proceeded along the same path, unravelling the intricacies of the turn-taking machinery in both medical talk (Atkinson and Heath 1981; West 1984; Ten Have 1987) as well as mixed-sex conversations (Zimmerman and West 1975). In fact, it was the distinctive way that the turns were ordered, with physicians asking all the questions, interrupting with equanimity or controlling how topics were introduced, developed, and closed off, which accounted, once again, for the constitution of social order; i.e. the organization of the conversation as a medical event, recognizably different than a chat between friends. Whereas the resources employed by participants might be different, this difference was, for the most part, not regarded as problematic.

I do not wish to dispute the claim that turn-taking is fundamental to the way conversations are ordered. Similarly, it is not my contention that formal descriptions of how it works in medical talk are incorrect. They are, however, hopelessly inadequate as descriptions of how institutional encounters are constructed. The impression is unavoidable that, having being taken a few miles along the road toward understanding what makes medical talk different than other kinds of talk, you are left sitting by the way-side. The most important questions have been touched upon, but left, tantalizingly, unanswered.

How, for example, can an account be provided of how consultations are 'opened' (Heath 1981) or 'topicality controlled' (Hughes

1982), without attending to the issue of the decidedly unequal conversational rights of the participants? Similarly, after having been shown how 'troubles-talk' disorders a service encounter (Jefferson and Lee 1981), how can we explain the remarkable persistence of patients in engaging in that very same activity anyway?

The problem is, quite simply, that these activities cannot be adequately analyzed without taking the institutional context as well as the respective social positions of the participants into consideration. The turn-taking system does not exist in a vacuum, but is embedded in relations of power. As Thompson notes:

> *which* interpretation is to count, and *which* participant is to succeed in holding or usurping the right to speak, depends on *who* the participants are and how much *power* they respectively have in the situation concerned (1984:116).

Descriptions of conversational openings or variations in the way turns are taken, interruptions made, utterances repaired and the like are incomplete unless they are considered in the light of who is allowed to do what and who is, in fact, doing it. Even the simplest procedures for organizing medical talk are embedded in the institutional arrangements of which they are a part and cannot be explicated separate from that context. By ignoring power and limiting themselves to the local context of the talk; i.e. the sequential ordering of the utterances, much of the current research on medical talk has ended up providing truncated and inadequate descriptions of the activities they set out to delineate in the first place.

In the interests of a truly empirical description of interaction in medical encounters, we can no more afford to leave power out than we can ignore members' commitment to the production of talk as orderly and meaningful or their ongoing display of an orientation to 'normal appearances.' Power needs to be treated as a central feature of talk, part and parcel of the analysis of topicality, sequential organization or any other conversational phenomenon.

Paying lip-service to the centrality of power in interaction, however, is not sufficient to being able to delineate it. It is to this issue that I shall now direct my attention.

Interpretation and power

Interpretation is involved in any description of social life, whether it is produced by laypersons or social scientists (Schutz 1962).

Descriptions of how power works in medical encounters are no exception. In order to investigate power, it will not be enough that the investigator assumes that it is there to find. She will also need to employ her 'interpretative competence' to link what she sees happening in the data to the relations of domination, which it serves to sustain.

This is, however, precisely where conversation analysis draws the line - at least in its original formulation. It is not so much that conversation analysis is not concerned with interpretation. The entire turn-taking system, in fact, presupposes members' interpretative activities. Members must ongoingly analyze one another's utterances, designing their own talk so that it can be understood and elicit an appropriate response. They must also display an ongoingly understanding of the talk. Although conversation analysts are willing to devote considerable attention to the careful and detailed description of how others engage in interpretation, they are noticeably silent when it comes to their own interpretative activities. Unlike ethnomethodology, which treats the investigator's subjectivity as a valuable resource, to be used and explicated as part of the analysis of any social phenomenon,<9> research in the field of conversation analysis refrains from discussion of how descriptions of social interaction are generated. The investigator remains a shadowy figure in the background. This is not simply a matter of oversight. In the interests of providing naturalistic descriptions of talk, conversation analysis seems to have dropped the concomitant ethnomethodological theme that these descriptions will inevitably entail the interpretative activities of the person producing them.

The main problem with this practice is that it doesn't work. As we have seen in current investigations of medical interaction, they are anything but neutral and objective accounts of how the talk is organized. As Mishler (1984:53) notes, they display a marked 'medical bias' and, more to the point, it is a bias which seems to have 'slipped through the back door' unnoticed. None of this is particularly surprising, of course. The notion that it is the physician who determines what happens in a medical consultation belongs to the dominant definition of what a medical encounter is all about. The question arises, then, why the investigator should not draw upon such stocks of knowledge in attempting to show how a medical interview is constituted. In fact, she will, in all probability, be making use of many of the 'shared understandings' which the participants themselves; i.e. the physicians and patients, draw upon in order to make sense of *their* interaction.

It is not this reproduction of professional perspectives in medical talk, which is reprehensible. The problem is rather that the descriptions simply aren't what they are 'cracked up to be' (Garfinkel 1967:vii).<10> Had investigators, mindful of their ethnomethodological heritage, explicated their own interpretative competence, the result would have been an accurate and undoubtedly valuable account of how a conversation between a patient and a physician comes to be constructed *as medical event* - nothing more, but also nothing less.<11> Unfortunately, by leaving out this explication, the result is obfuscation - the misleading presentation of the description as corresponding unproblematically to the actual interview.

This issue becomes even more serious in an investigation, like the present one, where the research problem is explicitly concerned with how asymmetrical power relations are being constructed in medical encounters between male GPs and female patients. It goes without saying that the primary aim here does not lie in the reproduction of professional or medical vantage points. On the contrary, the aim, as set out in Chapter 1, is to subject general practice consultations to some critical scrutiny from an - admittedly - feminist point of view. The focus, then, will not so much be on how the encounter looks to the professional, but rather its implications for the female patients. This means allowing the patient to take center stage in the analysis, paying careful attention to how she constructs the consultation as a medical event, but also how she may be undermining or otherwise influencing how the interview is being defined and organized. This will entail 'interrupting the voice of medicine' (Mishler 1984) or being on the lookout for gaps or discontinuities in the construction of the interview as a strictly medical enterprise. Special attention needs to be paid to ambiguities in the interactive behavior of physicians, to moments when they are not strictly oriented to professional schemes of relevance, and, most importantly, to the 'voice of the life-world' and the patient.

Just as contemporary investigations of medical interaction employ - albeit implicitly - the investigator's interpretative competence in constructing accounts of how that talk is organized, it will be necessary to use my own (feminist) interpretative competence <12> to retrieve the patient and uncover how she engages in a struggle for power in the course of a medical interview. Finding this, will involve looking for it from the outset of the inquiry. There is, however, more. As Thompson (1984) notes, interpretation is always a creative and ongoing process of constructing meaning. Discovering relations of power in medical talk, as well as delineating how they work, will require that the investigator

> ... *construct* a meaning which unfolds the referential dimension of discourse, which specifies the multiple referents and shows

> how their entanglement serves to sustain relations of domination. Reconnecting discourse to the relations of domination which it serves to sustain: such is the task of interpretation (Thompson 1984:138).

It is a task which will be part of the analysis, every step of the way. Whereas interpretation is an essential ingredient to studying power in medical talk, it is not sufficient. A theoretical framework for conceptualizing power will be needed as well.

Conversation analysis and theories about power

Up until now, it has been suggested that two revisions are necessary if conversation analysis is to serve as research method for the present inquiry. To begin with, the assumption must be made that medical interaction will involve power; i.e. that it will be there to find. Second, delineating how power works will inevitably entail interpretation; i.e. *who* is exercising *what* kind of power with *which* effect. It is neither possible nor desirable to ban the interpretative activities of the investigator from the description of how medical talk is organized. Whereas these two changes could provide the basis for an analysis of power at the level of interaction, they are, in and of themselves, not enough. A theoretical framework, which specifies how power relations involving domination and subordination are constituted, is required as well.

Conversation analysis stands on somewhat shaky footing with the business of theorizing. It is not so much that theorizing is to be abandoned altogether. Sacks (1984a), much along the same lines as Glaser and Strauss (1967) before him, advocated observation first, theories later.<13> In his view, theoretical constructs or abstract and hypothetical versions of how the world is or should be can only constrain what the investigator is able to discover in her data. Thus, he notes:

> ... when we start out with a piece of data, the question of what we are going to end up with, what kind of findings it will give, should not be a consideration. We sit down with a piece of data, make a bunch of observations, and we see where they will go ... if we pick any data, without bringing any problems to it, we will find something (Sacks 1984a:27).

He is advocating, then, that theories about social life need to be grounded in observations and naturalistic descriptions of members' actual social practices. It is this belief in the primacy of interaction as the locus for the constitution of social life - itself a theoretical commitment *par excellence* - which leads to his subsequent rejection of any other theorizing prior to analysis. Theories should not be formulated *in advance*. It is certainly not my contention to question the notion that theories should be grounded. Nor is the main thrust of my criticism the original formulation of a conversation analytic stance on theorizing by Sacks. It seems plausible that he, in attempting to develop a research program for the study of conversation, might find it expedient to put certain activities 'aside until later,' and theorizing may well have been one of them.

This becomes more problematic, however, in studies produced by latter-day conversation analysts. Whereas they have followed Sacks' injunction that theoretical constructs should not be formulated prior to describing conversational interaction, they proceed *plus royaliste que le roi* and forget that theories do, at some point, need to be formulated. Thus, their descriptions of talk remain disembodied, free of any link to theoretical considerations about the institutional arrangements or power relations, which are part of the encounters in which the talk occurs.

This is more a question of how the results of empirical studies of talk can be integrated into theories about the constitution of social life, however. My preliminary analysis, as well as current studies on medical talk where power is explicitly an issue, convinced me that theorizing is not only necessary, after-the-fact, but must also be done in advance. My reasons for this claim are two-fold:

1. *'Common sense' notions concerning power and domination can be wrong.*

   My preliminary analysis indicated that what I thought about the workings of power and gender in medical encounters did not fit what I actually saw happening in the consultations. To continue to look for overt instances of repression or straightforwardly discriminatory practices on the part of the GPs towards their female patients would have been a useless undertaking. It would have meant attempting to press my data into my preconceived notions about male domination and female subordination. As Sacks would surely agree, this would amount to nothing less than a constraint, blocking my vision for all of the 'something's' available in my data.

   If I assume, however, that power will be implicated in my data, just as it is in any encounter between the sexes or a professional and a layperson, then it makes sense to stop and take another look at my 'common sense' notions on power. Whatever the felicities of 'common sense,' rooted as it is in daily life experiences, it is not infallible. As a case in point, my own common sense clearly required some re-thinking.

This meant, in effect, taking the claim that theory must be 'grounded' and that it should be generated from empirical data rather than the other way around (Mehan and Wood 1975) more seriously than had previously been the case. This entailed some theorizing in advance; i.e. a time-out before continuing with the analysis of my data.

2. *Theoretical notions concerning power have far-reaching ramifications, some of which may not be intended by the investigator.*

Some conversation analysts *have* dealt explicitly with the issue of power in medical talk, most notably, the feminists. They have employed theoretical notions concerning power and gender in order to make sense of their data, much as I have done in this inquiry. These theories were not always explicated. In the case of West (1984), however, they were. She adopts, for example, a Parsonian perspective on power where patients and physicians have interlocking and asymmetrical roles, which constrain, but also produce the interaction in a systematic and characteristic fashion. Whereas Parsons regards such interaction as not only the way medical interviews generally *are*, but also how they *ought* to be, West takes a more critical stance. Based on her empirical findings, she questions the desirability of such asymmetrical interaction for the provision of quality health care.

Unfortunately, a Parsonian view, whether explicit or implicit, has certain side-effects for the study of power in medical interaction. One of these is that the patient, as she slips into the role of lay person, not only relinquishes any authority she might have had in the world outside the doctor's office, but she may find her position as competent member on shaky grounds as well. Moreover, there seems to be little room in such a conception for different roles or even different concepts of the same role being in play simultaneously.

Despite repeated assurances to the contrary, the interaction begins to lose some of its negotiated lustre; it seems to result from the ministrations of the professional, as the inevitable and unambiguous effect of role-based rights and obligations. Inevitably, a theoretical perspective influences what you are able to 'see' in your data. In the case of a theory of power *à la* Parsons, it becomes difficult to maintain an ethnomethodological commitment to interaction as the active and knowledgeable accomplishment of its members or delineate *how* asymmetrical relations of power are ongoingly negotiated in the interaction.

Even when investigators embark upon the study of medical talk with the explicit intention of linking the negotiated practices of participants in medical encounters with institutional structures involving power asymmetries as Fisher and Todd (1983; 1986) have done, some of the same problems manifest themselves. By focussing on the physician as dominant interactional partner, the view of the patient becomes obscured. Similarly, medical definitions of the situation are given a central role in the analysis, thereby making it difficult to uncover other, subordinate definitions - including, just parenthetically, definitions toward which the doctors themselves may be oriented. It is my contention that this is an 'unintended consequence' of the investigators' concept of power (and gender) relations (Davis forthcoming). In attempting to come to terms with structured forms of domination and oppression, feminist scholars of (medical) interaction have tended to treat women patients as primarily reactive or, in some cases, as the victim of the all-powerful male physician. Although the shortcomings of such an approach for understanding gender relations is currently subject to some debate within feminist scholarship (Rosaldo 1983; Leydesdorff 1984; Gerson and Peiss 1985; Cott 1986; Aerts 1986), it remains a complex and difficult task to develop feminist theories on power and gender, which can uncover the ways individual women attempt to exercise some degree of control over their situations without blaming them for structural inequities.

To this end, it is essential that a closer look be taken at the conception of power and domination being employed for studying relations between the sexes in a particular context. Any conception of power has empirical implications. In view of the results of current studies of medical interaction, these conceptions can require some theoretical elaboration.

CONCLUDING REMARKS

In the previous chapter, it was suggested that conversation analysis could provide the methodological tools for coming to terms with the present research problem. Not only did it offer a method for an in-depth study of social interaction, using actual conversations as data, but, after a decade of research in the field, a wealth of concepts and procedures are available for studying how talk is organized in various contexts, including medical encounters. Whereas issues like power and gender inequality were not commonly found in conversation analysis, I saw no reason why its 'conceptual scaffolding' could not be used to investigate them as well.

My initial confrontation with my data as well as the review of current studies in the field of medical interaction convinced me, however, that this would not be possible after all. It would be necessary, before embarking on my inquiry, to make at least three revisions.

First, power can not simply be added on as somehow separate from or different than ordinary talk. It belongs to any description of talk, particularly talk in an institutional or gendered context. In order to provide accurate descriptions of what is happening in the ongoing talk between male physicians and female patients, an account is required, which links the sequential ordering of the conversation with who is doing the ordering, how the institutional context is being constructed (and undermined) as well as the way this works to construct asymmetrical power relations between the participants.

Second, it is not possible to provide naturalistic descriptions of social interaction without accounting for the interpretative activities of the investigator. Attempts to do so, particularly where institutional interaction is concerned, can only lead to a reproduction of dominant; i.e. professional, vantage points, which are then misleadingly presented as objective accounts of The Consultation. In order to intervene in this practice, it will be necessary to account for how I, as investigator, am able to see something else in the consultation as well as using my (feminist) interpretative competence to link up what is happening in the data to the social relations of domination that are being produced and sustained.

Third, it becomes clear that being explicit about investigating power is not enough, but that the investigator's notions about what constitutes power need to be worked out in advance. Contrary to conversation analytic claims that theoretical constructs should be left until later, I discovered that, in the case of power, this will not work.

With these three revisions in mind, I shall now address the problem of conceptualizing power in keeping with the findings of the preliminary analysis. This theoretical framework having been explicated, I can embark, once again, upon the analysis.

NOTES

1. 'Common sense' is a term originally developed by Schutz (1962) to refer to the way members of society, including sociologists, make sense of the social world, the procedures used to construct this social world, as well as its factual properties; i.e. its sense of being ready-made and independent of the individual's perception. Common sense includes the stock of knowledge at hand, the natural attitude (sense of social structure), and practices of common-sense reasoning.

2. 'Theoretical sampling' refers to

> the process of data collection for generating theory whereby the analyst jointly collects, codes, and analyzes his data and decides what data to collect next and where to find them, in order to develop his theory as it emerges. This process of data collection is *controlled* by the emerging theory, whether substantive or formal. The initial decisions for theoretical collection of data are based only on a general sociological perspective and on a general subject or problem area (Glaser and Strauss 1967:45).

3. Role theory, in particular as developed by Parsons (1951; 1964), is not only extremely popular in the study of sex differences, but it is the primary theoretical perspective within the field of medical sociology. It continues to occupy this central position, despite the fact that it has been extensively criticized since its inception. See, for example, Wrong (1967); Cicourel (1972); Bloom and Wilson (1979); Wolinsky, (1980).

4. This review of current literature on medical interaction draws heavily on Mishler (1984), who provides a much more comprehensive discussion of the various research orientations than was possible within the scope of the present inquiry. In particular, he compares mainstream (i.e. quantitative) approaches to discourse analysis with 'alternative' approaches (i.e. conversation analysis) in terms of how they transform observations into data, their coding systems, and how the results are ultimately interpreted. He claims that the alternative approaches tend to be 'more respectful' of the structure and meaning of medical talk and are, for this reason, to be preferred over the more quantitative approaches (Mishler 1984:20). Since I share his view (see Chapter 2) as well as many of his criticisms, I have only provided a brief description of quantitative approaches to medical discourse, using one of the most well-known examples of this type (Byrne and Long 1976). For a more comprehensive review, the reader is referred to Mishler (1984:17-57).

5. There have been several collections of studies on medical interaction in recent years, most notably: Wadsworth and Robinson (1976), Atkinson and Heath (1981), Fisher and Todd (1983; 1986). Approaches to medical interaction from a specifically conversation analytic perspective are, for example, Heath (1981; 1982; forthcoming), Frankel (1983; forthcoming); West (1984), Ten Have (1987). These studies focus primarily on sequent-

ial ordering of doctor-patient conversations. Studies with a broader orientation (linguistic, pragmatic or interpretative sociological), which deal with substantive issues or specific interactional problems inherent in doctor-patient interaction include: Skopek (1979); Bliesener and Siegrist (1981); Conte (1981); Lacoste (1981); Maseide (1981); Northdurf (1981); Drass (1982); Hughes (1982); Fahrenfort (1984,1985); Mishler (1984); and most of the studies in Fisher and Todd (1983) and Wadson and Robinson (1976). This is, by no means, an exhaustive list, but will give the reader a general idea.

6. This point has been made by Silverman and Torode (1980) in their critique of ethnomethodology and conversation analysis. In their insistence on retrieving ordinary everyday life as a practical accomplishment for serious sociological scrutiny, conversation analysts have endorsed a vision of the world as 'the best of all possible worlds.' There seems to be no room for members who speak with a different 'voice' or look out at the world from a different vantage point. Social life is, however, as it turns out, not the 'best of all possible worlds' for everyone. I shall be returning to this point later on.

7. Murcott (1981) provides a sensitive account of this in her investigation of how cancer specialists construct a view of 'problem patients' as those who experience symptoms, but postpone seeking professional help. She shows that, by taking the 'problem of patient delay' not as an empirical fact, but rather as a topic of inquiry, a rather different picture emerges; i.e. the problem becomes a management difficulty for the specialists rather than that it says anything much about the patient.

8. See, for example, Fisher and Groce (1985) where a consultation between a Mexican-American woman and the resident of a family practice clinic is examined for how his cultural assumptions about her 'character' negatively influence the information and, ultimately, health care she receives. This is contrasted with a middle-class woman who does get the help she came for. She gets it, of course, because the doctor provides it *for* her.

   Upon closer examination, it appears that she does all the things patients and women are expected to do (asks questions, displays naivete in sexual matters, etc.). The other patient, however, talks about her motor cycle and does not appear particularly impressed with what the doctor has to say. At the end of the conversation, she treats him as an equal, conversationally speaking (Fisher and Groce,1985:356):

   D: 'It was nice meeting you.'
   P: 'Same to you I'll come back whenever I need to see a doctor ...'

   Whereas this patient did not get the medical help she came for, she did not relinquish her position as an autonomous adult either. In view of the increasing tendency to medicalize women's problems, this is not an unimportant event in terms of the negotiation of asymmetrical gender relations. This is true, even when the outcome is not in her favor.

It should be noted that the same authors do address the problem of the invisible patient in their later work and make an attempt to uncover her activities in the consultation (Fisher and Groce 1986).

9. An excellent example of an ethnomethodological study, which does this is Smith (1978). The topic of the paper is how a factual account may be constructed that someone is mentally ill. The investigator recovers the structure of the account by showing how she (and her students) 'recognize that that is what it is.' In other words, they use the same methods for establishing mental illness as a fact. By using her own subjectivity as a resource, an analysis can be provided for everyday 'labelling' practices.

10. Garfinkel suggests that the social scientist in no way inhabits a different world than ordinary members. It is the task of the sociologists to immerse herself within that world, uncovering all the varieties of ordinary reasoning procedures. He is explicitly against formulating 'remedies' for these activities 'as if it was being found out about practical actions that they were better or worse than they are usually cracked up to be' (Garfinkel 1967:vii).

11. Within such accounts of consultations as medical events, some provide notably 'more' information than others. For example, treating medical talk as unproblematic will provide a less comprehensive view of most medical interaction than a more critical (feminist) account might be expected to do. The problem of 'medical bias' (Mishler 1984) holds true for both, however.

12. Cicourel, Douglas, and even more radically, McHugh and Blum diverged from conversation analysts by picking up on the ethnomethodological concern with the scientist's interpretative activities. Of particular interest, is Cicourel (1972, 1980, 1982, 1983), who explicitly addressed the matter of the 'interpretative competence' of the investigator. In his view, her task to show exactly how she fits her findings back into the context of which they are a part. This involves explicating her sources of knowledge and showing precisely how she knows what she knows. Whereas Cicourel does not refer to feminism as a part of the investigator's 'interpretative competence,' a source of knowledge to be explicated as one goes along, I see no reason why it should not be included in the case of the present inquiry. See Stanley and Wise (1983) for a more detailed look at what such a feminist interpretative competence can entail.

13. Glaser and Strauss (1967) are also firm advocates of 'grounding' theories in empirical observations. Unlike Sacks, however, they assume that any investigator embarks on an inquiry with certain background assumptions about the world or 'sensitizing concepts.' These concepts, in fact, are useful for helping her to find her way into her data. Glaser and Strauss paid special attention to how this process might be facilitated and systematized. To this end, they developed a procedure for generating theories from data - something with which - sadly enough - Sacks *et al* did not concern themselves.

4

CONCEPTUALIZING POWER: A THEORETICAL FRAMEWORK

The preliminary analysis of my data as well as a critical review of current studies in the field of medical interaction convinced me that standard conversation analytic practices would require some revision for the present inquiry. A theoretical framework for conceptualizing power would be needed *prior* to embarking upon the empirical analysis. Its task would be to show how relations involving power and domination can be ongoingly produced and sustained at the level of medical interaction. Moreover, conceptual schemes are necessary, which can help me interpret some of the complicated and confusing phenomena encountered in my consultations as instances of power. And, finally, it has to be a framework which can counteract some of the pitfalls involved in investigating power from a feminist point of view; namely, the tendency to reify the physician as omnipotent villain and the patient as helpless victim.

This chapter will be devoted to presenting a theoretical framework, which will, hopefully, fit the bill; namely, Anthony Giddens' theory of structuration. Before delineating this theory and, in particular, how power is conceptualized within it, a few words are in order of a more general nature concerning theorizing about power.

PROBLEMS IN THEORIZING ABOUT POWER IN THE SOCIAL SCIENCES

Since the inception of sociology as a science, power has been rather standard fare for debate. Despite massive theorizing about the 'nature of the beast,' there has been surprisingly little agreement as to how it should be defined. Lukes (1974; 1979a; 1979b; 1982), himself a leading theorist on power, sums up some of the major differences in theories on power as follows:

- Power is something which is *possessed*; it can only be *exercised*; it is a matter of *authority*.
- Power belongs to the *individual*; it belongs only to *collectivities*; power doesn't belong to anyone, but is a *feature of social systems*.
- Power involves *conflict*; power doesn't necessarily involve conflict; power usually involves conflict, but doesn't have to.
- Power presupposes *resistance*; power is primarily involved in *compliance* (to norms); power is both.
- Power is tied to *repression* and domination; power is *productive* and enabling.
- Power is *bad*, *good*, *demonic* or *routine*.

Is it any wonder that the reader feels confused? As Lukes (1982:135) notes: when it comes to power, apparently 'anything goes.'

Fortunately, he does not leave it at that, but attempts to provide some order in the confusion surrounding discussions on power.<1> He begins with the question of why it has been so difficult within the social sciences to reach an agreement about what power is and how it should be defined. This is not simply a question of multiple-meanings; i.e. a semantic problem, but rather the result of the concept of power being 'essentially contested.'<2>

> By this I mean to suggest that any given interpretation or conception of it is inextricably tied to further background assumptions which are methodological and epistemological, but also moral and political assumptions, which themselves partly govern which reasons count in favour of one conception rather than another. Thus it is that there can be disputes about the proper use of the concept of power which are both 'endless' and 'perfectly genuine,' different uses serving 'different though of course not altogether unrelated functions' for different social groups, and contending positions 'sustained by perfectly respectable arguments and evidence' (Lukes 1979a:262).

This has several implications:

First, it means that a question, which might otherwise seem fairly straight-forward; namely, 'is this an exercise of power?' cannot be answered straightforwardly at all. Locating power empirically is not a neutral undertaking, solely dependent on what is 'out there' to find. What the investigator sees as a factual instance of power depends on the theory she starts out with. The identification of something as an instance of power, then, is *theory-relative*. It can only be seen with reference to a particular conception of power.

Second, the 'facts' are not the only things which are relative; so are theories. Theories are *value-dependent*, whereby the moral, ethical, and political background assumptions of the investigator determine her theoretical position on power. Whether or not a theory is considered adequate or satisfying as an explanation for how power is exercised depends on the investigator's view of society, what she sees as a desirable state of affairs in terms of the distribution of goods and opportunities, her conception of human agency and more.

Thus, a social scientist interested in how individuals with unequal rights or resources at their disposal resist subordination will probably not be satisfied with a conception of power, which emphasizes it as an 'essential and desirable component' of any highly organized society (Parsons). She may find herself flinching at a theoretical framework, which regards the individual as an 'effect' of power (Foucault) or as the mere 'representative' of class relations (Marx).

Third, if power as a concept is essentially contested, then the debate between contending views can never be resolved in any definitive way. There will always be good reasons for selecting one concept over another. In that case, the only option left to the investigator is to weigh the *methodological costs* involved in holding a particular view on power. This means examining various concepts for their suitability for coming to terms with a particular research problem and enabling the investigator to 'see further and deeper' into her material (Lukes 1979a:272). Ultimately, it will also entail assessing their acceptability with reference to her values and world-view. Selecting a theory of power is, in that respect, a political choice. In the final analysis, the dilemma of the 'relativity of power'(Lukes 1979a) is unresolvable. All the investigator can do is make a motivated choice for a particular theoretical framework.

In conclusion, Lukes' diagnosis of the debate around power confirmed the experiences I was having in the initial phases of my analysis and provided some reassurance to the effect that I was neither floundering alone nor was I doing so without good reason. If the conceptualization of power has stymied both philosophers as well as theorists in the social sciences ever since the beginning of this century, then it is hardly surprising that it has proved problematic in my own inquiry. Unfortunately, such reassurance does not help much when it comes to getting on with the analysis itself. I still needed a framework, however 'relative' and 'value-laden,' which would enable me to show how power was being exercised in my own data.

Ultimately, I discovered a theoretical framework for investigating power, which seemed to fit the purposes of the present inquiry: Giddens' theory of structuration. Like Lukes, Giddens also discusses the dilemmas inherent in the debate about power, both in classic social theory as well as in modern schools of thought. Although his treatment of the dilemmas, as well as his conclusions in terms of the methodological and political costs, resemble those of Lukes, he takes the discussion one step further. By attempting to resolve one of the difficulties, which has plagued social theory in general, and theories on power in particular, he is able to provide a way of conceptualizing power, which is not only sophisticated and comprehensive enough to help me analyze the phenomena I originally discovered in my data, but it is especially suited to investigating power at the level of interaction.<3>

In the remainder of this chapter, I shall be showing how the theory of structuration was developed, starting with the problematic agency-structure dualism. This will be followed by a consideration of how power is dealt with in the theory proper and how relations involving domination and subordination, particularly at the level of interaction, are conceptualized. Having set out this framework for delineating power, I shall evaluate its usefulness for the present inquiry. In other words, I shall be doing what Lukes has advised; namely, making the methodological, political, and ethical costs of one particular conception of power explicit. The final test, of course, will be the analysis of my data.

THE THEORY OF STRUCTURATION

Introductory remarks

Giddens' notions on power are embedded in his theory of structuration. Since power is part of a framework of interdependent and interlocking concepts, it cannot be tidily extracted from the rest of the theory. For this reason, it is necessary to begin by setting out the basic elements of the theory of structuration. Unfortunately, this is by no means an easy undertaking. The theory of structuration is part of a rather ambitious project. Beginning with a critique of the classics (Marx, Durkheim, Weber), as well as the more recent contributions to social theory (Goffman, Habermas, Althusser, and Foucault), Giddens attempts to uncover some of their inherent theoretical weaknesses while, at the same time, salvaging their respective contributions. This admittedly eclectic approach led to the development of his own theory of structuration, designed to serve as a kind of 'grand sociological theory.' The aim of his theory is nothing less than to provide a conceptual basis for a substantive analysis of all the 'concrete processes of social life' (1984:xvii)! Having developed an analytic framework, he then proceeds to tackle such subjects as the class structure of modern capitalism, the nation state, the industrialization of war and more. The theory of structuration has been the subject of a series of books produced so prolifically within the last decade that the reader barely has time to keep up.

In view of this, it should be clear that I have only been able to provide a very truncated description of the theory of structuration. I have confined myself to those features of the theory which seemed absolutely essential for the analysis of power at the level of interaction. This meant leaving out issues like the time/space dimension in social life, the nature of historical change as well as the reproduction of social institutions. All of these issues are central to the work of Giddens and, moreover, are by no means irrelevant to the study of power in medical settings. Due to the limited scope of the present inquiry, however, I have had to suffice with a cursory version of the theory of structuration as well as its central concepts. I have attempted to present it as clearly as possible, touching only very briefly on some of the problems involved in developing the theory and, for the rest, referring the reader back to Giddens' work.<4>

With these unavoidable limitations in mind, I shall now begin with a description of how the theory of structuration was developed, what the theory itself entails, and finally, how power has been conceptualized within it.

The Dilemma of Agency and Structure

The theory of structuration was founded on a constructive critique of interpretative (Giddens 1976; 1977), functionalist (Giddens 1976; 1979) and structuralist (Giddens 1979; 1987) schools of thought. The most important result of this critique was to pinpoint and clarify one specific dilemma - a dilemma, which has had pervasive and negative effects on the development of an adequate theoretical framework for coming to terms with the constitution of social life. This dilemma - the *dualism of agency and structure* - refers to the tendency within the social sciences to separate structural theories of social causality from subjectivist accounts, emphasizing the intentional activities of individuals.

Faced with the inability to reconcile agency and structure, most social scientists have opted for one pole, thereby closing their eyes to what the other had to offer. Thus, theorists have *either* focussed on the study of human conduct *or* on the study of social systems and institutions, *but not both*. However expedient this solution was,<5> it resulted in an inadequate theoretical account of social life and gave birth to a host of related, and equally problematic, dualisms as well (Giddens 1979:4-5; Giddens 1984:xx-xxi).<6>

The dualism of agency and structure manifests itself within interpretative, functionalist, and structuralist schools of thought in general, effecting, at the same time, the way they tended to deal with power at a theoretical level.

In the *first* tradition, which includes interpretative sociological schools of thought like ethnomethodology, symbolic interactionism, or phenomenology, social life is viewed as the product of competent actors. The underlying assumption is that human beings not only know a great deal about the world, but act purposively and reflectively. When asked, they can generally rationalize or account for their conduct.

Whereas studies within interpretative sociology have come up with some exceedingly sophisticated accounts of how humans behave and the means they employ to construct and sustain a meaningful social world, they have paid little attention to, or, perhaps found no way of coping with, structural features of society. Relations of domination, institutional control or social change are not favorite topics for investigation and power has, on the whole, received scarcely more than a passing reference. As a tradition, it is 'strong on action, weak on institutions' (Giddens 1985:31).

In the *second* tradition, that axiom can be turned around: it is 'strong on institutions, but weak on action'(Giddens 1985:33). Regarded in terms of the dualism of agency and structure, such otherwise divergent schools of thought as Marxism, functionalism, structuralism and post-structuralism as well as most of what we would call empirical sociology, focus their attention on the institutional or structural characteristics of social systems. Social systems are viewed as having an existence independent of the individual's activity. Structural features like social class or gender determine how individuals behave within a given context and place limits and constraints on that activity. This tradition has had a lot to say about institutions, social change, conflict, and, of course, relations of power. They have not, however, been able to come up with a theory of the subject, which can adequately account for human action and agency. People tend to appear as 'judgmental dopes' (Garfinkel), blindly driven by social forces beyond their control and comprehension.

Not surprisingly, the dualism between agency and structure does not just effect social theory, but it has taken its toll in the way power is conceptualized as well.

In the *first* tradition, the emphasis is on power as an intentional or willed phenomenon. Consider, for example, Weber's (1956) definition, which is, perhaps, the most celebrated of this type:

> Macht bedeutet jede Chance, innerhalb einer sozialen Beziehung den eigenen Willen auch gegen Widerstreben durchzusetzen, gleichviel worauf diese Chance beruht (Weber 1956:28). <7>

The conception of power as implicated in conscious, purposive action is shared by Hobbes and, more recently, by Russell and Dahl.

The difficulties with an intentional approach to power center around the fact that people regularly act under conditions, which are not of their own choosing and which they can only partially oversee. Moreover, they may intend certain outcomes, but there is no guarantee that these outcomes will, in fact, be obtained or, for that matter, that other, unintended ones will not be reached. Patterns of domination, extending over time and space, cannot be adequately taken into account by describing the conduct of individual agents. Power is not just a property of an individual, something to be exercised 'over' another. People often act in concert and, more importantly, with genuine consent.

The *second* tradition takes up such difficulties, resulting in quite a different treatment of power. Here, power has little to do with the individual or her ability to achieve certain outcomes. On the contrary, it is regarded as the property of the community or social collectivities. Power is 'generated' by the social system:

> a generalized capacity to serve the performance of binding obligations by units in a system of collective organization when the obligations are legitimized with reference to their bearing on collective goals (Parsons 1963:237).

Power, then, is the medium by which collective interests may be realized. This conception of power may be found in the work of Parsons, Durkheim, Arendt, but also in Marx, in which case, collective interests are class interests. These conceptions of power differ, among other things, in whether they regard power as asymmetrical, thereby involving potential conflict and struggle (Marx, Poulantzas), or whether they view power primarily as the capacity of a community for achievement, thereby implying consensus (Parsons, Arendt).

Another approach to power can be found in structuralist and post-structuralist thinking (Althusser, Foucault). Here power is neither a property of individuals nor collectivities, but rather an omnipresent and neutral force, implicated and implemented by means of discursive practices.

Foucault, for example, describes power as

> something which circulates, ... something which only functions in the form of a chain. It is never localised here or there, never in anybody's hands, never appropriated as a commodity or piece of wealth. Power is employed and exercised through a net-like organization. And not only do individuals circulate between its threads; they are always in the position of simultaneously undergoing and exercising this power. They are not only its inert or consenting target; they are always also the elements of its articulation. In other words, *individuals are the vehicles of power, not its points of application* (1980:98, italics mine).

Whereas structuralist approaches to power differ radically from both functionalist and Marxist treatments, they share a belief that power cannot be studied in terms of the practices of competent and, for the most part, knowledgeable social actors.

This second perspective on power is no less problematic than the first, although for different reasons. Power tends to become 'something out there,' a feature of institutions or a 'state of affairs.' Whereas is may affect the concrete, everyday activities of human beings, it is not investigated at that level. Whether it is considered a desirable and necessary element in the maintenance of the social order (Parsons), or as something which will miraculously disappear along with the division of class interests (Marx), or as something which is simply everywhere, 'co-extensive with the social body' (Foucault), it is not dealt with as part of the routine, daily life practices of social actors. This is, however, precisely where it needs to be located if we want to find out *how* relations of power are produced and, more importantly, undermined in various contexts of everyday life.

In conclusion, having delineated some of the difficulties resulting from a dualistic treatment of agency and structure in social theory, Giddens proposes a reconciliation.<8> The subject needs to be recovered as active and rational without lapsing into a subjectivist, voluntaristic position, which allows no room for social structure. This synthesis requires, first of all, a re-working of both the concept of agency as well as that of structure. Once this has been accomplished, they can be set in *relationship* to one another. In this relationship, the *duality of structure*, agency and structure presuppose one another rather than standing in opposition, thereby paving the way for a more comprehensive social theory and more adequate account of power.

I shall now describe how this synthesis is put forward in the theory of structuration.

Re-defining agency

To begin with, Giddens assumes that people *are* agents, active participants in the construction of social life and, at least to some extent, masters (*sic.*) of their fate. Inherent in the term agency is the possibility that a person at any given point in time 'could have acted otherwise;' i.e. she can always exert some influence, however minimal, over the course of events. This means that the world doesn't hold a pre-determined future for the actor (Giddens 1976:75). By taking this stance as starting point, a break is made with deterministic social theories where human activity is viewed primarily as the product of social forces beyond the control of the individual.

People are, however, not only agents; they are *knowledgeable* ones as well.

> All social actors know a great deal about the conditions and consequences of what they do in their day-to-day lives (Giddens 1984:281)

They are never 'cultural dopes' or mere 'supports' of social relations. In this vein, both Althusserian marxism and functionalism can be rejected for their shared blindness to the evidence of our everyday experience; namely, that we do, in fact, have considerable awareness of our social environments.

Although the structuralist and post-structuralist position is honored that the subject in social theory requires some 'decentering' in the sense that consciousness is neither a reliable nor an autonomous source of knowledge, but is made possible by structures not immediately available to it, Giddens is quite adamant in his rejection of the theory of the subject, which accompanies this claim (Giddens 1987:86-89). Subjectivity becomes a discursive phenomenon, a product of the internal play of signifiers within a text. The subject is reduced to a grammatical form, an 'I,' which only exists as sign constituted by virtue of its being different from other signs ('you', 'we', 'they', etc.). The author of the text, whether the object of social analysis or the investigator herself, becomes more or less irrelevant to our understanding of it. This 'evaporation of subjectivity into an empty universe of signs' (Giddens 1984:xxii) can never provide an adequate account of the situated practices of human beings.

On the contrary, what is needed is:

> an interpretation of the agent, rather than the subject, and of agency rather than subjectivity alone. 'Subjects' are first and foremost agents (Giddens 1987:98).

This entails, however, developing a more sophisticated account of agency than has hitherto been provided. Drawing upon ethnomethodological notions on agency and knowledgeability, the traditional distinction between activity which is not accessible to the actor (unconscious) and activity which is (conscious) is extended to include two separate senses of consciousness.

The first is *discursive*, referring to what actors are able to say about social conditions and particularly about their own actions. This is an ability possessed by all competent members, allowing them to stay in tune with why they do what they do while they are doing it. If asked by others, they can usually supply reasons or accounts for their actions (Giddens 1984:374;376). All members exhibit some degree of discursive 'penetration' of their environments, although this is not necessarily displayed in propositional form. Members often display understanding of authority relations, for example, by means of humor, sarcasm or irony (Giddens 1984:290).<9>

The second form of consciousness is *practical*. This refers to the taken-for-granted stocks of knowledge upon which we routinely and unproblematically draw in the course of our everyday life. It is what enables us to know 'how to go on' without being able to put that knowledge in discursive form. Most of what we know about the world falls under this category. It is what is known without being articulated as such. Such knowledge could, of course, be made explicit and incorporated into our discursive consciousness - a consideration which has important consequences not only for social research, but for the possibility of social change as well.

Practical consciousness belongs to a conception of action as an ongoing flow of interventions-in-the-world, initiated by the actor (Giddens 1976:75). It is not a matter of discrete 'acts,' guided by clear-cut intentions on the part of the actor. On the contrary, it is a continuous flux. Actions are purposive in the sense that actors routinely and tacitly monitor what they are doing, how others react to what they are doing, and the circumstances under which they are doing it. This *reflexive monitoring of action* is the crux of practical consciousness and forms the cornerstone of agency according to Giddens.<10> By providing a more complex account of the reflexive capacities of the actor, it becomes possible to include knowledgeability of a practical, on the spot variety.

Whereas this conception of agency as well as the practical, processual character of action owe an unmistakable debt to the ethnomethodology, Giddens does not accept the concomitant stance that social life is *nothing more than* the skilled performance of knowledgeable actors.<11> Society is not made under circumstances of actors' own choosing. Moreover, social activities take place under conditions lying outside actors' self understanding. Action is bounded by 'unacknowledged conditions' and 'unintended consequences'(Giddens 1984:9-14).

This, among other things,<12> is where social structure enters the scene. Much of what social actors do consists of highly repetitive activities, structured as institutional practices over time and space. Conventional forms of conduct, but also much of what occurs in an institutional setting, would fall under this heading. It is what makes, for example, a medical consultation recognizable as such, involving all the routine activities belonging to 'what one does at the doctor's.' Regularised or habitual activities will have regularised consequences as well. This is the case even when they are unintended by those engaging in them.

In conclusion, a conception of agency <13> is proposed, which makes it an integral feature at every level of social life. The notions of intentionality and action are extended to include ethnomethodological understandings of knowledgeability as a routine, practical, and processual affair, largely a matter of knowing what to do rather than being able to put it into so many words. An attempt is made to avoid the problem of assuming that members unrestrictedly embark on the construction of social life or that social life is explainable solely in terms of human agency. Although the production of social life cannot be analyzed without reference to human agency, it is not reducible to it.

Re-defining structure

Structure is a well-worn concept within the social sciences, generally invoked to account for those aspects of social life, which cannot be explained with reference to the intentional activity of individuals. It refers to those features of society, which stretch through many decades and over large or fixed domains. Dissatisfied with the way the concept has often been used in social theory and analysis, Giddens attempts to clarify some of the issues involved, while, at the same time, giving the definition of structure a novel twist.

To begin with, functionalist versions of structure are rejected, which treat it in a static, mechanical way, as an almost visible pattern of social relationships 'akin to the girders of a building or to the anatomy of a body' (Giddens 1982:33). This kind of imagery carries with it the implication that structure exists outside the individual actors, 'causing' them to behave in specific ways. Such a conception of structure entails, almost by definition, constraint. Structures 'determine' what people do. Whereas this conception can be employed descriptively to account for certain observed regularities in social practices, it falls short as an explanation for *how* structures come into being or are re-constituted and sustained in everyday life (Giddens 1976:59-65).

A more useful way of conceptualizing structure may be found in structuralist (Saussure, Lévi Strauss) and post-structuralist (Althusser, Foucault, Derrida) writings. Using language as an analogy, structure is compared with the absent corpus of semantic and syntactic rules, which is drawn upon in the production of discourse. Like language, it has no actual existence except as it is known and used by a community of speakers (Giddens 1976:118). Speech, on the other hand, does exist - as a situated activity, involving the actual presence of at least two speakers. When a speaker produces an utterance, she will draw upon the structural properties of language. These rules make it possible for her to construct or generate the utterance in the first place. By producing a syntactically correct and meaningful utterance, she is, at the same time, engaged in reproducing the language itself.

Analogously, interaction is constituted in and through the activities of individuals, just as social structure has a virtual existence. It is drawn upon by actors in the course of their interaction and, in that process, reproduced or transformed. Prior to structuralist thought, social scientists tended to treat structure in the Durkheimian sense of 'social facts;' i.e. external, objective properties of society, limiting or curtailing the individual's scope of action. By linking structure with constraint, the individual became pitted against society, her only vestiges of freedom residing in some nature-bound realm beyond (Giddens 1984:174). Such a conception proved inadequate, however, when it came to accounting for a highly complex and fluctuating reality. It could not come to terms with such messy social 'facts' as one person's constraint being another's opportunity or the refractory quality of repression opening up new avenues for action and even rebellion.<14>

In structuralist thinking, structures are not equated with constraint. Whereas they can place limitations on an individual's activities in a particular situation or be tied up with repression, structures are also what make that activity possible in the first place. Social structure is *enabling as well as constraining.* An adequate explanation of social phenomena will, therefore, entail taking both dimensions into account and showing how structural constraint and enablement 'intersect' in various social and material contexts (Giddens 1984:179).

Despite his unmistakable debt to structuralist thinking on the subject of social structure, Giddens' conception differs in two important ways. To begin with, his conception of structure is always and everywhere linked to social *practices*; i.e. the activities of agents. In order to link agency to the production and reproduction of social structure, structure is reformulated in terms of *rules* and *resources*. Dipping into the conceptual arsenal of ethnomethodology once again, structures are described as sets of 'rules and resources, recursively implicated in the reproduction of social systems'(Giddens 1984:23).

Rules refer to the 'dazzling variety of procedures and tactics used by agents in the constitution and reconstitution of encounters' (Giddens 1984:23). Members behave in accordance with countless rules, which, though rarely expressed in any tacit way, constitute their practical consciousness. They are often informal, involving the mobilization of normative rights and obligations. Even without being formal prescriptions for conduct, they can be extremely compelling. Rules are embedded in struggles concerning rival interpretations about what the situation is about as well as participants respective positions within it. For this reason, they cannot be conceptualized in isolation from resources.

The concept of resources has a double meaning, referring to the

> media whereby transformative capacity is exployed as power in the routine course of social interaction, but they are at the same time structural elements of social systems as systems <15> reconstituted through their utilization in social interaction (Giddens 1979:92).

Taken together, rules and resources are the constitutive elements of social structure.

The second crucial way that Giddens' concept of structure differs from structuralist renditions of structure is that it is not limited to discursive phenomenon. Although language is employed as a useful analogy for how social structures work, social life is, in no sense, equated with language or 'discursive practices' (Giddens 1976:118).<16> On the contrary, social structure needs to be analyzed in terms of three dimensions or modalities. These modalities are always implicated in social life and mediate both interaction and social structure.

Represented graphically (Giddens 1979:82):

INTERACTION	communication	power	sanction
(MODALITY)	interpretative scheme	facility	norm
STRUCTURE	signification	domination	legitimation

In order to communicate, actors draw upon interpretative schemes in the course of their interaction, which are organized at the level of structure in symbolic systems of signification. When they apply sanctions, they make use of norms, which can be analyzed at the level of structure as moral rules or systems of legitimation. Power in interaction involves employing facilities to achieve certain outcomes. These facilities are the rules and resources comprising structures of domination. The distinction between these modalities is obviously more analytical than substantive. In actual social practices, they overlap and interweave.<17> All levels are implicated and belong to an analysis of social life.

In conclusion, Giddens re-works the concept of structure in such a way that historically stable, regularized institutional patterns can be described, without resorting to mechanical causal explanation. Our attention is not specifically focussed on *what* features belong to social structures, but rather on the connection between them and especially *how* they come to be constituted through concrete social practices. Structures cannot be analyzed separately from the activities of social actors, where they are being instantiated in diverse, flexible and, most importantly, routine ways. Whereas social structure is linked to constraint and domination, it is not synonymous with it. Structure is also implicated in the 'degrees of freedom' of human action. It can even enable activities, which challenge the social order or disrupt existing hierarchies.

Linking structure and agency

Having re-thought both the concepts of agency and structure based on a constructive critique of functionalist, interpretative and structuralist traditions in the social sciences, the *theory of structuration* is an attempt at synthesis. Instead of the dualism between subject (human agent) and object (society, social institutions) which has proved so problematic for social analysis, a *duality of structure* is proposed where agency and structure are not regarded as an opposition, but rather analytically presuppose one another.<18> The duality of structure refers to the

> essentially recursive character of social life: the structural properties of social systems are both the medium and outcome of the practices that constitute those systems (Giddens 1982:36-37).

All human action is structured: it does not 'just happen.' Actors need to make use of what is already there to know what to say, what to do, and even how to begin in the diverse situations of their everyday lives. At the same time, however, it is only by virtue of this activity that social structures come into being in the first place. They are always and everywhere instantiated and sustained in the social practices of actors. It is this dialectic connection between production and reproduction, which accounts for the 'recursive' character of social life.

Structuration refers to *how* social life is produced and reproduced over long periods of time and over large domains of space. It involves processes of social reproduction, the 'conditions governing the continuity or transformation of structures, and therefore the reproduction of systems' (Giddens 1979:66). Any social system can exist in stable, more or less regular forms by virtue of the fact that we tend to do today what we did yesterday. People living in a particular locality or involved in a specific type of encounter rely on conventions for conduct, which have been followed by others before them. Structuration, then, is what endows social interaction with its systematic character over time and space. It puts the actor back in historical practices of change and continuation. It is precisely the task of social analysis to account for *how* systems are produced, maintained, and transformed (Giddens 1985:39).

In summary, a cursory look at the most central concepts in Giddens' theory of structuration has been provided. I began by showing how he tries to come to terms with one of the most enduring problems in the development of social theory; namely, the relation between action and structure. After clarifying some of the issues involved in conceptualizing both, he proceeds to set them into relationship with one another, delineating how the activities of social actors are related to the institutional and structural features of the societies of which they are a part. In a series of interlocking concepts, a basis is provided for the analysis of how actions are 'structured,' as well as how structures are ongoingly reproduced in the practices of knowledgeable social actors.

This synthesis is an analytical one and there has been some debate as to whether it has been entirely successful. The criticism has been raised that the theory of structuration does not reconcile agency and structure in any definitive way. This is attributed to various features of the theory: the primacy given to the subject over structure (Smart 1982; Callinicos 1985), the inadequate conceptualization of structural constraint (Urry 1982; Thompson 1984), and the lack of empirical grounding (Hirst 1982; McLennan 1984). I shall be returning to some of these issues at the end of this chapter. Despite these criticisms, however, the theory of structuration remains one of the most sophisticated, comprehensive, and consistent attempts to tackle the relationship between agency and structure. As such, it can provide a theoretical basis for reformulating a series of important sociological concerns, including the analysis of power. It is to this issue that I shall now direct my attention.

GIDDENS ON POWER

According to Giddens (1984:283), 'there is no more elemental concept than power.' It is not something to be 'tacked on' to other more basic concepts as, for example, in functionalism, where power is secondary to 'norms' and 'values,' or in marxist theory where 'class interest' is central and power an appendage that will disappear together with class divisions. On the contrary, social theories that disregard the centrality of power in social life will inevitably prove inadequate. Taking the position that power is central to the study of social life does not mean, however, that it automatically becomes more central than other concepts. He rejects the tendency of Foucault and others who, under the influence of a 'Nietzchean radicalization of power,' give power primacy over everything else. Power appears as some 'mysterious phenomenon, that hovers everywhere and underlies everything' (Giddens 1984:226). The result is a reductionism every bit as faulty as the previous normative and economic reductions had been.

In his theoretical framework, he tries to overcome what he regards as a 'disembodied dialogue' (Giddens 1976:113). Power is treated as *one* of several primary concepts, each essential for the analysis of social life. They are all clustered around relations involving action and structure and cannot, therefore, be explicated without reference to the duality of structure.

I shall now address the way power is conceptualized in the theory of structuration. In dealing with the central notions on power, I shall, however, be engaging in a bit of 'methodological bracketing.'<19> In view of the requirements of the present inquiry, I am particularly interested in *power at the level of interaction*; i.e. how actors draw upon rules and resources in order to influence the course of events in a conversation. The following features will be dealt with:

1. *Power as integral to social interaction.*
2. *Power as intrinsic to human agency.*
3. *Power as relational concept, involving relations of autonomy and dependence.*
4. *Power as enabling as well as constraining.*
5. *Power as process.*

After discussing Giddens' conception of power, I shall evaluate the implications of using such a framework for the present inquiry.

Power as integral to social interaction

Power is an integral feature of social life. Whether we are talking about institutional contexts, structural inequalities based on class, race, or gender, or an intimate relationship, power will be involved. It is never something 'out there,' but part and parcel of any encounter, implicated even at the micro-level of interaction. Although power cannot be eliminated from a study of interaction, this does not mean that it is sufficient for explaining how social interaction is produced. The production of any instance of social interaction will always involve three basic elements: its constitution as *meaningful*; its constitution as a moral or *normative* order; and its constitution as the operation of relations of *power* (Giddens 1976:104-113).<20> These elements are 'subtly, but tightly' interwoven, making it impossible to deal with one without taking the others into account.

Power involves the skills and resources, which members bring to and mobilize in the production of interaction, thereby directing or influencing its course. It is implicated in the production of meaning in the sense that actors have capabilities to make certain 'accounts count.' It is also involved in ensuring compliance to moral claims. Members exercise power by enacting or resisting sanctions (Giddens 1976:83).

Whereas the production of meaningful and appropriate interaction is a collaborative endeavor, the contribution of participants is rarely an equal one. Not only do they have access to different resources and skills, but these resources are asymmetrically distributed. Thus, members are seldom really peers in the interaction game.

> *The reflexive elaboration of frames of meaning is characteristically imbalanced in relation to the possession of power*, whether this be a result of the superior linguistic or dialectical skills of one person in conversation with another; the possession of relevant types of 'technical knowledge;' the mobilization of authority or 'force,' etc. 'What passes for social reality' stands in immediate relation at the distribution of power; not only on the most mundane levels of everyday interaction, but also on the level of global cultures and ideologies, whose influence indeed may be felt in every corner of everyday social life itself (Giddens 1976:113)

In conclusion, if power is implicated at *all* levels of social life, then it will also be part and parcel of even the smallest and most insignificant face-to-face encounter. This means that an analysis of power relations can quite legitimately and with an hitherto unfamiliar equanimity take social interaction as its starting point. Analyzing power involves showing how it is involved in both the production of meaning, as well as the constitution and maintenance of normative orders concerning how the situation and the respective social positions of the participants are to be defined.

Power as intrinsic to agency

The connection between power and agency is, first and foremost, a logical one. In its most general sense, power is 'the capability of the actor to intervene in a series of events so as to alter their course' (Giddens 1976:11). This basic 'can' or 'is able to' forms the cornerstone of any debate on agency and the nature of action. Action entails, by definition, a transformative capacity. Any agent possesses a range of causal powers, which she deploys in the course of her everyday activities. This does not mean that interactants share equal resources, but rather that, even in asymmetrical interaction, the weaker party can exercise some power. She is, at any given moment, always able to 'make a difference' or do otherwise than she perhaps actually did. When this is no longer the case - as in death or extreme forms of coercion - she ceases to be an agent and we can no longer speak of power.<21> Taking this view has two immediate consequences:

First, by positing a logical connection between agency and power, the possibility is rejected that actors can ever be completely determined by their circumstances. Even when their activities are severely restricted or where the power imbalance is extreme, there is always a sense of *could have done otherwise*. Sometimes this means doing nothing at all or 'acting powerless.'<22> It also means that actors can display outward compliance with even the most oppressive contingencies of their situations without the conclusion being warranted that they have been 'driven irresistibly and uncomprehendingly by mechanical pressures' (Giddens 1984:15). Frequently, compliance is the result of a decidedly rational assessment of the situation and viable alternatives. In other words, compliance does not automatically entail agreement.<23>

Second, power cannot be considered in isolation from agency. It is the means by which all social life is 'made to happen,' the product of the activities of knowledgeable social actors. Thus, power becomes *logically prior* to various other issues inhabiting current sociological debates. For instance, if it is a regular and routine feature of action, it cannot be linked to a particular type of activity. Any activity will involve the exercise of power. It becomes unimportant whether the action was intended or desired by the actor (Weber) or whether power is exercised in order to accomplish sectional; i.e. class, race or gender-based, interests (Lukes). Conflict or struggle, whether overt (Dahl) or covert (Bachrach and Baratz), does not have to be involved. Whereas power is contingently related to all of these features in much of what we know of as social life, there is no logical or necessary connection between them (Giddens 1976:71-80).

In conclusion, by introducing agency so strongly to the study of power, an attempt is made to redress a perceived imbalance in social theory. Without denying the structured nature of power relations, the claim is put forth that power can not be analyzed without taking agency into account. In its most general sense, we are reminded that power *is* agency. It is the 'can,' mediating between the desired or intended outcomes of social actors and the actual realization of these outcomes in their daily social practices. It is also the 'could have done otherwise,' which is implicated in every situation, even the most restrictive and oppressive. By linking power to agency, it becomes essential to uncover the subtle mix of what actors do (and refrain from doing), what they achieve (and fail to achieve) and what they might have done (but didn't).<24>

Power as relational concept, involving relations of autonomy and dependence

Power is more than a general transformative capacity; i.e. the actor's capacity to 'get things done.' In its more restricted sense, power is a relational concept. Agents can exercise 'power over' other agents by influencing the circumstances under which their action takes place. 'Power may be harnessed to actors' attempts to get others to comply with their wants' (Giddens 1976:93). In order to do this, resources and skills are mobilized by the actors involved in the course of their interaction.

Members do not have equal access to resources for affecting the outcome of their interaction. Resources are asymmetrically distributed in accordance with structures of domination. Power, then, intervenes conceptually between transformative capacities for intervening in the events-of-the-world and domination as the asymmetrically structured access to certain skills and resources. <25>

Although power relations involve a certain asymmetry between participants, they are always reciprocal. This means that no matter how limited a person's resources or how subordinate her position *vis-à-vis* another actor or actors, she is never completely without resources for influencing the activities of her superiors. In the same vein, even the most powerful do not have total control over the activities of others. The exercise of power always involves a two-way relationship, expressing autonomy and dependence in both directions (Giddens 1984:39).

This conception of power breaks with simplified notions of power as a matter of 'have's' and 'have not's' (the so-called zero-sum model), which have led to an overestimation of the power of the powerful. Our eyes become closed to the 'chinks in their armour.' Equally regrettable is the tendency to miss the myriad ways the less powerful have for exercising some control over their lives, even in situations where stable, institutionalized power relations are in operation.

By re-formulating power as a relation of autonomy and dependence, our attention is focussed on power as a struggle, involving a *dialectic of control* (Giddens 1984:16). The emphasis shifts from the question of *who* 'has' the power to *how* each party attempts to exercise control over the other's activities and the kinds of skills and resources they employ in the process.

In conclusion, a conception of power is proposed, which locates it in a *relationship* between actors drawing upon rules and resources to accomplish their respective ends. Whereas they do have unequal access to skills and resources and, consequently, unequal opportunities for influencing the course of the interaction, their relationship is always a reciprocal one. As in any relation, each party will be to some degree dependent on the other and will display some autonomy of action. Examining power involves uncovering how power relationships are dialectically constructed.

Power as enabling as well as constraining

In many of the contexts of everyday life, power does go hand-in-hand with structured forms of domination. Members do not simply intervene in the course of events, but they also try to exercise control over one another. This is accomplished by means of sanctions, which are structurally available. By means of sanctioning processes, some actors can restrict the activities of other actors or get them to do things that they might not have done under other circumstances. Whereas this sanctioning can be accomplished by means of threat, force or violence, that is the exception. More common or mundane sanctions entail disapproval, criticism or simply the absence of any response at all (Giddens 1984:175).

Sanctions are not only restrictive, however. They can be enabling as well, inducing actors to engage in specific activities. In fact, sanctions are the very means by which social interaction comes to be constituted as orderly and 'normal'; i.e. the way it should be.<26>

In conclusion, Giddens borrows Foucaultian notions of power as enabling as well as constraining, thereby rejecting the tendency within the social sciences to treat power as something inherently negative. <27> Despite their 'inherent social association,' power cannot be equated with coercive forms of domination. It is not an 'inherently noxious phenomenon,' the 'capacity to say no' (Giddens 1984:32). It is also implicated in the constitution of social life, making it productive, enabling, and even positive. Any study of power will involve taking *both* dimensions into account and showing how power is connected to constraint and enablement in any specific instance of social interaction.

Power as process

Power, like every other central feature of social life, will reflect the duality of structure. Conceptually, it can be viewed as intervening between agency on the one side (as the actor's ever-present capacity to influence the world around her) and structure on the other side (domination as a structural feature of society). This relationship is not a static one, but ongoingly reproduced by means of structuration processes. Power 'only operates as such through the utilization of transformative capacity as generated by structures of domination' (Giddens 1979:92). It is exercised as a process, part and parcel of the ongoing flux of situated practices of social actors. Structured relations of power involving domination and subordination are produced, reproduced, and transformed through these practices.

Such a conception of power precludes studying power in terms of outcomes or end products. Whereas it is undoubtedly easier to establish empirically whether or not a particular outcome has been accomplished, this does not tell us very much about power or how it works within a specific context. Power as process becomes subject to structuration. Just as members routinely and ongoingly construct the 'normal appearances' of their everyday contexts, they will, simultaneously and in similar ways, be constructing power relations. Power belongs to these 'normal appearances,' as part of the ongoing flux of everyday life. It will be exercised as members reflexively monitor their interaction and employ skills and resources in flexible and on-the-spot ways to gain control over their situation.

In conclusion, Giddens conception of power focusses on the 'how's' of power; how members ongoingly and routinely construct, maintain, but also change and transform their power relations. In this way, structures involving domination are reproduced as power practices in everyday contexts. The question becomes not so much whether or not domination is part of a specific situation, but rather how it is involved.

In summary, an attempt has been made to set out some of the basic features involved in Giddens' conception of power. It is a conception, which focusses on power as a process. Relations of power involving domination and subordination are constructed ongoingly in the course of interaction by means of the same sorts of reflexive procedures employed by actors to sustain and maintain any situation. Power becomes integral to any social encounter, even the most casual or intimate one. It is exercised at the micro-level of interaction as well as being implicated in institutions or social systems. The exercise of power is not a matter of straightforward, top-down, or repressive forms of control. Far from being openly authoritarian or coercive, it can be enabling or productive. Power has a Janus-face. The exercise of power is simultaneously routine, processual and multi-dimensional.

A PRELIMINARY EVALUATION OF GIDDENS' THEORETICAL FRAMEWORK FOR THE PRESENT INQUIRY

Having set out some of the central concepts in the theory of structuration as well as how power is conceptualized within it, the next step is to evaluate its usefulness as theoretical framework for the present inquiry. I shall now be directing my attention to three issues. First, I will consider some of the *general advantages* of using a type of theory like Giddens' theory of structuration in an inquiry like my own, with its alleged aim of providing an empirically grounded feminist critique of the health care system. Second, I will be taking a *critical* look at his theory, indicating where it falls short as theoretical framework for coming to terms with power and gender in medical encounters. Third, I shall begin to *link* the *theoretical notions* about power, which have been set out in this chapter with the results of my preliminary *analysis*. In other words, having engaged in this brief 'time-out' from my empirical analysis, I shall return to my data, constructing my theory toward it as I go along.

Some general theoretical, methodological, and political considerations

Giddens' conception of power is inseparable from his theory of structuration and, for this reason, it makes sense to evaluate the usefulness of this theory as a whole for the present research. The following considerations will be of a general nature, concerning the type of theory it is as well as its applicability for an investigation of power and gender at the level of face-to-face interaction. I shall also deal with its applicability for research being done from feminist perspective.

Advantages of conceptualizing power within a general social theory

As Lukes (1979b:634) has noted, conceptions of power are never simple, but are intricately connected with a whole series of interlocking notions and values concerning society, as well as opposing points of view about what is necessary and possible in the course of human life. The kinds of decisions that the investigator makes about what constitutes power or what is relevant in analyzing its workings in any given situation depends on the general social theory to which she subscribes.

Of course, not every theorist on power, let alone those engaged in doing empirical research on the subject, can be bothered to develop a full-fledged social theory. Giddens, however, has attempted exactly that. His project, in fact, entailed developing a general social theory as framework for studying and illuminating the 'concrete processes of social life' (Giddens 1984:xvii). In other words, it is 'grand theory' at its best.

Whereas there are certainly objections to be made with respect to the adequacy of any grand theory for explaining social life (Mills, 1959) - a matter to which I shall be returning shortly - there are some definite advantages as well, particularly where a concept like power is concerned. Within the structuration theory, power as one of several central concepts becomes embedded in an elaborate framework, where it is linked to the other central concepts and issues of the theory as a whole. For this reason, it becomes possible to examine the ramifications of the conception of power in a much more comprehensive way than would have been possible had Giddens been 'just' a theorist on power. In other words, his theory of structuration provides not only a satisfyingly complex way of conceptualizing power, but it enables me to identify some of the 'methodological and political costs' (Lukes) involved in employing such a conception. Because they are explicit, I can make sure that they square with the requirements of my particular research problem and method, as well as with my explicit intention to provide a feminist critique of a specific medical context.

Advantages of conceptualizing power within a heuristic theory

Not only is Giddens explicit about what his theory of structuration is; he is explicit about what it is *not* as well. It is not his intention to provide a 'set of deductively related laws,' to be 'tested' in the course of an empirical investigation. Nor does he wish to propose a 'theory as explanatory generalization' (Giddens 1984:xviii-xix). The theory of structuration is a heuristic theory; a theory which will provide a set of 'conceptual schemes,' enabling the investigator to interpret what she sees happening in her data.

As previously noted, there has been relatively little research done on how power works at the micro-level of face-to-face interaction. In those cases where it did become a topic, it was either operatio-nalized in accordance with functionalist conceptions of asymmetrical 'roles' between physicians and patients (West) or with an exclusive focus on the dominance of male physicians, thereby obscuring the power of the patient (Fisher and Todd). Despite their merits, these approaches proved inadequate for understanding how power worked in the consultations like those of my own data, where the interaction, despite the asymmetrical power relation between the participants, distinguished itself primarily by virtue of its elusive and friendly quality. Even more important, however, is the fact that such models preclude discovering power struggles between the participants- assuming that such things do exist - or of finding ways that women might be resisting or otherwise undermining the oppressive features of medical encounters with male doctors. These questions were, however, precisely the issues which interested me most at the outset of my inquiry.

Thus, both my preliminary analysis as well as my dissatisfaction with existing research indicated that I would, in any case, need a different framework for conceptualizing power. It would have to be a framework, which, above all, did not restrict my view to 'what we already know about power' in medical encounters between male doctors and female patients. Moreover, it would need to offer conceptual support for understanding some of the confusing features of the consultations of my data and help me look at my data in a fresh and more comprehensive way. In short, a theory was in order, which could 'clear up queries' and attempt a 'contextual' (rather than generalizing) explanation (Giddens 1984:xviii) for some puzzling instances of how gender and power relations are constituted in the specific context of a general practice consultation. Seen in this light, a heuristic theory like the theory of structuration has obvious attractions.

Advantages of conceptualizing power within a critical theory

Power as an issue is 'eradicably evaluative' (Lukes). It is never free of considerations of a more political or ethical nature. The present inquiry is obviously not aimed at delineating how power works in medical encounters for strictly academic reasons. On the contrary, it is meant to supply a relevant and, hopefully, useful contribution to current feminist scholarship on asymmetrical gender relations, particularly as they are produced and reproduced in the health care system. I therefore need a theory, which does not prevent the possibility of such a critique.

The theory of structuration is not only an attempt to alleviate some of the conceptual difficulties in contemporary social theory in order to establish a more comprehensive framework for empirical research, but it is intended as a critical theory. Since social research involves social critique and the 'structuration theory is intrinsically incomplete if not linked to a conception of social science as critical theory' (Giddens 1984:287), some guidelines are suggested for the form such critique might profitably take.

To begin with, empirical research must always take the situated practices of actors as its analytical starting point. This involves delineating the various activities in which they are engaged in order to construct and sustain institutional contexts. Social analysis will entail sorting out the often-very-fine line between how individuals' activities are systematically and specifically bounded by things for which they, if asked, could provide an account and by things of which they are not aware. Since most activity is of a practical, taken-for-granted variety, putting it into discursive form - which is, in effect, what any research report does - will have the immediate effect of 'de-routinizing' it. It becomes accessible to both the actors themselves, as well as to other social scientists, in a new form, thereby opening up possibilities for reflection and, at least potentially, alternative courses of action. Moreover, by describing the ways actors' conduct is bounded and specifying how this 'bounded-ness' is related to the institutional orders of which the conduct is a part, the investigator introduces a 'moment of critique.' She can show how actors' expression of common sense beliefs about 'the way things are' may be invalid or inadequately grounded (Giddens 1984:327-343).<28>

As recipe for social critique, this proposal is a decidedly undramatic and praxiological one. It essentially involves taking a close look at what people are already doing in the everyday contexts of their lives, making this 'discursively available' and specifying where and in what ways these activities are structurally bounded.

The question now becomes: does this approach to social critique have anything to offer an investigation like the present one, which is aimed at elaborating the feminist critique of how power is being exercised in consultations between male GPs and female patients? It is my contention that it does, and for two reasons:

1. The focus of feminist scholarship, including most of the investigations of how women fare in the health care system, has traditionally been the analysis of global or general forms of domination and subordination in gender relations. Relatively little attention has been paid to what this means in actual encounters between men and women. For example, we know very little about *how* women's problems are 'medicalized' (Riessman, 1983) or the 'control over their bodies usurped' (Ehrenreich and English, 1979). In a similar vein, we are virtually in the dark about the kinds of resistance or de-routinization *already* taking place in medical

encounters between women and male physicians. There may well be a gap between the strategic critique put out by the feminist health movement and what 'ordinary' women patients are doing right now 'on their own'.<29>

The present inquiry is an attempt to take the feminist critique of the health care system as a starting point and fill in the gaps by investigating actual instances of interaction between male GPs and female patients. It is precisely this 'unchartered territory' where the theory of structuration in its function as social critique can be employed. It enables the investigation of the most routine and everyday forms of domination, the minuscule power struggles of daily encounters, and the microscopic attempts of social actors to resist or otherwise undermine relations of authority in institutional contexts. As such, it can provide a useful starting point for an inquiry like the present, with its express aim of developing a 'grounded' theory of gender relations in medical encounters.

2. There has been a tendency in feminist theory and research to treat women as the victims of male supremacy with men inevitably cast in the role of the ill-meaning oppressor. Gender inequality appears as a kind of 'sexist conspiracy.' In recent years, this perspective has born the brunt of considerable criticism, the claim being that it is out-dated, over-simplified, and methodologically untenable (Chapter 3).

   Nevertheless, it remains a thorny problem to show how women are actively involved in giving shape to their lives without having to resort to a 'blaming the victim' stance; i.e. if they are oppressed, then it's their own fault. As we have seen, one approach to the problem, particularly within the feminist literature on medical interaction, was to concentrate primarily on what the doctor is doing. In an attempt to uncover relations involving the subordination of women in medical settings, the focus became the dominant party, the male physician, at the expense of the female patient as agent.

   Another 'solution' to the problem has been to abandon the notion of agency altogether. Following in the footsteps of their (post-)structuralist forebears, some feminist scholars maintain that neither women nor men are 'masters of their fates.' Or, more to the point, agency like power, femininity, and gender, is simply a 'discursive phenomenon,' to be investigated as construction.<30>

   I do not find either approach a satisfactory alternative for the problematic exaggeration of the intentionality of men and the helplessness or victimization of women. My first reason is *substantive*. Such a perspective closes off areas of research - at this point, most notably, my own - which are valuable for understanding the ways women (and men) produce or cope with the inevitable contradictions of their everyday lives (Grunell, 1986). It is difficult to imagine subscribing to a view that people are not (really) agents and then turn around and investigate their activities, which is, after all, what a study of interaction in the final analysis is all about. Whereas this is not the only level of social life worthy of our attention, it does seem to be one which should not be left out of critical feminist research.

My second reason is one of *preference*. I would personally rather treat the people I am investigating as basically competent, as knowing - at least to some extent - what is going on in the situations in which they find themselves. Their concrete activities, as well as how they account for the situations in which they find themselves, are of particular interest to me, even when - or, perhaps especially when - their conduct seems to stand in opposition with what their 'objective interests' might be.<31> Perhaps, ultimately, this is one of those 'essentially contested' (Lukes) aspects of any social theory, the bottom line which makes it satisfying (or not) for the investigator.

In conclusion, the theory of structuration can be of use as a general and heuristic theoretical framework for investigating how power works at the level of interaction. Moreover, it can provide a starting point from which to construct a critique of how asymmetrical gender relations are being produced, sustained, but also ongoingly altered in medical encounters. The theory of structuration, however, is nothing more than a starting point. As soon as an attempt is made to apply it to one of the 'concrete processes of social life,' (Giddens 1984 xvii) it becomes clear that there are quite a few missing pieces. It is to this issue that I shall now turn.

Some critical remarks concerning the theory of structuration

Giddens' theory of structuration, including his conception of power, are highly general and abstract. Although he professes to have developed a framework suitable for analyzing social life at various levels and in a variety of contexts, the framework itself is curiously devoid of any connection to concrete social contexts, whether institutional or mundane. Similarly, his emphasis on the centrality of situated social practices for any social analysis remains strictly theoretical. At no point does he provide more than the most gratuitous of references to actual, flesh-and-blood individuals or groups of social actors. It might be argued, of course, that a conceptually coherent theory is all that is needed for the investigator, who must then prove herself enterprising enough to apply it to the study of some particular area of social life. I do not believe, however, that we can let Giddens off the hook so easily. In particular, when it comes to investigating power in gender relations, two separate, but related, problems emerge. The first concerns the inevitable contextuality of power; the second, that gender is not accounted for in his theory. I shall deal with each issue in turn.

The contextuality of power

Giddens' conception of power is empty; i.e. it has no content. It suggests that actors - any actors - are apparently exercising power as a matter of course, routinely, anywhere and everywhere. No mention is made of *which* actors we are talking about, although any actor worth her salt will also have a gender, a race or ethnic background, and will belong to a particular social class. This information is not ancillary, but relevant to any study of power.

Similarly, we are informed that rules and resources are drawn upon by these actors to influence the course of events or achieve certain outcomes. Social life, of course, is full of rules and there are all sorts of resources that can be employed by actors. Clearly, however, some rules will be more essential than others for understanding, say, how asymmetrical gender relations are being constructed. Within Giddens' framework, we have no way of assessing *which* rules and resources are important for understanding how power works in a particular institutional setting or how they are implicated in specific structures of domination. In other words, there is no 'criterion of relevance' (Thompson 1984:158) in the theory of structuration for deciding what to include in our analysis of power.

And, finally, we are told that resources are structured, and, more to the point, asymmetrically so. This means that some groups are systematically and differentially affected by their limited access to these resources. The knowledge that this *is* so does not, unfortunately, provide us with insight into *how* it is so. We do not know which structures affect specific groups in which ways.

> What is at issue is the fact that the restrictions on opportunities operate *differentially*, affecting unevenly various groups of individuals whose categorization depends on certain

> assumptions about social structure; and it is this differential operation or effect which cannot be grasped by the analysis of rules alone (Thompson 1984:159).

Power relations are always and everywhere contextual. They are embedded in *specific* contexts, involving *specific* actors or groups of actors drawing upon *specific* rules and resources which are organized in *specifically* structured ways. Power, along with structures of domination, is implicated in concrete situated social practices.

Although Giddens would be the last to deny the above, his theory of structuration does not provide anything beyond a series of generalized features describing what power is all about. More importantly, it makes no mention of how power is to be located and analyzed within these social practices. This, however, is precisely what needs to be done. There simply can be no analysis of power, which does not take the contextual particulars into account.

Because it is devoid of contextual grounding, Giddens' conception of power can never be anything more (but also nothing less) than a starting point for the analysis of power. In order to show how power works, particularly in the production and reproduction of structures of domination, *empirical grounding* is required; i.e. the theory must be embedded in a specific context and set of social practices.

Conceptualizing gender

With the exception of the occasional gratuitous reference,<32> Giddens is notably silent on the subject of gender. The theory of structuration does not deal with gender as theoretical construct, nor is there any reference to gender relations or even power relations between the sexes. Whereas the same criticism could (and should) be made with respect to any number of specific social groups, the fundamental problem is that we have no way from within his theory to derive an analysis of gender or, more to the point, how gender and power are related.

Gender, like race and class, has a troublesome tendency to collide and become hopelessly conflated with power at all levels of social life. It pops up in the course of a conversation between the sexes, but it is also implicated in the reproduction of any number of social institutions. A case could also be made for considering gender as a kind of general structural property, which constitutes, but also limits the opportunities and resources available to the sexes (Wolffensperger forthcoming). Within feminist scholarship, gender has been analyzed at all levels of social life - in terms of *individual* men and women (their behavior, attitudes, gender identity), in terms of *social structure* (gendered divisions of labor or social activity), and in terms of *symbolic orders* (how we think masculinity and femininity). As Harding notes, it is the analysis

> through which the division of social experience along gender lines tends to give men and women different conceptions of themselves, their activities and beliefs, and the world around them (1986:31).

We might expect that any theory, which claims to be a framework for investigating 'all the concrete processes of social life,' would provide some clues for conceptualizing something as pervasive to social life as gender. This is, however, not the case in the theory of structuration. It lacks any account of gender as structural feature of contemporary society.

No theoretical basis is provided for establishing structural identities or differences of any kind, whether they are based on class, ethnicity or gender. This means that we have no way of linking together the experiences of various women (or men) under the common banner of gender. By the same token, we cannot determine what differentiates asymmetrical gender relations from relations between other dominant and subordinate groups. There simply are no criteria for deciding which structures of domination are specific for gender relations and, more to the point, which ones are even relevant for understanding how power works in encounters between the sexes.

If we assume that gender is part and parcel of social structure, then we don't know where to locate it in Giddens' framework. Does gender entail, for example, a series of resources, organized in a specific structure of domination ('gendered power')? Or is it best conceptualized as a structural property, a kind of organizing principle for relations between the sexes (Meyer 1983; Wolffensperger forthcoming)? Perhaps this is much too limited and we should be thinking more in terms of gender as a system in and of itself, the very foundation of how any culture is organized (Rubin 1975)? Giddens, by focussing solely on the 'how's' of power and the processes of system reproduction, leaves questions like these unanswered. The various levels of structural analysis have an irritating tendency to fall apart, leaving us empty-handed when it comes to explaining structural arrangements and constraints based on gender.

It is not my contention that gender *is* the most useful theoretical starting point for analyzing asymmetrical power relations between the sexes. In fact, I shall be arguing in the course of the present inquiry that it is *not*. I do think, however, that any social theory, which pretends to provide the basis for analyzing power, cannot afford to ignore the problem of how gender and power are related. This relationship or, at least, how it can best be conceptualized needs to be addressed in *any* theoretical account of power.

This problem can be alleviated, in part, by anchoring the generalized properties of power, as set out in the theory of structuration, in the situated social practices of actors in concrete everyday contexts. This is, however, only part of the solution. The relationship between power and gender is not simply an empirical matter; it is a theoretical one as well. Thus, Giddens' conception of power is not only in need of empirical grounding, but it could use some *theoretical* grounding as well.

These reservations having been voiced, the task of my own inquiry becomes clear. I shall now direct my attention to the necessary activity of writing the theoretical considerations on the workings of power at the level of face-to-face interaction 'towards' my data. In this way, a start will made in the process of constructing a theory of power and gender in medical encounters, which is grounded, both empirically and theoretically.

Returning to the preliminary analysis of the data

An attempt will now be made to consider the results of the preliminary analysis as described in Chapter 3 in light of the conception of power set out in this chapter. As the reader will recall, the exercise of power in general practice consultations between male physicians and female patients had proved to be anything but a straightforward affair. In fact, it was subtle, elusive, and, at times, even confusingly benevolent. My brief foray into current studies on medical interaction was not particularly helpful in supplying a model for conceptualizing power of this sort. If power was dealt with at all, it tended to be regarded as a primarily top-down and repressive affair. Having elaborated these notions of power, however, my data can be re-examined and emerges in a somewhat different light.

Power and interaction

My research problem, i.e. how power is exercised in consultations between male GPs and female patients, already implies that power and interaction *are* somehow connected. It assumes that the process of diagnosing and treating patients' complaints is more than an interactional achievement between peers. The process will also reflect the power relationship between the physician and patient as well as the socially structured gender inequality. These assumptions were made at the outset of the inquiry on the basis of the - by now - extensive social and feminist research on gender relations, as well as critique of the health care system (Chapter 1).

In order to investigate this particular research problem, it is not surprising that I turned to interpretative sociology for a suitable method of analysis. After all, it is a research tradition, which has specialized in studying how actors go about constructing their everyday encounters. Whereas it has provided many useful tools for analyzing face-to-face interaction, it has, unfortunately, not been particularly helpful in describing how power works in that interaction.

In an inquiry like my own then, it is obviously of central importance that I can weave these two strands together. I need to be able to treat power as a structural feature of medical encounters or gender relations as well as to analyze the consultations, using methodologies taken from interpretative sociology.

The theory of structuration provides for this possibility, at least at a *conceptual* level. It takes the problem of linking structural forms of domination and inequality with everyday practices as its starting point. By introducing the duality between agency and structure, power becomes integral to even the most personal or fleeting encounter. More important, it suggests that no study of interaction is complete unless power is taken into consideration. This has two implications for the present inquiry:

First, it provides a rationale for even embarking on an investigation like mine in the first place. If power is implicated in social life at all levels, then it will inevitably be involved in any encounter between a GP and a patient as well. I neither have to abandon my research problem, nor treat power relations as a structural issue, to be 'tacked on' later when I have completed my analysis.<33>

Second, if power is part of any encounter and, therefore, the consultations of my data, it will be involved even when it, at first glance, does not seem to be. In other words, I can rest assured that power will somehow be implicated, even in the very nicest and most intimate of my consultations. With patience and perseverance, I will be able to find it.

Patients as agents

Pivotal to the theory of structuration as Giddens' notions on power is the skilled and knowledgeable actor. Even when a social context seems to be structured in entirely predictable ways, well beyond the control of the individual, the investigator is warned never to lose sight that there will also be many aspects of that situation, which are made to happen by the actors involved. Like the ethnomethodologists before him, Giddens refuses any stance that people are 'cultural dopes.' Their activity will be bounded, of course, but it is precisely the task of social analysis to specify those bounds, uncovering how unintended consequences of the activities in question are connected to what agents do intentionally (Giddens 1984:294).

This conception of agency enabled me to avoid one of the pitfalls, which has plagued conversation analysis in institutional settings; namely, the ubiquitous professional vantage point. In most medical interviews, physicians do seem to be in charge. They have more interactional control over which topics are discussed, the form the talk will take, as well as the last word on what is wrong with the patient and what should be done about it. This can occur with or without the cooperation of the patient. Concern with these empirically observable regularities of medical power has deflected attention away from what patients are actually doing in consultations. The fact that patients do not win interactional power struggles does not mean that they go down without a fight.

Patients also exercise power in the unfolding consultation in various subtle and not-so-subtle ways. This can include, for example, microscopic attempts to 'flout convention' or to exhibit behavior which does not fit habitual or appropriate modes of conduct for institutional encounters. Patients may 'distance' themselves from medical routines or the GP's attempts to define their problems in a particular direction. This does not have to be in-so-many-words, but can be tellingly revealed in humor, irony or just plain silence. Opposition does not necessarily take the form of an outright confrontation; it can also be a 'continuous nagging' or a 'picking at the bases' of institutional authority (Giddens 1984:291).

Even when patients go along with aspects of the encounter, which seem to be blatantly oppressive or against their best interests, this does not necessarily mean that they are in agreement with what is happening. Compliance may be a matter of habit, embedded in the practical consciousness of going on as before without stopping to think about it. Often coping behavior occurs as a result of an accurate and rational assessment of the situation and the lack of viable alternatives. If we accord actors some degree of 'discursive penetration' (Giddens 1984:289-293), then we do not have to resort to explaining compliance in terms of the actor laboring under 'false consciousness' or having had the ideological wool pulled over her eyes.

Taking this view means treating the patient as an active participant in the encounter, competently engaged in constructing the consultation along with, but sometimes at odds to, the GP. It becomes possible to uncover not only what she is saying and doing, but also the resources she employs in the interaction. In this way, I can sort out if and where she manages to intervene in the course of events, where she displays resistance, and at which points she offers nothing but compliance. These activities have not been the focus of research done from the vantage point of the physician and cannot be delineated unless the patient is, first of all, installed as an agent in the ongoing interaction. By developing an explicit theory of agency, it becomes possible for me to elaborate current feminist studies on medical interaction to provide a more comprehensive account of how power works in encounters between female patients and male practitioners.

Nice GPs

In addition to restoring the patient to the medical encounter, Giddens' conception of agency enables a more realistic view of what GPs are doing as well. Human conduct is not only skilled and knowledgeable; it is also bounded. Our activities take place under conditions of which we are unaware or only partially aware. This can be due to unconscious motives or our behavior can have effects, which we were not able to foresee and/or did not want. The conditions under which the interaction takes place are not always known to the actor in all of their ramifications. In short, what we do is always a mix of intended and unintended activity.

In the last chapter, I described my initial confrontation with my data as one of confusion. The GP's did not fit my stereotypic views of medical professionals. Whereas they were exercising power in some of the ways criticized by feminists, they were doing so in a decidedly nice and friendly way. I could not get around the fact that they were displaying benevolence at every turn of the conversation and undisputedly interested in providing quality care for their patients.

I did not want to reject this finding as an exception or a fluke just because it did not fit my preconceptions of how doctors are supposed to be. It seemed that if asymmetrical power relations are, indeed, part and parcel of everyday encounters between men and women, then they should be visible here as well; i.e. even when GPs are 'nice.'

By assuming that events are never simply the product of the actor's conscious intentions, it becomes possible to view a GP's behavior as, for example, contributing towards the trivialization of the patient's problems, while, at the same time, sincerely having 'her best interests at heart.' Or, he may consciously believe and propagate the risks of prescribing too many psychotropic drugs, while, simultaneously, coercing the patient into taking those very pills.

The notion, prevalent within the feminist critique of the health care system, of the physician as perennial villain, is in need of revision. GPs are not inherently evil nor do they have to seem so just because we are investigating how they exercise control over their patients. They can be doing this with the very best and most benevolent of intentions.

A correct-the-doctor-solve-the-problem approach does not alleviate the injustices faced by women in medical encounters. In order to alleviate these problems, a more accurate view of the injustices themselves is required; i.e. both in terms of how they emerge, as well as how they may be eliminated. If many of the problems encountered by patients in consultations are not consciously intended by GPs, then they cannot be overcome by a simple declaration of good will on his part either. Inequality based on institutionalized relations of authority, as well as on gender, is not completely within the control of the individual male physician.

In conclusion, a re-conceptualization of agency *à la* Giddens enables a more comprehensive, flexible, and, ultimately, realistic explanation of the conduct of both the GP and the patient. I can examine what both are doing without having to resort to stereotypic extremes of omnipotent doctors and passive patients. This involves taking *both* of them more seriously. In this way, a better view of not only the patient's possibilities, but the physician's limitations can be achieved.

Power and the dialectic of control

Having established that individuals are always agents, it is only a small step to the claim that power is exercised as part of a relationship. For Giddens, power is a two-way, reciprocal affair, involving relations of autonomy and dependence in both directions.

There can never be one party, who has all the power, any more than the other party is ever completely powerless. Even when the resources are asymmetrically distributed (as they are in institutional encounters or meetings between men and women), the relationship between parties is a fluctuating one, requiring ongoing negotiation. It becomes the task of the analyst to uncover this dialectic of control; i.e. to show how the less powerful (the patient) manages to exercise some control even in an established power relationship (Giddens 1984:374).

Current literature on medical interaction is rarely dialectical. By presenting the physician as absolute authority, 'having' all the power, we run the risk of depicting the patient as passive, incompetent, misinformed or - for the more politically-minded - as the object of medical (and male) domination.

The conception of power the investigator has will constrain what she 'sees' in her data; i.e. empirical facts are theory-laden (Lukes). If her perspective on power is a matter of 'have's' and 'have not's,' it is very likely that she will 'discover' physicians engaging in 'elicitation practices' and patients 'responding' to them. Moreover, if the investigator assumes that patients and doctors are primarily engaged in working cooperatively toward the accomplishment of some shared conception of 'normal appearances,' that is also what she will 'see' in her data.

I assumed at the outset of this inquiry that general practice consultations would not be the harmonious co-productions they are cracked up to be in current studies on medical interaction and that patients and physicians would, at least occasionally, be engaged in a power struggle. The feminist critique of the health care system made it possible to expect that patients might not always be in agreement with the ways their complaints were being defined, diagnosed, and treated. In that case, I anticipated (and hoped) to be able to uncover instances of resistance or, at the very least, mildly subversive behavior.

Whereas I did not know for sure what form this behavior might take, I certainly did not want to miss it, should it occur. Since I want to delineate potential struggles within general practice consultations, I need a way of conceptualizing power, which will allow for this *possibility*. In other words, I do not want to obstruct my view at the outset of my inquiry. This requires, at the very least, assuming that medical interaction, including how power and control are being exercised, involves a relationship between the GP and the patient. There can be no struggle if there is only one person doing the struggling. For, just as 'it takes two to tango,' a power struggle takes (at least) two as well.

In Giddens' conception of power, members draw upon rules and resources as structural properties of the social systems of which they are a part in order to produce, reproduce, and transform these systems in systematic or regularized ways. These structures are never simply constraining, but enable social activities as well. The aim of social analysis is to examine the inevitable mix of constraint and enablement within a given situation in order to determine how an individual's action is limited and where the degrees of freedom lie for achieving desired outcomes.

Power in medical interaction - when it appears at all - tends to be equated with repression and constraint. Physicians' questions 'limit' patients' possibilities for action to an 'answer,' which is, subsequently, 'interrupted' when he has enough information to accomplish a diagnosis (West 1984). They usually 'decide' what is wrong with the patient and which treatment is to follow (Fisher 1983; Todd 1983). This is not necessarily in accordance with medical criteria, but can also be based on negative cultural assumptions about how women should be (Fisher and Groce 1985).

Whereas I did find an occasional instance of this kind of straight-forward, authoritarian control in my data, it was, by far, the exception. Rather than being characterized by coercion or restraint, the most pervasive feature of my consultations was a quality of friendly intimacy. They bore more resemblance to a chat between friends than a chilly institutional encounter headed by a harsh, authoritarian physician. Moreover, many of the resources employed by the GPs in my data for gaining control over the consultations were not very different from those available to all members for sustaining orderly and harmonious social intercourse. For example, anyone can tell a joke or cluck sympathetically in response to a story about troubles.

These utterly mundane resources can prove decidedly problematic, however, when employed in a medical setting where the outcome is the trivialization of the patient's complaint, the reduction of her problems to something 'in her head' or the establishment of her continuous need for medical surveillance. In order to come to terms with this kind of control, I obviously need a way of conceptualizing power, which entails more than constraint. Knowing *that* power is being exercised in medical consultations and, in fact, even knowing what the outcome of the struggle will be, is not enough. Discovering, for example, that a patient leaves the GP's office with a prescription for Valium does not explain *how* this was possible. In order to answer this question and others like it, the exercise of power must be viewed as a process, constructed in the course of a consultation using ordinary and often benevolent conversational resources. Power is implicated both in the exercise of control as well as in friendly or intimate conduct. <34>

CONCLUDING REMARKS

An attempt has been made to discuss the advantages of Giddens' theory of structuration for coming to terms with how power is exercised in the way women's complaints are defined, diagnosed, and treated in general practice consultations with male GPs. Not only does his theory provide a conceptual link between power and interaction, thereby justifying a study like mine in the first place, but it offers some guidelines for delineating how power works at that level. His conception of power is relational, complex, and subtle enough for me to deal with the elusive and confusing phenomena, which the preliminary analysis of my data unearthed. Moreover, his notions on agency provide a needed corrective for some of the pitfalls encountered in the medical interaction literature. The theory of structuration is a useful heuristic device for uncovering how power is exercised in general practice consultations and developing a feminist critique of how asymmetrical gender relations are being produced and reproduced in that context.

Whereas I have attempted to show the advantages the theory of structuration presents for coming to terms with my research problem, it is by no means a recipe for studying how power is exercised in consultations between male GPs and female patients. It is a highly abstract, general social theory, whose advantage resides primarily in its sophisticated and relatively cohesive attempt to overcome some of the conceptual problems encountered in contemporary theories on power. However, it says nothing about how power is actually exercised in a specific context, nor what structured forms of domination and subordination look like in institutional encounters like general practice consultations.

Moreover, Giddens has nothing to say on the subject of gender relations. Gender as social category or 'structuring principle' does not appear in his framework. Nor does he specify the structures of domination, which are systematically linked to gender, i.e. the specific and asymmetrical availability of rules and resources to men and women in all areas of social life. These shortcomings are by no means inconsequential. It is my contention, however, that they are more a sin of omission than commission. What is primarily missing in Giddens' theory of structuration, as well as his conception of power, is *grounding* - empirical grounding in the situated practices of social actors and theoretical grounding in an elaborated feminist critique of asymmetrical gender relations in the concrete contexts of social life.

In this chapter, I began this process of grounding by writing Giddens' notions on power 'towards' the phenomena encountered in the preliminary analysis of the data. This is, of course, only the beginning. It will be the task of the remainder of the investigation to delineate precisely how power can be exercised in general practice consultations, using the theoretical framework set out here. I shall be employing the framework to help me find a suitable way for operationalizing the research problem, as well as in the final analysis of power practices in consultations between male physicians and female patients. Ultimately, I hope to take us a step in the direction of developing a grounded theory of power and gender in medical encounters, thereby showing how asymmetrical gender relations can be constituted, sustained, but also undermined in one specific context.

NOTES

1. Lukes (1979a) does this by delineating the 'contests' involved between contending positions. The first entails power as a form of *structural determinism* as opposed to *power being tied to human agency*. The correctness of the holder's view depends on whether she believes that the person with power could have done otherwise; i.e. is her behavior completely determined by structural limitations or is there always some degree of choice or freedom of action? The second contest concerns the *nature of power*. Whereas everyone would agree that the exercise of power involves A affecting B in A's but against B's interests, there is disagreement about what constitutes the effect (does it have to be observable at the level of overt behavior?), as well as the notion of 'interest' (might B have acted differently, had she known better?). The third contest concerns whether theories of power should focus on *coercion* or domination or on normative *consensus*; i.e. the 'essential and desirable component of a highly organized society.'

2. The notion that there are concepts, which are inherently subject to dispute, was originally formulated by Gallie (1956).

3. Lukes also has developed a sophisticated theory of power, but, ultimately, I would agree with Giddens (1979:89-91), as well as other critics (Komter forthcoming), that he tends to leave some threads dangling. He also sees the problem of agency and structure as central to the conceptual confusion surrounding theories on power, but he does not succeed in resolving the dualism at a conceptual level. Instead he introduces

additional concepts like 'structural determinism' or 'interest' to explain the boundaries to human action. Structure, thus, becomes primarily a matter of constraint, something which Giddens has managed to overcome. Finally, Lukes' conception is not readily amenable to investigating power in interaction, where, for example, the notion of 'interest,' latent or otherwise, is less relevant (because inaccessible to the investigator) than *how* members go about exercising power.

4. For explicating the theory of structuration and Giddens' notions on power, I have used, particularly Giddens (1976,1979,1984), with occasional elaboration taken from Giddens (1981, 1982, 1987). For an overview of the rest of his project, the reader is referred to Munters *et al* 1985.

5. An even more expedient solution was to reject the possibility or desirability of developing consistent social theories at all. Particularly in the field of applied social research, there is an unfortunate tendency to forget about theories and go straight to empirical investigations.

6. Examples of some of the classic dualisms in social theory are: *individual* and *society*, *voluntarism* and *determinism*, *conscious* and *unconscious* modes of cognition, *moment* (in which a specific action occurs) and *totality* (of social structures implicated in the activity, but extending beyond it in time and space), *contemporary* social life and *history*, and *social stability* and *social transformations*. Instead of treating these relationships as dualisms, Giddens proposes a conceptual synthesis in which they become connected, presupposing one another.

7. There has been some controversy around Weber's definition of power. In English translations, 'Chance' is sometimes translated as 'capacity' and other times as 'possibility,' the latter being somewhat less individualistic than the former. He, also, by no means limits the exercise of power to the individual. For a more detailed discussion, see Bendix (1960).

8. Other contemporary social theorists who have attempted to reconcile this dilemma in order to provide a more sophisticated framework for conceptualizing power are: Bachrath and Baratz (1970); Lukes (1974); and Wrong (1979).

9. This kind of 'penetration' is left out of much research reports and is missing altogether in discourse analysis, which limits itself to the study of propositional statements. For a good critique of the inadequacy of such methods for investigating ideology see Thompson (1984:Chapter 3 and 7).

10. In both the case of 'practical consciousness' as well as 'reflexive monitoring of conduct,' Giddens reveals his (sometimes unacknowledged) indebtedness to ethnomethodology. 'Reflexivity' is an old ethnomethodological stand-by, proposed in response to structural-functionalist accounts of behavior as governed and regulated by norms in pre-defined ways. In Garfinkel's view, the only 'given' in human activities is members' commitment to the production of some semblance of 'orderliness' in their dealings with one another. Aside from that, they tend to take it as they go along. Human

conduct does not consist of a series of discrete acts with clearly specifiable intentions, but rather is a continuous flux of activity: flexible, practical and, above all, subject to on-the-spot revision. The constitution of the situation as a particular type of situation, as well as the actors' motives and reasons for doing what they do, are all embedded in the activities and unfolding circumstances themselves. Or, as Garfinkel notes, in his plea for a reflexive account of action:

> the activities whereby members produce and manage settings or organized everyday affairs are identical with members' procedures for making these settings 'account-able'. The 'reflexive,' or 'incarnate' character of accounting practices and accounts makes up the crux of that recommendation (Garfinkel 1967:1).

Giddens' notions on how relations of power are produced and reproduced draw heavily on ethnomethodological understandings of reflexivity.

11. Giddens (1976:53) criticizes interpretative sociology on four counts:
1. Its concept of action is tied only to the construction of meaningful interaction rather than to the realization of interests.
2. It fails to recognize the centrality of power in social life.
3. It does not account for the fact that norms or rules are subject to differential interpretation, something which lies at the heart of power struggles.
4. It has little to offer when it comes to problems of institutional change and history.

12. The other boundary is imposed by unconscious sources of motivation. This is of obvious relevance for studies of a more psychological nature or ideology critique. I have not paid much attention to it here, since my own inquiry deals with what is said (discursive) and what is done (practical) and the fluctuating and (hopefully) permeable line between them.

13. In diagram form, Giddens'(1979:56) conception of agency yields a stratification model, whereby a kind of homeostatic loop is set in progress:

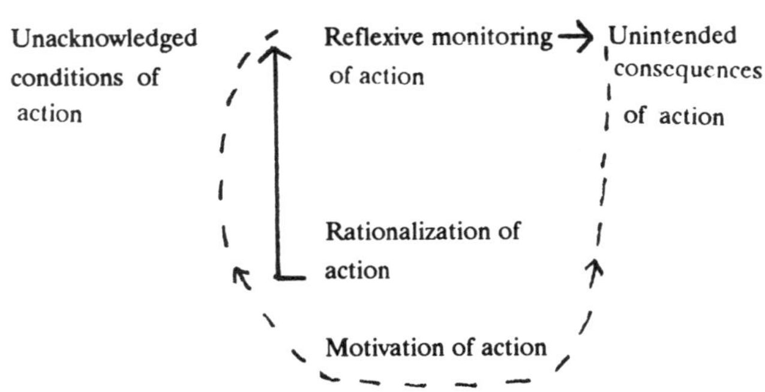

The diagram incorporates the elements of Giddens' notion of agency as a continuous process, part of the *duree* of everyday life. It includes the three features of agency, whereby 'rationalization of action' is the product of discursive consciousness; 'reflexive monitoring of action' is the product of practical consciousness and 'motivation' is tied to the unconscious. Action can be intentional, but, at the same time, bounded by 'unintended consequences' and 'unacknowledged conditions.' Any social analysis will need to be concerned with how these elements interlock and influence one another in the process of constituting and reproducing social life.

14. An example of this is the fact that economic constraint can force many full-time housewives to look for outside employment in order to make ends meet. What began as a limiting dimension, may provide opportunities as well. Not only do women discover new ways of shaping their lives, but economic independence can give them self-confidence and an edge in power struggles on the home front. Any analysis of the significance of outside employment for women will obviously need to take both dimensions into account. My thanks to Marianne Grunnell, who does exactly that in her analysis of women, who married and started a family after World War II (1984), for this example.

15. Giddens (1979:66) distinguishes between system and structure, whereby system refers to 'reproduced relations between actors or collectivities, organized as regular social practices.' Unlike structures, systems do exist in time and space. A medical encounter, then, is an example of a system, which is 'structured' by virtue of the rules and resources drawn upon by physicians and patients in the course of their interaction.

16. The other side of the coin is, of course, his rejection of (post-) structuralist conceptions of language, which display a unified tendency to detach language (the 'text') from the situated social practices of the speakers. It is difficult to see how a satisfactory account of language or signification can be given without showing how it is used in everyday life, as a 'medium of living in the world' (Giddens 1987:94). For a similar perspective on language, from a feminist point of view, see Cameron (1985).

17. Consider, for example, a social practice as elementary as a greeting ritual. In order to interpret a raised hand or a nod as an attempt to initiate interaction, members will need to draw upon systems of signification. Interaction, however, is not simply a matter of producing mutually shared meanings. Once a greeting has been made, the next person is compelled to 'greet back.' Failure to do so constitutes a normative breach of conduct; something requiring legitimation and, potentially at least, subject to sanctions ('What's the matter, cat got your tongue?'). Whereas both participants may be oriented to what constitutes the 'correct' response to a greeting, they may well have differential access to sanctions, should this response not be forthcoming. The cleaning woman who snubs her boss by not greeting back is probably in bigger trouble than would be the case, were the roles reversed.

18. This is somewhat misleading, as one Giddens-exegist has pointed out (McLennan 1984:126), since he seems to be advocating a duality of structure *and* agency.

19. A *methodological* distinction can be made between the analysis of strategic conduct

(the study of how actors draw upon structural elements in their social relations) and institutional analysis (the study of rules and resources as chronically reproduced features of social systems). These are the two entrances into the study of social life, whereby the investigator places a methodological epoche on one level of analysis. It is important, however, to remember that this is strictly a methodological bracketing and not an epistemological or ontological one (Giddens 1979:80).

20. Interpretative sociology has focussed on the first two elements: interaction as meaningful and interaction as constructing the normative order. It was Giddens' contribution to add power as a third basic elements.

21. Giddens suggests that even in the most constraining of all situations, as for example, in prison, where a person is gagged and bound or placed in solitary confinement, there is some degree of autonomy possible; i.e. refusal or, ultimately, suicide (Giddens 1981:63). He has been criticized on this point by Thompson (1984:169), who claims that it is quite difficult for an individual *not* to be an agent in Giddens' framework. This does not seem to be very satisfactory when it comes to analyzing situations where a person's choices are extremely limited.

    In medical encounters, there are obviously more degrees of freedom involved, making this distinction of less relevance.

22. That the 'weak' also have some degree of control over the world around them is illustrated by such common sayings as 'Behind every great man, there is a woman' and many more. Various authors have addressed themselves to the subtle and covert forms of power exercised by the 'weaker sex', notably Cloward and Piven (1979), Janeway (1981), Honegger and Heintz (1981), and Wise and Stanley (1987).

23. This approach is sympathetic in that it precludes a 'blaming the victim' stance so often found in social research. I am, for example, frequently warned that patients 'want' to have the doctor tell them what they have, presumably making any discussion of whether they also 'want' to be treated with disrespect or have their complaints defined as trivial, superfluous.

24. In his discussion of Willis (1977) research on working-class school boys, Giddens demonstrates convincingly how important it is to analyze relations of power as the production of (knowledgeable) agents. Even when the outcome of the struggle runs a predictable course, it is important to know in what ways actors both undermine *and* help to perpetuate the very conditions, which set limits on their situations in the first place (Giddens 1984:289-304). Not only does this provide us with a much more thorough picture of how power is exercised, but it shows more realistic possibilities for uncovering alternative courses of action. By making 'discursively' available what actors do without 'knowing' that they are doing it, the investigator provides them with information, which can change their starting position 'next time around.' This is the crux of social research having an emancipatory function, something I shall be turning to later on.

25. There has been a tendency within the social sciences to equate power with dominati-

on, as though we were automatically talking about the same thing. This is not surprising, since power, in the sense of domination, seems more relevant to us, particularly when we have some sort of critical pretension. Giddens also gives domination a central place in his theoretical framework, but he conceptualizes the relationship between power and domination as a contingent one. Structures of domination are an inherent feature of social life, so that even in the most causal encounter, actors will draw upon them, just as they draw upon structures of signification or legitimation. Power is, however, not the same as domination. It is contingently and inherently connected to the *totality* of structures, including but not reducible to domination (Giddens 1976:111-112).

26. An obvious instance of this is the adjacency pair (Schegloff and Sacks 1974)as one of the most basic ways that orderly conversation can be organized. At the same time, however, it places the 'next speaker' under constraint to respond in the appropriate way. If this response is not forthcoming, its absence may warrantably be treated as an absence and, consequently, require some accounting on the party of the speaker.

27. Giddens (1981) rejects Foucaultian tendencies, however, which equate power with the production of truth, thereby making all forms of discourse (including radical or alternative ones) just as normative or disciplinary as the traditional discourses. See, also, van Aalderen (1986:51) for a similar critique of such theorizing in connection with the patient movement. He disagrees that alternative groups don't do anything different than the medical establishment; i.e. that they also 'define' people's problems, engage in medicalization, etc. Such theorizing not only eliminates the question of whether or not some discourses are 'better' than others, but also ignores the 'creative potential that people have for resistance' and the fact that they are, time and time again, prepared to jump on the barricades.

28. To this end, Giddens distinguishes between 'mutual knowledge' and 'common sense.' The former is what members, including social scientists, need in order to make their way through life. It is involved in describing what members do tacitly as a result of their practical consciousness. It does not stand to be corrected, because it is what accounts for being able to have findings at all (Giddens 1984:336).

    'Common sense,' however, involves propositional beliefs, which, though founded and implicated in day-to-day activities, are fallible. My own 'common sense' notions on power are a case in point (Chapter 3) with the present inquiry being an attempt to 'correct' them or, at least, provide them with some more 'adequate grounding.'

29. I do not mean to suggest that feminist activists are not 'ordinary' or that 'ordinary' patients might not have feminist proclivities. However, much of the feminist literature criticizing current health care practice does provide guidelines for the patient, which are of a prescriptive nature. See, for example, *Our Bodies, Ourselves* (The Boston Health Collective 1971), one of the first handbooks put out on women and health care, where women are given suggestions on what to expect from the doctors and how to get what they want (Chapter 15). These suggestions can be helpful, but they do not pay attention to what women patients are doing already - activities, which may not

only be quite different, but also of use for developing feminist strategies for change.

30. For a good example of this kind of thinking, applied to power, see Aerts (1986). She, along with other 'constructivists,' should not regard agency as something which exists, but rather as an 'effect of power.' In other words, the research question becomes not what people are doing and how this is implicated in power relations, but rather the appearance that they are doing things and their erroneous belief that they are agents is what needs to be investigated. For a good criticism, see Komter (1986).

31. By this, I do not mean to suggest that there are such things as 'objective interests' in the sense that they can be definitively assigned by the social scientist. I am referring to the often puzzling experience, familiar to most investigators, that their research subjects can be seen doing things which seem, to the outside observer, to be 'wrong' or have all sorts of negative results. Such occurrences may be 'explained away' or ignored, but they can also be made into a topic in their own right, which is what I intend to do in this inquiry.

32. See, for example, Giddens (1985:121-122) where he stresses the importance of the feminism as contemporary social movement, as well as the reprehensibility of sexist pronoun usage, but is doubtful about whether gender fits into his theory at all. He suggests that feminists might do better to apply a concept like 'affective individualism' to relations between the sexes: every woman is responsible for her own well-being, ergo she can change traditional relationships and her own subordinate position within them.

33. This is, for example, what I did in an earlier inquiry (Davis, 1984b; 1986) where my aim was to show how a client's problems with her situation as full-time housewife and mother were transformed in the course of a therapy session into a purely psychological problem, something 'wrong' with her. The analysis itself was limited to the process of problem re-formulation as an interactional accomplishment. What this meant for women as a group or in terms of power imbalance between a professional and client was dealt with in the introduction as *raison d'être* for the research problem, as well as in the final chapters concerning the relevance of the findings. This led to some rather unexpected and unfortunate difficulties. For example, feminists tended to limit themselves to just reading the beginning and end of the book, whereas professional helpers and traditionally-minded social scientists occupied themselves with the process of problem re-formulation without thinking too much about what was wrong with it for women as clients. This fuelled my conviction that a link between levels of analysis would be essential for any adequate feminist critique of interaction between men and women in professional service settings. (Davis, 1983; 1984a).

34. It is worth mentioning that I am not concerned with the enabling features of medical interaction *per se*. My research interest has, from the outset, been to uncover the workings of power in terms of how relations involving domination and subordination are being constructed. Because my data showed such confusing entanglements of enabling and constraining elements, I needed a conceptualization of power which did not limit itself at a theoretical level to constraint or limitation. It is this entanglement, then, which interests me rather than sorting out the 'enabling' side of things. I am indebted to Aafke Komter for her useful insights on this matter.

II

LOOKING FOR POWER:

THE RESEARCH PROBLEM OPERATIONALIZED

5

CHOOSING ANALYTIC CONCEPTS AND CATEGORIES

INTRODUCTORY REMARKS

Up until now, the research problem has been dealt with in a general way. I began with a discussion of the background of the present inquiry, paying particular attention to the relevance of how women's complaints are being defined, diagnosed, and treated in consultations with male GPs. The advantages and disadvantages of conversation analysis as a suitable research method were discussed and, based on the findings of a preliminary analysis of the data, as well as the review of current studies, a theoretical framework was discussed for investigating how power works at the level of interaction.

I am now ready to move from a theoretical consideration of the research problem to the question of how power may best be analyzed, using transcripts of consultations as data. In other words, I need to find a way to operationalize the research problem, which does justice to the previously mentioned methodological and theoretical considerations.

The first step entails finding a suitable *analytic category* (Lofland 1971). By this, I mean a category, which can be used to order, explain or summarize the concrete details of the conversations, making them amenable to further analysis. As is the case in most qualitative research (Schwartz and Jacobs 1979), I shall be moving back and forth between the particulars of the data and more general or theoretical concepts. Depending on the phase of the inquiry, the categories necessary for the analysis will be more descriptive or more abstract. At this point, categories are necessary, which will provide a preliminary way of organizing the data.

This proved a rather complicated task and, in fact, the next three chapters will be concerned with how I went about it. In this chapter, I shall be showing how I selected an analytic category as starting point for the present inquiry. Three sources were used for this. To begin with, a category was required, which fit the preliminary analysis; i.e. a category which would enable me to investigate power in terms of agency as conceptualized in Chapter 4. Once my ideas had begun to crystallize concerning the type of category needed, I proceeded to explore *current studies* in interaction for potentially useful ways of organizing my data. This led to the selection of two promising analytic categories as starting point: troubles-telling and stories (about troubles). Although these categories have not been employed to any significant extent for the analysis of medical talk, I found them interesting possibilities for approaching my own research problem. The final step involved returning to my data to make sure that these categories, having been selected, were *empirically* available as well.

This chapter will be devoted to delineating this process, showing how each source was used in order to arrive at an analytic category suitable for coming to terms with the research problem.

SOME THEORETICAL CONSIDERATIONS INVOLVED IN OPERATIONALIZING THE RESEARCH PROBLEM

An analytic category for organizing the data in a preliminary way was required, which would meet the following requirements:

1) It would organize the data in such a way that I could show how women's complaints are being defined, diagnosed, and treated in the course of a general practice consultation.
2) It would facilitate an analysis of how power is implicated in this process.
3) It would reflect my theoretical framework concerning how power must be conceptualized at the level of interaction.

Beginning with the research problem itself, there are various possibilities for showing how a patient's complaints are transformed in the course of a consultation into a medical diagnosis. Much of the current literature on medical talk is concerned, in fact, with precisely this process (Atkinson and Heath 1981; Fisher and Todd 1983). In what seems to be a rather standard format, the patient's problems enter the scene already pre-packaged as 'complaint'<1> and proceed to make their tortuous way along the steps of the physician's medical 'agenda' (history taking, diagnosis, examination and/or treatment suggestions). Analytic categories for describing the conversational activities of the participants can generally be 'fit' back into this scheme and, in fact, serve to display how it came to be organized that way in the first place. Thus, for example, patients are described as 'giving information' (Conte 1981) or 'answering questions' (West 1984), while physicians do the asking or, more elegantly, engage in 'elicitation acts' (Drass 1982). Whereas the patient may produce the occasional 'initiative' (Coulthard and Ashby 1976; Bliesener and Siegrist 1981), as the term itself suggests, this rarely gets beyond the most fragile beginning.

It is by no means my contention that such activities cannot be found in medical talk, nor that the categories themselves are incorrect as descriptions of patient and physician behavior. In fact, my own material yields ample evidence of 'question-answer sequences' as well as that 'undesirable' (Bliesener and Siegrist 1981) phenomenon, the patient initiative. Whereas I might have organized my data using such categories, they proved inadequate for the purposes of the present inquiry. This is, quite simply, because, by using them, it is nearly impossible to see how:

- the patient has a specific and possibly even coherent contribution to make to the ongoing interaction
- the physician is anything but in complete control over how the encounter is being organized
- the consultation does not unproblematically coincide with (dominant) medical definitions.

None of these issues is particularly relevant as long as the investigator is interested in showing how the consultation is being constructed as a medical event. They become so, however, when she is interested in showing how the patient is involved in the process of transforming complaints into a diagnosis or a treatment decision.

Such a perspective would depend upon a definition of power as relational, involving at least two participants, who are actively engaged in exercising it. Analyzing power would then entail showing how both draw upon rules and resources, which are (asymmetrically) available to them, in order to influence the course of events. It is precisely this dialectic of control, which must be the focus of any study on how asymmetrical relations involving gender and power can be constructed, sustained, and, of course, undermined.

Thus, in order to come to terms with power struggles involved in the business of defining and treating women's complaints in the context of a general practice consultation, I needed to approach my material in such a way that the patient was viewable as agent. In other words, analytic categories were required, which did not transform her activity *in advance* to a response to the GP's ministrations. If I do not wish to preclude the *possibility* that she is actively and competently engaged in constructing the consultation, as well as resisting potentially oppressive aspects of it, I will need analytic categories which leave her activity intact. This means taking her activity as starting point for the analysis and treating it as a coherent and specific contribution in the context of an ongoing consultation.

Finding an analytic category, which would restore the patient to a position of agency, thereby enabling an analysis of power relations in medical encounters, is a step in the right direction. It is, however, not enough. My preliminary analysis indicated that the exercise of power was not a simple matter of medical authority and, in fact, the consultations did not always seem like medical encounters at all. Whereas patient and doctors are undoubtedly oriented to medical definitions of what the situation is about, they also appear to be attuned to alternate realities as well.<2> Subordinate definitions of the situation may well be essential for understanding how asymmetrical gender relations are being constituted and sustained in medical encounters. Thus, I needed a way of organizing the data, which at least allows these realities a 'voice' (Mishler 1984), making them available for further examination.

In conclusion, these considerations led to the conviction that I could not approach my data using categories found in much of the current literature on medical interaction. Another analytic category was required, which would do justice to the present research problem, enabling me to uncover the patient's 'voice' in the process of defining, diagnosing, and treating her problems in a consultation with the GP. It would have to be a category, which takes the patient's activity as starting point. Moreover, this activity could not be reduced at the onset of the analysis to a medical event ('complaint'), a mere discourse fragment ('initiative') or a reaction to the physician's activities ('answer,' 'response,' etc.).

Thus, my search began for an analytic category, which would help me organize my data in a way suited for the study of power, focussing on the patient's activity as specific and, at least potentially, coherent contribution to the interaction.

CURRENT STUDIES ON TROUBLES-TALK

Introductory remarks

Taking troubles-talk as possible analytic category for organizing my data was, by no means, an arbitrary choice. Just on the basis of common sense, it seems to be a promising starting point for the present inquiry. To begin with, patients will presumably be seeking the services of a GP, because they have some troubles, bodily or otherwise. They do not go to the doctor unless 'something is wrong.' This means that doctors and patients will be talking about troubles, at least for part of the time.

An investigation of troubles-talk will have as its focus the person with the trouble. It will be the teller, who has a vested interest in getting her trouble told, requiring her to perform various interactional tasks for its accomplishment. Whereas other participants' activities as recipients will also be involved for its successful production, it will be the teller of the trouble, who must orchestrate the activity; she will have center-stage in the analysis. Thus, by investigating troubles-talk in consultations, I can automatically ensure that the patient be viewed as active participant in the encounter.

And, finally, there has been considerable research on troubles-talk, both as interactional activity (Jefferson and Lee 1981), as well as a specific kind of story with characteristic features (Quasthoff 1979; 1980; Ehlich 1980). Studies show that troubles-talk is not only a common occurrence in everyday conversations, but that it may be found frequently in institutional encounters as well (Flader and Giesecke, 1980; Rehbein 1980; Wodak 1980; Bennett and Feldman 1981). Although troubles-talk has only rarely been investigated in medical settings,<3> current research can provide considerable insight into the kinds of interactional difficulties, which can arise when troubles-talk is done in professional service settings. Since general practice would indisputably fall under the heading of a professional service,<4> it seems reasonable to assume that these findings could be relevant for that particular setting as well.

In short, a closer look at research in the field of troubles-talk, particularly in institutional settings, seems to be warranted. In order to do this, I shall be drawing upon two distinct, but related research traditions. The first deals with how troubles are told in conversations; i.e. *troubles-telling as interactional activity*. This work is primarily of a conversation analytic variety, with Gail Jefferson as its primary proponent. The second tradition is more concerned with the *structural features of stories about troubles*. More linguistic in orientation, these studies are concerned with what distinguishes one kind of story from another, the stages stories go through, and their characteristic features. Recently, attention has also been devoted to why people tell stories and how this effects the situation in which they find themselves. Of particular interest is the work of Uta Quasthoff on storytelling in institutional settings.<5>

I shall now turn to each of these approaches to troubles-talk and discuss what they have to offer in terms of providing a suitable analytic category for the present inquiry.

Troubles-telling

Within mainstream conversational analysis, it is certainly Jefferson's work which provides the most comprehensive treatment of how members discuss trouble in 'ordinary' situations. Following her work on the sequential organization of storytelling within conversations (Jefferson 1978b), she addressed herself, together with Lee (1981) to how members tell their troubles to friends, acquaintances, and family members. Whereas they were primarily interested in routine troubles-talk, they also collected a smaller corpus of taped troubles-tellings taken from a service encounter - specifically, an emergency ambulance service. This material was used to elaborate some of their findings concerning ordinary talk about trouble.<6>

Upon examining their material, the authors found that talk about trouble was a 'robust phenomenon' (Jefferson and Lee 1981:395) with a coherent organization and specific membership categories and activities.<7> Whereas it displayed an observable order, it was an order of a rather gross sort. Troubles-telling seemed to be a particularly messy conversational activity, subject to constant disordering. They set out to discover why this was so.

It appeared that, even in ordinary conversations, troubles-telling required special management (Jefferson and Lee, 1981). In a service encounter, it became even more problematic, however. Troubles-telling did not seem to 'fit' at all. By lack of fit, the authors were not referring to any incompatibility of topic. Ostensibly, the client and the service provider were talking about the same topic - her trouble. Nor was the problem that troubles-telling as conversational activity was so flexible that it could easily be disrupted (and therefore was). On the contrary, it appeared that two separate conversational activities were being performed simultaneously and that these activities were themselves, in part, mutually exclusive (Jefferson and Lee, 1981).

In a *troubles-telling*, the person with the problem has the floor. Her concern is to share her trouble with another party and elicit some kind of emotional reciprocity like 'how terrible!' or 'that happened to me once, too' or 'well, try to look at the bright side.' The person and her experience are the focus of the talk, rather than the trouble itself. Troubles-talk constitutes something members 'normally' do together, during coffee breaks, on the telephone, or in the pub. Bad times need to be shared and the sharing often has the effect of drawing the teller back into the community. Her lot may be a hard one, but she is still one of the group and is expected to behave accordingly; i.e. go back to work after the coffee break. One could say, then, that the aim of a troubles-telling is *continued function, regardless* (Jefferson and Lee 1981:417).

In *giving service*, the service provider has the lead. Her main concern will be dispatching a specific task. The task may involve providing information, offering advice or performing some activity. There is always a 'problem to be solved.' Whereas the problem 'belongs' to the person requesting service, once she has presented it, the focus of the talk is on the problem, rather than her experience with it. Having separated the person from the problem, the teller as the person with the problem makes her departure from the conversational arena. What remains is the task of finding a suitable way to eliminate or, at least, alleviate the problem. Put simplistically, service encounters are about *maintenance and/or repair* (Jefferson and Lee 1981:417).

It is not surprising, considering these differences, that the topical coherence provided by the same trouble might not be sufficient to prevent the occasional interactional snag when members engage in both forms of talk within the same encounter. Or, as the authors put it, attempts to combine the two produce a 'misbegotten hybrid (Jefferson and Lee 1981:419).' Rather than shunting back and forth between the two, they advise opting for one or the other and making the best of it: 'recognizing and enhancing the deeply remedial potential of emotional reciprocity' or 'recognizing and enhancing the intensely relational potential of the instrumental colloquy (Jefferson and Lee 1981:421).'

Before returning to the problem of finding a suitable analytic category, I shall touch briefly on what the work of Jefferson and Lee can provide, which is of use for the present inquiry. In addition, I shall indicate where they fall short, necessitating a further step before my analysis may begin.

To begin with, applying the category troubles-telling to what patients are doing in consultations with GPs enables me to view their activity in a way which does not automatically reduce it to a mere response to the GP. The patient can be displayed as someone with her own agenda for the consultation. Her tasks, as well as the difficulties she encounters in performing them, can be uncovered.

The service encounter itself - and general practice is an instance of such an encounter - becomes an event where more than one conversational activity may be occurring at any one time. It becomes plausible that parties engaging in these different activities might run into difficulties. Since I am specifically interested in how power works within the actual interaction between male GPs and female patients as they go about determining what is 'wrong' with the latter, it seems sensible to begin by locating those points in the ongoing talk where struggle occurs. Jefferson and Lee provide some useful leads for doing this.

Whereas they help us understand *how* conflict is inevitable when troubles-telling and service encounters converge, they leave us dangling when it comes to the question of *why* such problematic talk is produced in the first place. In other words, why would patients persist in engaging in troubles-talk, when it is so manifestly doomed to failure? Or, why do GPs insist on providing service, if all their patients 'really' want is a friendly ear for their troubles? Even their suggestion for recognizing and enhancing the respective forms seems to indicate that such a thing is possible - if participants were only aware of what they were doing.

Aside from skirting the empirical fact that clients in a wide variety of institutional settings do apparently engage and continue to engage in troubles-telling, despite its unwelcome reception (Ehlich 1980), this stance neatly avoids the issue of interactional power and possible asymmetries between the interactants. In other words, we cannot come to terms with how participants deal with their interactional difficulties unless we understand what makes the phenomenon of troubles-talk so irresistible to patients and so problematic for physicians. This is more than a problem of incorrect sequential organization.

I shall now turn my attention to an inquiry, which can serve to broaden the way troubles-talk in institutional settings might be conceptualized.

Stories (about Troubles)

Whereas troubles-telling has received rather selective attention as research topic within mainstream conversational analysis, storytelling as conversational activity has been the focus of considerable research. This effort has not only been international, but it has straddled various disciplines and covered a wide variety of aspects of the phenomenon itself.

Whereas I shall be dealing with this research in more detail in the next chapter, I would like to turn briefly to one representative taken from the German research tradition on storytelling - the work of Uta Quasthoff. She provides a useful way of expanding troubles-telling as analytic category to meet the requirements of the present research problem.

Quasthoff (1980) has dealt extensively with how people tell stories in everyday conversations. She goes beyond the sequential approach employed by Jefferson and other conversation analysts to include the psychological and interactive functions storytelling may have. Whereas her research is primarily concerned with stories as they are told in ordinary settings, she has also investigated storytelling within institutional, personal service contexts - in particular, a social welfare office (Quasthoff 1979). Obviously, any stories being told by clients in this context will be about some sort of trouble-inadequate housing, financial difficulties, conflicts with authorities, etc., and might, therefore, be regarded as a form of troubles-talk.

Like Jefferson and Lee, Quasthoff also found that client's stories about their troubles conflicted with the requirements of the institutional situation. Social workers need to get relevant information from their clients as quickly as possible so that they can reach a decision about how to help them (Quasthoff 1979:107). Whereas a story about a trouble does provide this information, it does so in a round-about way, taking up more time than strictly necessary and providing considerable extraneous material. It is not surprising that, given a choice, social workers might prefer a simple report to a story about the client's troubles.<8>

Whereas many clients do oblige by presenting their difficulties in report-form, others, apparently, do not.<9> When Quasthoff examined those instances where clients engaged in storytelling, she found the same types of interactional difficulties as those mentioned by Jefferson and Lee. Two forms of talk were being produced within the same conversation. What interested Quasthoff particularly was not so much *that* parties seemed to be conversing at cross purposes, but rather why this might be so. In other words, why do social workers seem to dislike (and therefore obstruct) storytelling and, more important, why do clients persist in telling them anyway?

Quasthoff's (1980:11) starting point is that people like and, in fact, even *need* to tell stories. She does not regard storytelling as an exceptional or in any way disturbing activity, but rather what most people, given the choice, would prefer doing when engaged in face-to-face encounters. Storytelling is *the* standard form for communicating everyday experiences (Ehlich, 1980). Thus, when confronted with a situation where storytelling efforts of certain members were being blocked or obstructed, she assumed that relations between participants must be hierarchical, therefore meriting further attention (Quasthoff 1979:71).

In order to come to terms with what she saw occurring in her data, she examined some of the possible functions, which storytelling might have in a specific situation.<10> In other words, why do members engage in storytelling and what sorts of effects does it have on the ongoing conversation? In addition to myriad psychological and interactional functions for both the storyteller and the story recipient(s),<11> Quasthoff suggests that storytelling is a means of defining the situation itself. By telling a story, a member can exercise control over the type of encounter it will be. Storytelling establishes the interactional *frame* of the situation.

Since this particular function of storytelling is of special interest for the present inquiry, I shall deal with it in more detail.

Storytelling as Conversational Frame

The notion of frame was originally developed by Goffman (1974). It refers to the way understanding is organized in face-to-face encounters. The frame is the way members recognize a set of concrete occurrences as belonging to or constitutive of a specific type of situation. It does not concern itself with the meanings themselves, but rather the 'principles of organization which govern events - at least social ones - and our subjective involvement in them (Goffman 1974:10-11).' A frame, then, is a kind of grammar or syntax of social life.

Frame has the advantage over the sociological standby 'definition of the situation' in that it does not simply exist independent of members' activities, as a set of features or characteristics of an encounter to which participants are oriented. The notion of frame includes the activity of defining the situation as well.<12> According to Quasthoff, the conversational activity of storytelling can serve to establish a specific frame for the ongoing interaction. Stories are produced routinely and appropriately in a particular type of situation. By the same token, the activity of storytelling can serve to construct the encounter as belonging to that type of situation as well. Stories tend to occur in situations with the following features (Quasthoff 1980:74):

- The encounter is *informal*, *private*, and *non-goal-oriented*. Business transactions or meetings are rarely conducive to storytelling, whereas a luncheon date or a party would be.
- Participants tend to have a relationship, which is characterized by some degree of *intimacy*. This may be due to emotional ties or temporary circumstances as, for example, finding themselves sharing a train compartment for a few hours. Perfect strangers or persons in a relationship involving authority do not launch easily into storytelling.

- There must be enough *time* and *leisure* to be able to tell stories. It is not appropriate when driving in rush-hour traffic or when there are more pressing concerns to be dealt with.

Taken together, these features belong to what we might call a *friendship frame*.<13> Storytelling is not only a regular feature of a chat between friends, however. If frames are the ways that members make sense of their experience, as well as constructing an encounter as making sense, then the activity of storytelling can also serve to construct an encounter as a friendly one. In other words, friends are likely to tell one another stories and, at the same time, by telling each other stories, they are becoming friends.

Having established storytelling as a conversational frame, the next question becomes: if storytelling is something that is part of a friendship frame, why do clients or patients engage in such an activity in an institutional service setting? It would seem to be a highly irregular and inappropriate activity for a conversation, which can hardly be construed as an informal meeting between friends. Since the clients in her data-base (conversations in a social welfare office) did, in fact, persistently engage in storytelling, she considered some possible explanations for this 'dysfunctional production of stories (Quasthoff 1979:112-113).'

The first possibility was that the clients were *confused* as to the character of the conversation, did not realize that stories were inappropriate there or mistook the social worker for their friend. The second was that they *did not accept the professional definition of the situation* and were interested in forcing the social worker to engage in a personal conversation with them. And, finally, *stories were being used to define the situation in an alternative way* not intended to change the dominant professional frame, but rather to enhance it. In other words, the friendly frame was being constructed alongside the service one. <14>

It seems unlikely that clients would not be oriented to the professional nature of the service contact. Research has indicated that common sense notions are not separate from institutional practices (Stimson and Webb 1975). Moreover, it will not be in their own interest to break with the professional frame altogether, as they are presumably also in need of assistance for their problems. Thus, the third explanation seems, at first glance, the most likely. Assuming, then, that clients attempt to produce an additional definition to the ongoing situation, the next question becomes why they would want to construct a friendship frame.

To begin with, friendly encounters are generally preferable to unfriendly ones. It will be in the best interests of both the client and the professional helper that their conversation proceed as harmoniously and pleasantly as possible. For the client, however, there are additional advantages. In a personal conversation, she will be able to share her experiences as she sees them and not have to 'package' them to fit external requirements. Telling a story prolongs the contact. Most important, by defining the conversation as a friendly one, she will, at the same time, define the other participant as an equal, someone who may be expected to react with sympathy or, at least, involvement. It is not surprising that these conditions might be attractive for a person with a trouble.

According to Quasthoff, there is much more at stake here, however. Storytelling and, along with it, a friendship frame, are not simply one *possible* way of defining the ongoing encounter. They constitute, in fact, the *only* interactional form available to clients by means of which they can actively and with some degree of autonomy determine how an institutional encounter is to be defined (Quasthoff 1979:118). Establishing friendship frames in service encounters is linked to the moment-by-moment struggle for power. Just how essential this link is for understanding how power works in medical consultations will become clear in the analysis to follow.

In conclusion, Quasthoff's work on storytelling in institutional encounters indicates that in a context where clear constraints exist concerning the procedures and routines, which should be properly followed, storytelling may be the only way a client (or patient, defendant, student, etc.) can 'break the frame' - at least temporarily - and establish another kind of social relationship. Whereas a patient will rarely have access to some alternative professional frame, and in the medical frame her experience is subordinated to medical expertise, she can speak in a dominant voice by engaging in storytelling. This is possible because stories about trouble fit, however imperfectly, into the context of a professional service encounter.

Seen in this light, troubles-talk or storytelling in institutional settings cannot be reduced to a simple matter of 'mismatched membership categories' (Jefferson and Lee 1981). There is much more at stake than that. Storytelling is implicated in the way the situation itself comes to be defined, the kind of encounter it will be, and the positions to be taken by participants within it. Interactional control over how a situation is defined is irrevocably connected with power.<15> This means that if storytelling is taking place in an institutional encounter, we would expect it to reflect and give rise to the occasional power struggle as well. For this reason, the patient's storytelling activities might be a useful place for locating and investigating potential power struggles between her and the professional over how the situation comes to be defined and, as such, a useful analytic category for the present inquiry. <16>

Before reaching this conclusion, however, I would like to address some considerations of a more empirical nature.

SOME EMPIRICAL CONSIDERATIONS CONCERNING THE USEFULNESS OF STORYTELLING AS ANALYTIC CATEGORY

Up until now, some of the theoretical problems involved in selecting an analytic category for the present inquiry have been discussed. Troubles-talk and, in particular, troubles-telling and stories about troubles, have been considered as potential categories for organizing my data for further analysis. Before embracing these categories with open arms, however, the question arises whether they are empirically available in my data. In other words, if stories are not being told by the patients in my material or are being told only very sporadically, one could reasonably object to the adequacy of storytelling as analytic category for this inquiry.

The question is obviously relevant since, as the reader will recall, I did not begin my analysis with the intention of examining the consultations as potential instances for patients' storytelling endeavors. On the contrary, storytelling emerged as a possible analytic category in the course of the analysis itself, mid-way between my preliminary analysis of the data and my grapplings with the theoretical ramifications of the research problem. It had, to speak with Glaser and Strauss (1967:36-39) all the makings of an 'emergent category,' which, however, having emerged, required examination for its empirical usefulness for the present inquiry.

I shall be addressing this issue from two different and, at first glance, contradictory angles. First, I shall concern myself with the possibility that the stories discovered in my data are exceptions. Second, the actual empirical frequency of stories in my material will be discussed.

1. *Stories as exceptions*

   It could be argued that the frequent occurrence of storytelling in the consultations comprising my data is what makes it a worthwhile topic for investigation. In other words, if a phenomenon seems to be such an important part of a particular setting, further examination is justified. Whereas this is certainly a *raison d'être* for many scientific inquiries, it is not the most relevant criterion for my choice of storytelling as analytic category for the present inquiry. Here my considerations are primarily of a theoretical nature and have to do with the type of research problem I am dealing with. Or, put in another way, even if storytelling was very rarely done in general practice consultations, it could still make sense to examine those instances where it did occur.

As the reader will recall, the GPs in my sample are in some ways a-typical of the more general population in that they are younger, have had some specialist training, and all expressed willingness to cooperate with a project geared toward improving training programs (Chapter 2). If they display differences *vis-à-vis* their colleagues, it seems safe to assume that they will be exceptional in a positive sense; i.e. better educated and possibly more in tune with some of the progressive developments in the field of general practice than the run-of-the-mill GP might be expected to be.<17>

This was in keeping with the theoretical considerations outlined at the outset of the inquiry. It was not my intention to limit my study of gender and power inequality in medical encounters to the most extreme examples of medical misogyny. On the contrary, I am assuming that it will be part and parcel of the most ordinary and, in fact, pleasant encounters. Since my focus is on asymmetrical power relations as an integral feature of medical encounters or meetings between sexes, it makes good theoretical sense to look for it first where it might be least expected. In other words, if even the nicest and best-educated GPs engage in sexist practices, then the extreme cases begin to look less exceptional as well.

Applied to storytelling, the following claim might be made: If patients are engaging in storytelling in the course of a consultation, then it is also because they are being allowed to do so. It will always require some degree of cooperation from the physician. In the face of institutional constraints, which make such activities dysfunctional, it is safe to conclude that the GPs are giving their patients an unusual amount of conversational space by allowing them to do troubles-talk. If I can demonstrate how relations of dominance and subordination are being constructed, even in those consultations where the GPs are being so manifestly empathetic and generously allowing patients talk about their problems in their own way, then a case may surely be made for power as a structural feature of medical talk. Thus, whether or not storytelling occurs in the majority of the consultations, it still can serve as a suitable analytic category for the purposes of the present inquiry. At the risk of contradicting myself, I shall now take a look at the actual frequency of storytelling in my data where - as it turned out - stories appeared often enough to warrant a closer look anyway.

2. *Empirical frequency*

In the course of my preliminary analysis, my first impression was that the patients spent considerable time talking during the consultations. This struck me, undoubtedly, because it ran counter to my pre-conceived notions that this would not be the case. I had not expected that women would have so much conversational space to talk about their problems with male physicians.<18> This was, of course, nothing more than an initial observation. However, it was enough to make troubles-telling or storytelling seem like a possible candidate for coming to terms with patient activity in my analysis. This was further supported as I took a closer look at *how* the women in my consultations were talking about their troubles to GPs. In other words, it resembled the activity described in the current literature on troubles-talk.

Once I had selected storytelling as tentative category, I returned to the larger collection of 315 taped consultations (see Chapter 2) and investigated them for instances of storytelling. I wanted to see whether the 20 tapes of my preliminary analysis were in any way exceptional. At this point, my concern was in getting a *rough* idea whether storytelling was occurring at all and, if so, to what extent. I say rough, because I hadn't completely worked out the category of storytelling (about troubles) yet.<19>

The result was as follows:<20> there was some evidence of storytelling activity, either in the form of completed stories or interrupted stories, in about 3/4 of the consultations. Whereas the majority of the patients appeared to be making storytelling attempts, there was also a marked sex difference. Of the consultations where no storytelling attempt could be located; i.e. where the patient provided nothing beyond a short answer to the GP's question, more than twice as many of these patients were male. The GPs, on the other hand, only very rarely engaged in storytelling. There were only 13 examples of physician-told stories, five of which were told by the same GP.

It is not my intention to say anything conclusive about the frequency in which storytelling occurs in general practice consultations in general. Nor do I wish to suggest that women as a group have the tendency to tell more stories than their male counterparts.<21> These findings indicate, however, that storytelling occurs often enough in my data to make it a good starting point for the present inquiry.

CONCLUSION

In this chapter, an attempt has been made to link my empirical observations to some theoretical considerations. After rejecting some of the categories currently in vogue within studies on medical interaction as being contaminated by medical or professional vantage points, I turned to another focus, which would support what I found in my data and take the patient's activity as its starting point. This led me to consider troubles-talk as a possible way of organizing my material for further analysis. To sum up, my reasons are as follows:

1. As starting point, troubles-talk (which encompasses both troubles-telling as well as storytelling, when the stories are about trouble) provides a way of investigating how a patient's initial presentation of her problems may be transformed in the course of a consultation without automatically relegating her to a respondent position. Her activity is left intact. She becomes viewable as someone with her own interactional agenda, which may or may not fit in with the GP's notions concerning what the consultation is about.
2. Having made the patient visible as active participant with -at least, potentially - her own agenda, it becomes possible for the first time to investigate the consultation as not immediately and unproblematically synonymous with the GP's agenda. The analysis, even in the face of the dominant voice of medicine, can be geared toward uncovering the resources and practices employed by both parties constructing the encounter.
3. Whether troubles-talk is regarded as an interactionally divergent activity within a service encounter (Jefferson and Lee) or as a separate conversational frame (Quasthoff), its occurrence within institutional settings suggests that conflicts are just around the corner. Both studies indicate that clients may initiate and continue to produce troubles-talk in spite of considerable opposition from the professional. Functionalist understandings of medical situations with complementary roles neatly distributed and more or less imperfectly internalized prove inadequate for understanding this event. Or, put in more ethnomethodological terms, members appear to be not only sustaining institutional interaction as an orderly and harmonious

endeavor, but also constructing it as an arena of struggle (Molotch and Boden 1985).

4. Whereas Jefferson and Lee provide categories for locating potential points of conflict at the level of interaction between troubles-telling and the constraints of a service encounter, they do not enable us to see these conflicts as much more than an interactional mishap. By including Quasthoff's analysis of storytelling, the patient's activities can be viewed as contributing to the construction of the situation in which the conversation occurs. Storytelling becomes a vehicle for exercising power over how the situation is defined. Since it may often prove undesirable from the point of view of the professional, it might be expected that patients cannot proceed with their stories unobstructed. It will be precisely those points in a storytelling where the patient runs into obstacles, that an analysis of power might profitably begin.

Based on these considerations, I decided to take troubles-talk as analytic starting point for my analysis. Although it has not been employed with any regularity in the analysis of medical talk, I saw no reason why this could not be done. A brief look at my material convinced me that troubles-talk does occur in general practice consultations, just as it does in other service encounters.

NOTES

1. In the original formulation of the research problem, the term 'complaint' was used. This already represents a kind of medical pre-packaging of what the patient is talking about; i.e. the way the patient presents her problem to the GP and the way he will be likely to hear it. Whereas the participants themselves are oriented to the business of constructing complaints, it seems important that the researcher avoid applying a medical category to the patient's activity before other possibilities have been considered.

   This is what I shall be doing in this chapter, ultimately leading to a slight reformulation of the research problem.

2. I am taking a phenomenological stance on the matter of 'reality' here. This assumes that there is not just one world out there for everyone to find, but rather that members subscribe to and produce multiple versions of reality. Thus, for any given encounter, there will be more than one definition of what is going on.

   For an interesting account of some of the interactional difficulties, which 'multiple realities' can entail, the reader is referred to Pollner (1975). His analysis is also relevant as it demonstrates how asymmetrical power relations are implicated in the ultimate determination of which reality is more 'real' in a given situation.

3. There are several exceptions, but for various reasons they are not dealt with extensively here. Bliesener (1980) analyzes how patients attempt to 'tell stories' during hospital rounds. His contribution - the only one of its kind in an anthology on storytelling in various contexts (Ehlich, 1980) shows how patients' stories are actively prohibited. In fact, they never get beyond the most fragmentary initiative. Ten Have (1987) mentions troubles-talk as one of the three main forms of talk occurring during general practice consultations. His own research, however, focusses on what he calls the Ideal Sequence of the consultation, which is the usual medical definition. Troubles-talk is referred to as a disturbance.

4. See Goffman (1961) for a structural analysis of personal service encounters as a type of social relationship with its own particular features. He applies it to a number of encounters ranging auto repair shops to psychiatric counselling. In each case, there is a client, who puts herself in the hands of a professional server. In exchange for a fee, she receives help for some troubles. Whereas Goffman is particularly interested in the structural similarities between such events, he does make the distinction between those relationships in which the client can leave her malfunctioning object with the service provider (Workshop Complex) and those where she, herself, is the object in need of repair. Obviously, general practice consultations belong to this latter category.

5. Just for the record, a few conceptual distinctions need to be made. I shall be employing the term *troubles-talk* to refer to common sense understandings of the everyday activity in which members talk about their problems with one another. *Troubles-telling*, on the other hand, will refer to the work of Jefferson and Lee, who showed how this social activity is sequentially ordered within a conversation. *Story-telling* and *stories* are concepts taken from both conversation analytic as well as linguistic traditions of discourse analysis. The distinctions between them will be worked out in the next chapter.

6. Although mainstream conversational analysis is typically interested in ordinary talk as opposed to what happens in institutional contexts (Chapter 2), it is interesting that here, taking a look at what happens to troubles-talk in a service encounter could provide insight into ordinary troubles-talk as well. One might say that the institutional setting magnified the 'ordinary' interactional difficulties which such talk entails.

7. Sacks (1974b) has coined the term 'membership categorization' for the apparatus we build for linking various activities to categories of people known to 'do' them. This is not an empirical concept, but rather the way we construct and interpret meanings. As example, he presents a child's attempt at storytelling:

   The baby cried. The mommy picked it up.

   Although this is not said in-so-many-words, we 'hear' the baby as belonging to the mommy in the story. Since 'baby' and 'mommy' are both members of the same 'family'-category, they will be heard as belonging together. This meaning is automatically constructed by the recipient of the story. Moreover, both activities mentioned; i.e. baby crying and mommy picking it up, are category-bound activities. Everyone 'knows' that babies cry and when they do, mommies pick them up. Thus, we hear that what the mommy picks up, is, in fact, the baby and not something else. There are myriad membership categories, as well as category-bound activities, available for anything which needs to be talked about.

8. The distinguishing features of *reports* and *stories* will be discussed in Chapter 6.

9. Quasthoff (1979:106) discovered that in 75% of the conversations occurring in the welfare office, no storytelling took place. In those cases where stories were told, the following features were present: clients were the only ones who engaged in storytelling; when the stories were told, social workers rarely provided sufficient affiliative response making it difficult for the client to tell the story at all; in those cases where they did react, it was with reticence; i.e. at no point did the social workers go into the clients' story in any deta

10. By 'function', Quasthoff does not mean a static, causal relationship between the story and the goals or desires of the speaker. On the contrary, it is

    > der Zusammenhang von intendierter und beabsichtigter (bzw. nicht beabsichtigter) Wirksamkeit einer Handlung, die gemäss konventioneller Handlungsmuster vollzogen wurde und damit interpretierbar ist (1980:145).

    Her typography of functions is based on an empirical study of a large corpus of stories, collected in a variety of contexts.

    I shall not be using this typography in the present inquiry, since I am not interested in the reasons people engage in storytelling so much as how it can be used to define a situation.

11. Functions are typified in terms of being related to the person telling the story, to possible recipient(s) of the story or to the context in which the story is told (Quasthoff 1980:146-169). For example, a story enables a person to make sense of her experiences, particularly if they have been frightening or troublesome. The actual storytelling can have a cathartic effect, enabling her to live through what happened and come to terms with it. Since the storyteller is often the only person present personally involved in the events, she may use the story as a vehicle for presentation of self; i.e. to display herself in a way that she would like others to see her. In terms of the interaction, stories may be told for entertainment or simply to pass time when people get together. They may, however, be put to all sorts of additional interactional tasks. Stories can be used for purposes of illustration, edification, justification or persuasion. In short, stories are a highly versatile device for communicating experiences in social settings. This accounts, no doubt, for their frequent occurrence in everyday conversations.

12. It is not surprising that I have selected a frame-oriented approach to interaction for my analysis. The advantage of 'frame' over 'definition of the situation' is that it bridges the gap between accounts of social structure or 'the world out there' as produced by individuals engaged in face-to-face interaction and the actual business of defining or constructing the encounters. The latter, of course, has always been the focus of ethnomethodological studies. The notion of frame operationalizes Giddens' conception on the relationship between social structure and interaction as well.

13. It should be noted that Goffman does not talk specifically about friendship frames. His own model is a dramaturgical one; social life is depicted as a stage where members perform for the benefit of an audience. He is interested in how their activities are 'staged.' Since frames refer, however, to structural principles for organizing experience, I see no reason why a substantive substitution cannot be made. See, also, Tannen (1984), who uses a frame-perspective for analyzing talk among friends.

14. Quasthoff deals exclusively with storytelling as something clients engage in opposition to professionals. My own data indicate that professionals are not necessarily or unambivalently engaged in obstructing storytelling, but may be utilizing it for their own ends as well, often to the detriment of the patient.
This point will be dealt with in the course of the analysis.

15. For a comprehensive discussion of the connection between defining a situation and power, the reader is referred to Berger and Luckmann, who write:

> The confrontation of alternative symbolic universes implies a problem of power - which of the conflicting definitions of reality will be 'made to stick' in the society (1966:126).

And, later, as if in answer to the question implicitly posed here:

> He (*sic.*) who has the bigger stick has the better chance of imposing his definition of reality (1966:127).

This conception of power is itself, of course, a bit simplistic, as well as being overly optimistic, as I hope to show in the analysis to follow.

16. At this point, I am not concerned whether or not storytelling is an *effective* way for clients to exercise interactional control over institutional encounters. Quasthoff is, in fact, quite skeptical about the advantages of this particular conversational strategy. Whereas it does provide some degree of control over the situation, it brings clients, at the same time, into a no-win position which can, ultimately, be degrading. This is because of the non-reciprocal nature of the encounter, whereby professionals do not engage in storytelling themselves nor set themselves up as friends. Thus, the very reciprocity, which the story is designed to ensure, is doomed to failure (Quasthoff 1979: 119).

17. It is not my intention to say anything conclusive about the GPs in my data compared to other GPs. I do not have access to material from practices where the physician has had no vocational training. Nor do I have any way of knowing what the physicians themselves think about their work or their patients. My impression, however, based on my experience with my data, was that the GPs allowed their patients considerable opportunity for engaging in troubles-talk. Moreover, I recognized various therapeutic techniques, which I know belong to the usual arsenal of training programs for GPs. Since the above deals with the possible advantages or disadvantages of a positive bias, the speculative nature of these observations does not seem problematic.

18. This was based on the feminist critique of how medical men treat women. I assumed for example that 'not taking women seriously' would also mean not giving them a chance to explain their difficulties or that 'medicalization' would entail transforming women's initial complaints with a minimum of muss and fuss into a prescription for tranquillizers.

    Whereas I do not reject the critique, my material, even at this initial stage of the inquiry, indicates that further research is warranted concerning how these various social 'facts' may occur in actual interaction. In other words, my naive assumption that GPs were not letting women talk or abruptly chucking them out of their offices with a prescription for Valium proved far too simplified.

19. My preliminary definition of storytelling was:

    -that it take more than one utterance to produce
    -that it concern some experience of the person telling the story
    -that it contain emotional or evaluative elements ('It really hurt.')

    In Chapter 6, I shall be dealing with storytelling in more detail; i.e. showing how this definition was derived, how stories differ from other narrative forms, the organizational features of storytelling in conversations, etc. Since there is considerable controversy in the field as to what a story actually 'is,' I shall be discussing some of the points of contention as well.

20. This was done with the help of Yen Ni Yap and Netteke Schouten-van Meeteren.

21. There are, as it turns out, reasons for assuming that this is true. See, for example, Wodak (1981), who discovered that in a group therapy setting, men and women chose different forms of communication for talking about their problems. Men provide reports or descriptions of the circumstances, whereas women seem to provide narratives of their life histories.

6

TALKING ABOUT TROUBLES: STORYTELLING IN GENERAL PRACTICE

INTRODUCTION

In order to operationalize how patients talked about their troubles in general practice consultations, I began by taking a look at the more general conversational practice of storytelling, assuming that troubles-telling, as a specific kind of storytelling, would share many of the same interactional features. If, in my initial optimism, I had expected to find a straightforward description of what constitutes a storytelling, particularly when the story is about a trouble, I was soon to be disillusioned. Upon examining the current research on the subject of stories and storytelling, it became clear that there were nearly as many definitions as there were studies. These definitions range from Sacks' (1978) pragmatic formulation of stories as 'ways of packaging experience' to more elaborate treatises on the structural elements, which must be present if we are to talk about stories at all (Rummelhart 1975; Labov and Waletzky 1967). Stories are viewed as literary texts, linguistic units or social activities, depending on the discipline and vantage point of the particular researcher.

In this chapter, an attempt will be made to find a way through this confusing collection and come up with some useful working definitions for the present inquiry. I shall begin with stories and storytelling, showing how they distinguish themselves from other forms of talk and discussing some of the interactional tasks involved in their accomplishment. This will be followed by a closer look at one particular kind of story, stories about troubles, and the specific tasks and problems involved in a troubles-telling. And, finally, some guidelines will be formulated concerning story- and troubles-telling in general practice consultations, which will serve to get me started on the analysis.

STORIES AND STORYTELLING AS GENERAL CATEGORY

Some preliminary distinctions

Before defining storytelling, a few preliminary distinctions need to be made. The first concerns *stories as literary objects* and *stories in everyday situations*. Stories have originally been the domain of literature; i.e. written texts produced by professionals using aesthetic standards. The kind of storytelling produced routinely by people in everyday life was considered too trivial to merit serious attention. It is a relatively recent development that linguists and social scientists have concerned themselves with non-professional storytelling.<1> Since then, storytelling has been examined in a variety of mundane contexts, ranging from friendly gatherings to institutional settings like courtroom hearings, therapy sessions, social welfare consultations, and visits to the hospital ward (Labov and Waletzky 1967; Polanyi 1978,1983; Quasthoff 1980; Ehlich 1980).

Another important distinction can be made between stories, which are *elicited* by the investigator and stories *occurring naturally*. Just because investigators became interested in everyday stories, this did not mean that they were prepared to wait around until the stories were told spontaneously by their subject(s). For example, in a pilot study on storytelling by Black adolescents in Harlem, Labov (1972) actively encouraged the production of stories by means of questions like 'Were you ever in a situation where you were in serious danger of being killed?' or 'Were you ever in a fight with a guy bigger than you?' This, together with the occasional prompting like 'What happened next?,' proved sufficient for eliciting a full-fledged storytelling.

Whatever the advantages of this technique, it is not useful for the investigator, who does not have direct access to subject(s). Moreover, it suggests that all stories might have the same form as those actively elicited by the investigator. Recent studies on stories emerging spontaneously in specific contexts, however, indicate that stories vary according to the context (Ehlich 1980).

A third and related distinction concerns viewing stories as *linguistic units* or as *social activities*. Generally speaking, this distinction reflects a linguistic or sociological approach to storytelling. In the former, stories are expected to have a standard or ideal format, which proceeds along chronological lines. Labov's (1972:363) scheme is a good illustration where stories are supposed to be replete with abstract, orientation, complicating action, evaluation, result or resolution, and coda. Whereas the possibility is taken into account that there may be the occasional omission or exception, it remains exactly that - a deviation from the norm.<2>

Sociologists, particularly within the field of conversation analysis, have been less interested in stories *per se*. Their interest has been in storytelling as a social activity (Sacks 1970, 1973, 1978; Jefferson, 1978b; Ryave, 1978). If a definition is provided at all, it tends to be of a common sense variety like 'a way of packaging experience' (Sacks, 1978). Assuming that everyone knows *what* a story is, they proceed to show *how* stories come to be told, what kinds of interactional tasks are involved, and how various social activities can be accomplished by means of a storytelling. This yields a *praxiological* definition of storytelling; namely:

1. Stories take more than one utterance to produce (Sacks 1970:1).
2. Stories have to be 'fitted' into the conversation; they don't 'make their own place' (Sacks 1978:261). This 'fitting' has to be negotiated.
3. Stories are interactional accomplishments, requiring specific tasks of both the teller and the recipient. It is the teller's business to 'tell the story with respect to its import for him' and the recipient's to 'display his understanding of it' (Sacks 1978:261).

Returning to the aims of the present inquiry, it will be clear that I am not interested in stories as the literary productions of professional storytellers. My focus will be on everyday storytelling, in particular, as it occurs in the context of a medical encounter. While patients, when asked, will provide stories (Webb and Stimson 1976), I shall be looking at storytelling, which occurs spontaneously in the course of a consultation. Furthermore, my interest in storytelling is sociological rather than linguistic. I am not concerned with delineating the structural features of patient-stories in general. On the contrary, my aim is to show how patients and physicians engage in storytelling during medical consultations and how this storytelling functions in an ongoing struggle for power. Since I am assuming that this activity will meet with some obstruction, I will be exploring how participants go about negotiating these obstacles. It is my hope that these negotiations will provide a good starting point for investigating how asymmetrical power relations are being constructed. Thus, in my inquiry, storytelling is a means to an(other) end: a good analytic category for a study of power at the level of medical talk (Davis 1985).

Although the praxiological definition of storytelling produced by conversation analysts provides a starting point for delineating the social activity of storytelling, there are, in fact, other related ways of 'packaging experience.' For example, narratives, descriptions, and reports not only involved similar conditions for their production in conversations, but they could be found in my own data as well. This raises the question of what it is exactly, which makes a story a 'story.' After examining several alternative forms of talk, I shall provide a working definition for the present inquiry, which includes the specific quality of a story. This will be followed by a more detailed look at the interactive tasks involved in storytelling.

Storytelling vs. other narrative forms

To being with, stories need to be distinguished from *narratives*. In many languages, they are used interchangeably, where one term applies to both.<3> In English, 'narrative' is a general concept, including, but not limited to, stories. The Oxford dictionary defines it as a 'tale,' 'story,' or 'recital of facts.'

Although a story and a recital of facts both fall under the heading of narratives, common sense tells us that they are different and it is to this difference that we must turn in order to discover the most essential feature of a story.<4>

To this end, two narrative forms will be examined, which entail the provision of a factual account; namely, descriptions and reports. This will be followed by another look at stories. Despite the similarities between these forms, it will be demonstrated that there are significant differences as well. These differences involve the subject matter, the way the speaker organizes the recipient's attention, and the kind of situation, which is appropriate for each narrative form (Rehbein, 1984).<5>

Descriptions

A description is about the visible, spatial or surface aspects of an event, an object or a state of affairs (Rehbein 1984:74-89). Its aim is to provide the recipient with sufficient information about the object in question that she can form a perceptual impression of it. Obviously, there is very little which does not lend itself to description.

Descriptions are rarely produced 'for their own sake;' i.e. for the enjoyment of the person doing the describing or the person on the receiving end.<6> It is much more likely that a description will be elicited by someone, who needs certain information in order to orient herself. It is geared toward providing as full and accurate an account as possible, enabling the recipient to find her way about. Thus, descriptions are provided, following a query about how to get from one place to another in a strange city. During a job interview, the applicant will normally receive a description of the kind of work the position entails. In medical consultations, descriptions are frequently elicited and provided by patients concerning the location of their pain, how it feels, etc.

Descriptions are the only narrative form, which is considered, almost by definition, to be objective. Since the aim is to provide an accurate and complete account of the surface dimensions of reality, it is naturally assumed that the final product will be, more or less, identical with that reality. What the person doing the description thinks or feels about the object in question, is not relevant for the description itself. Whereas there is some disagreement as to whether it is possible to produce descriptions free of subjective 'contamination,'<7> it is not surprising that parties, interested in obtaining an accurate picture of what is 'out there,' might prefer to have it provided in the form of a description.

Reports

Unlike descriptions dealing with the surface aspect of reality, a report is about a series of events or occurrences, which took place at some time in the past (Rehbein 1984:89-104). Its aim is to provide the recipient with an overview of what happened. She does not necessarily have to be able to visualize the events in question, as long as she can reconstruct them.

Reports, like descriptions, are not produced for pleasure, but rather to provide information. Frequently presented in written form or with a standard format, they are couched in the technical language, specific to a particular institutional context. The speaker does not have to have been directly involved in the events in question and, in any case, her involvement is not relevant for the report itself. She functions as a neutral witness or a transmitter of information. By reconstructing events in a logical and orderly way, the recipient will be able to follow what happened and, ultimately, reach some kind of decision. Some familiar examples of reports are: school reports, research reports, weather reports, and sales reports. In medical settings, reports are a familiar occurrence. Results of lab tests or medical examinations are typically communicated in report-form. A medical history may be written up as a report, whereby the patient's symptoms and the course of her illness are documented in such a way that a diagnosis is facilitated.

Reports are not, strictly speaking, true. Much of the reporting which appears in the newspaper might be criticized on that count. They are, however, invariably presented as factual accounts of what happened. In fact, a report suggests that reality is sometimes in need of some professional editing; i.e. it must be organized in order to make it useful for other, often institutional, purposes.

Stories

Stories, like reports, involve a recounting of something that happened, usually an event or series of events at some time prior to the telling. As in a report, these events are, often, but not necessarily, organized chronologically.<8> Stories will frequently include descriptions of the people involved, the location or other surface information required for understanding what happened. This is, however, where the resemblance ends (Rehbein 1984:104-113).

Stories are not factual accounts of the events in question. The events may have actually happened as, for example, in a True Story,<9> but the emphasis is not on describing them accurately or objectively. On the contrary, stories focus on the unusual character of what happened. There will always be a twist, a complication or something unexpected, which transforms even the most ordinary happenings into a story-worthy item. Stories are about something *especially* saddening/sickening/frightening/amusing/shocking, etc.

Stories may provide information, but this is more likely to be a side-effect than its *raison d'être*. They are much more likely to be produced for purposes of entertainment, establishing social contact or presenting the storyteller in a particular light - for instance, as hero, victim of circumstances or a Good Samaritan.

Stories are the only narrative form, which is admittedly subjective. Whether or not the storyteller is the protagonist, <10> she will be involved emotionally in the events in question. This is why she is telling them to begin with. Stories are interactional productions in which, by re-enacting what happened, the storyteller attempts to draw her audience into the experience so that they can share her feelings. In other words, recipients are expected to feel sympathy, express surprise or laugh. In fact, the story will be organized in such a way that the significance of the events is continually being underlined, thereby increasing the teller's chances that she will obtain the appropriate response from the recipient(s).

In conclusion, a cursory look at several narrative forms - descriptions, reports, stories - indicates that there is considerable overlap. Reports may contain descriptions. Stories will inevitably display segments resembling both reports of events or descriptions of characters and places.<11> Particularly in conversations, which are, after all, rather messy affairs, we could expect that these forms can rarely be separated with any precision.

Nevertheless, it should also be clear that stories have one element which distinguishes them from other kinds of talk; namely, that they are admittedly subjective. No other narrative form has this quality. Stories make no pretense of being objective or factual accounts. They must, in fact, be more interesting than normal life to be worth telling at all. In short, it is the evaluative and subjective quality of a story which, ultimately, makes it a 'story.'

A Working Definition of Stories and Storytelling

Having examined some of the narrative forms available for 'packaging experiences,' an attempt will be made to provide a working definition for the purposes of the present inquiry. By 'working definition,' I mean exactly that: a definition that works. Its function will be to help me locate stories, occurring in my own data. This means that the definition will need to be flexible enough to account for stories, which emerge spontaneously as well as those which are the result of elicitations on the part of participants in the course of a consultation. Since these stories will presumably be subject to some disruption (Chapter 5), the definition must be broad enough to account for even the most fragmented of storytellings. In other words, it must include disorderly stories or stories, which, in the Labovian sense, are missing some of their parts.

To this end, I propose taking Sacks' and cohorts' definition of storytelling as a social activity in conversations a step further. Since it does not pay attention to what makes a story distinctive from other ways of talking about experience; i.e. the evaluative point, or take into account of the way asymmetrical power relations between participants effect how stories may be told, some elaboration is required. This brings me to the following definition:

1. *Stories are about the storyteller's experiences*. The storyteller is not necessarily the protagonist of the story, but she will have some involvement in the events. Stories may not be identical with what 'really happened.' They do not have the

pretention as do reports or descriptions of being objective representations of some series of events.

2. *Stories have an evaluative point*. There is always a reason for a story being told and that reason will involve the fact that the events in question are in some way unusual - in any case, unusual enough to make them story-worthy.

3. *Stories are interactional happenings*. The storyteller and story-recipients both have specific tasks in a storytelling. The former is responsible for telling the events with respect to what is important for her. The latter are responsible for displaying understanding, particularly of the story's point.
   Since stories always take more than one utterance to produce, negotiation will be required for them to be told. These negotiations will reflect and construct relations of power between participants and will be analyzable as such.

Whereas this definition proved sufficient for locating stories in my data, it was only the first step. My primary interest is not the stories themselves (although I will be discussing them as well). In particular, I am concerned with *how* they are being told and, more importantly, what happens when difficulties arise in the course of their telling. In order to uncover potential obstacles to storytelling, it makes sense to take a closer look at the kinds of tasks required for a story to be told. It is to the interactional work involved in a storytelling to which I shall now direct my attention.

Some interactional features of storytelling

To begin with, telling stories in conversations will involve more than recounting what happened or accounting for why the story needs to be told. It will have to be linked to the ongoing talk and, once completed, provide participants with a way of getting back to that talk. Or, as Jefferson (1978b), who has focussed particularly on the business of beginning and ending a story, notes: stories are *locally occasioned* by the talk and *sequentially implicative* for it.

Stories as locally occasioned

Conversations are essentially orderly affairs. Utterances will not be produced randomly, but must be made accountable in respect to the ongoing talk. They are related to the preceding talk in terms of topic, as well as activity being performed. Stories are no exception. They will not be abruptly dropped into the conversation, but must be derived from the talk-thus-far and fitted into it. Stories are 'triggered' (Jefferson 1978b:220); i.e. they must make sense both in terms of the ongoing talk and their narratibility or story-worthiness.

Stories as sequentially implicative

Not only are conversations topically coherent, but they are organized on a turn-by-turn basis. In other words, at least one, but not more than one party will speak at a time in a single conversation (Sacks, Schegloff, and Jefferson 1974). This turn-taking rule has consequences for a storytelling as well. A story will invariably take longer than one utterance, i.e. the storyteller will be embarking on a protracted stretch of talk. Thus, a story constitutes a disruption of the normal turn-by-turn organization of conversation.

This means that other parties will need to be oriented to the fact that a story is going to be told, so that they can temporarily relinquish the conversational floor and become aligned as story recipients. Moreover, they must be alerted to the story coming to a close in order to re-engage in turn-by-turn talk. Thus:

> The elaborated turn-by-turn emergence of a story may be methodically constructed, not only by reference to its appropriate, locally occasioned occurrence, but by reference to its projected reception (Jefferson 1978b:228).

On the basis of the above, it would seem that for a story to be 'methodically constructed,' at least the following tasks need to be accomplished:
- getting the floor
- recounting what happened
- making the point
- relinquishing the floor

I shall now deal with each task, showing what it entails for the storyteller as well as the story recipient.

1. *Getting the floor*

Since a story takes more than one utterance to produce, it will automatically entail an interruption of normal, turn-by-turn talk. For this reason, the first task facing any storyteller is getting the conversational floor long enough to tell her story. This involves alerting the recipient(s), preferably in the first utterance, that a story is about to be told.

Not only does the storyteller have to get the floor, but she needs to ensure that recipients are going to listen to what she is saying. This involves mobilizing their interest or, as Sacks (1970) puts it, not only getting the floor, but getting their 'ear' as well. Thus, the recipients, recognizing that a story is about to follow, will need to indicate a willingness to give up their turn at talking (temporarily), so that the storyteller may develop her story.

Potential interactional difficulties at this stage involve having the floor contested. Recipients may have other ideas about what the talk is about or even want to engage in a little storytelling themselves. Moreover, getting the floor is only the beginning. The storyteller must see to it that she holds the floor - at least long enough to tell her story.

2. *Recounting 'what happened'*

Stories are told about events or experiences. They involve various characters, as well as details about the time and place in which the events occurred. Whereas stories told in ongoing talk do not necessarily tidily proceed in stages (Labov 1972; Rummelhart 1975), the events do need to be recapitulated in such a way that the recipient is able to follow what happened (and what happened next).

The recipient will be required to demonstrate that understanding has occurred as the story progresses, enabling the teller to revise or embellish her tale as needed. Should the recipient fail to provide appropriate signs of alignment, the storyteller will attempt to elicit them. Thus, she will need to ensure that the story is an orderly achievement, making sense every step of the way.

3. *Making the point*

Stories, unlike reports and descriptions, do more than recount past events or states of affairs. The teller has a reason for telling the story. This reason will be that the events she (or someone else) has experienced are unusual enough to warrant being shared with others.

A story always has to be worth telling (Polanyi 1979).

Unusualness can mean particularly frightening, amusing, upsetting, etc. In any case, the events constitute some deviation from what the teller felt entitled to expect 'under normal circumstances.' Her own, often emotional evaluation of this unusualness is the story's point.

Unfortunately for storytellers, their point is not always immediately evident to other parties. It may even become the subject of considerable negotiation in terms of its unusualness, or reportability (Sacks 1978; Ryave 1978). 'So, what's so terrible about *that*?' the recipient may ask, thereby placing the storyteller under constraint to account for the story being told. Moreover, there is always the risk that the point will be missed altogether.

For this reason, the storyteller will need to establish the fact that the events in question constitute some violation of commonly-held norms about the way things should be. To this end, she will need to make use of shared background beliefs and knowledge. The recipient must provide recognition of the point in the form of an affiliative response, along the lines of 'Oh, really?,' 'How terrible/wonderful/crazy,' etc. Another commonly employed means of providing affiliation is 'Yeah, that happened to me once, too,' leading, more often than not, to a 'series-of-stories' (Ryave 1978).

4. *Relinquishing the floor*

If the floor must be obtained and held during the course of a storytelling, it must also be given away. At some point, the story will have to be brought to a close so that normal topic talk may resume. Not only is this a task facing the storyteller, but recipients will also be monitoring the story presentation for signs that it is completed or for possible transition relevance points (Jefferson 1978b), when they can return to turn-by-turn talk.

In conclusion, storytelling in everyday conversations requires considerable interactional work on the part of both the storyteller as well as the recipient. Stories are not told, monologue-fashion to brick walls, but are collaborative achievements.

Having described some of the general features of storytelling as well as how they are told in ongoing conversations, I shall now turn my attention to a particular type of story; namely, stories about troubles.

STORIES ABOUT TROUBLES AND TROUBLES-TELLING AS SPECIFIC CATEGORY

Stories may be told about a wide variety of events and experiences. In examining storytelling as it occurs in general practice consultations, however, it might be expected that patients will not be telling the same kinds of stories that they would tell at a family gathering or a meeting between friends. We might assume that if they engage in storytelling, then the story will be about some kind of trouble, physical or otherwise.

Whereas I did find an occasional example of other kinds of stories, the majority of the stories in my data were, in fact, about troubles.<12> For this reason, it makes sense to take a closer look at the research done on this particular type of storytelling. I shall be concentrating on two sources. The first deals with some of the structural features of stories about suffering (Rehbein 1980); the second with the extra interactional work required when troubles are discussed within any conversation (Jefferson 1980; Jefferson and Lee 1981).

Stories of suffering ('Leidensgeschichten')

According to Rehbein, there are three general types of stories: *heroic tales* (where the protagonist is confronted with a series of dangerous events and manages to overcome them), *stories about strange occurrences* (for example, UFO's or oddly-behaving people who produce feelings of puzzlement in the person relating the story),<3> and *stories of suffering.*

His own work deals primarily with the last story-type, based on stories told by second-generation Pakistani immigrants to social welfare workers in East London. Whereas the stories deal with the kinds of suffering one might expect to hear about in this particular setting (housing problems, difficulties with the children, confrontations with various authorities), he has attempted to delineate a general structure for the story of suffering as genre. As such, this structure might be applied to other contexts where such stories are likely to be told.

I shall now describe some of the features belonging to stories of suffering, which go beyond the general elements of storytellings and which distinguish them from stories about other kinds of happenings. An attempt will be made to relate Rehbein's findings to stories of suffering, which might be told in a medical setting.

Stories of suffering have a *characteristic structure*, which may serve as an underlying pattern for their production and interpretation. This pattern is not the same thing as a step-by-step account of how such stories would appear in an actual conversation, but rather a kind of interpretative scheme.<14> It has the following elements (Rehbein 1980:69):

1. The actor or protagonist of the story has been behaving normally.
2. The actor becomes a victim. Her victimization is caused by a direct attack from some offender, possibly directed at her because of the very 'normal-ness' of her actions.
3. The actor attempts to remedy the situation.
4. All attempts to remedy the situation meet with failure. Everything gets worse.
5. It becomes clear to the actor that she is a victim.
6. As the situation progresses, the actor begins to feel more and more helpless/powerless.
7. The actor seeks help from a person, who is perceived to be 'on her side'.

There are two aspects of this particular kind of story, which deserve particular attention. The first concerns the *circular or repetitive quality* of the storytelling; the second the *double-edged* quality of the *point*. I shall deal with each separately.

Repetition in stories about suffering

Unlike stories about other kinds of experiences, stories about suffering do not concern events, which can be relegated firmly to the storyteller's past. The events leading up to the suffering may be over, but the suffering itself is not. The storyteller still experiences the distress at the onset of the storytelling.

This unfinished quality of the events effects the way the story is told. Instead of proceeding in linear fashion, stories of suffering have a repetitive or circular quality, which is quite unlike other storytellings (Rehbein 1980). If the experience of suffering goes 'on and on,' it is not surprising that a storytelling, in trying to capture that experience, will also display a tendency to go on and on. This can be seen in the way storytellers repeatedly circle back to the emotional significance of the distressful event or link various current troubles into one continuing 'tale of woe,' seemingly without beginning or end. Whereas some authors regard this aspect of stories about suffering as something, which can be potentially bothersome for an ongoing conversation,<15> it seems more reasonable to regard it as part of the kind of story being told. The unresolved quality of the experience is something that the storyteller will need to convey. It is a crucial element of what the story is about.

The double point of stories about suffering

To begin with, stories about suffering, like any other story, concern an experience, which is considered unusual or abnormal enough to warrant being told. The storyteller/protagonist suddenly finds herself in a situation, which is different than what she would 'normally' expect. Because the events cause suffering, the point of the story will be something like 'This is *really* terrible/pain- ful/upsetting/frightening ...'

Several responses are appropriate for acknowledging the point of a story about suffering. The recipient may display *sympathy* or *commiseration*. 'How awful for you' or 'I know just what you mean-that happened to me once, too' are examples of this type of response. Or, the point may be acknowledged by disagreeing with it. 'Oh, come on, it's not so bad' and, a bit less affiliatively, 'Yes, you and a million others' are examples of such acknowledgement. By reformulating the grounds for the storyteller's suffering, the person with the trouble may be reassured. Her situation is bad, but not *that* bad (Sharrock and Turner 1978).

This is, however, only part of the point of a story about suffering. The second element of the point concerns that fact that the sufferer is a *victim of circumstances*. This is what Rehbein (1980:79) calls the 'skandalon.' The person with the problem or affliction has been struck down by circumstances beyond her control. Not only is she blameless, but she has done everything possible to help herself.<16> At the outset of the storytelling, she has reached the end of her tether. She is powerless to help herself. Thus, the point is also something like 'I need help.'

There are several ways that a story recipient might respond to this aspect of the story's point. One possibility would be to acknowledge or *protest* the unfairness of the situation. For example, 'Why did this have to happen to *you* of all people?' would highlight the injustice of the storyteller's particular predicament. Another extremely common response, of course, would be to attempt to *remedy* the situation. Providing some advice would be a typical reaction to the storyteller's display of powerlessness and inability to help herself.

It is this double-edged point, which not only distinguishes stories of suffering from other types of stories, but accounts for some of the difficulties involved for the recipient in acknowledging it appropriately. It is not as straightforward as the points of other stories.

Stories about suffering in medical encounters

Based on Rehbein's (1980) description of stories about suffering as well as my preliminary analysis of the data, I can begin to have a sense of what stories of suffering might look like in the context of a medical encounter - a context where the suffering is related to a complaint brought in by the patient. A patient's story of suffering would probably proceed along the following lines:

1. The patient/storyteller was, prior to the story's onset, healthy.
2. She becomes the victim of illness. Despite all efforts to the contrary (i.e. proper diet, keeping bundled up in the winter, getting enough sleep), she has been 'attacked' by a virus/flu/stomach-aches/insomnia.
3. The patient attempts to remedy her illness.
4. All attempts at curing herself fail. She keeps getting more and more ill.
5. It becomes gradually clear to the patient that she is *really* ill this time.
6. As her illness gets progressively worse, she begins to feel helpless and powerless about doing anything about it.
7. Finally, she seeks help from someone she thinks will be able to do something about this problem - her GP.

Patient-stories, resembling other kinds of 'Leidensgeschichten,' would also be likely to have a circular or repetitive quality. This would involve telling the same experiences over and over again or, perhaps, presenting one troubles-telling after another. In either case, the stories would be connected to the patient's pain or distress.

Patient's stories would also have a double point. On the one hand, the patient would need to stress that her illness or problems are unusual enough to warrant being told. Statistically speaking,<17> it is not unusual to be ill. Nevertheless, most people seem to feel that being ill is a deviation from the way they 'normally' expect to feel; namely, healthy and, more recently, happy (Illich 1976). Whereas this expectation may be unrealistic, it does provide the basis for considerable storytelling about illness and other problems of living.

On the other hand, patients, in addition to establishing their troubles as unusual and deserving of sympathy, would also have to establish their inability to help themselves. Illness, like other tragedies of fate, is something that 'happens' to the person without her being able to do anything about it.

For this reason, we might expect a patient to go to considerable trouble to establish that she has done everything she can to help herself before coming to the doctor with it. Asking for help is a potentially delicate business. In a culture where helplessness is not a particularly positive characteristic, it carries face-threatening <18> dimensions to indicate that one is in need of it (Goffman, 1967).<19> Thus, the patient would do well to establish herself as a competent person, prior to displaying her helplessness in the face of her illness. The recipient can then respond in kind, reassuring her that, despite her illness, she will remain an autonomous adult, entitled to the same conventional treatment and respect to be meted out to everyone else.

In a medical encounter, the issue of help will presumably be even more complicated than it is when a story of suffering is related to a friend. In a chat between friends, the recipient may want to lend a helping hand to the sufferer, but she is not under constraint to do so by virtue of her professional expertise. Presumably, both the storyteller-patient and the recipient-physician will expect the latter to provide help and somehow alleviate the suffering. Moreover, it may be easier in a conversation between peers to treat the patient as a 'member of the fold' than in a professional encounter with its various structural asymmetries.

In conclusion, we would expect stories about suffering told by patients to resemble Rehbein's (1980) 'Leidensgeschichten' in terms of their content and structure. Presumably, they will involve some repetitiveness and display a double-edged point; i.e. that the suffering is (unusually) terrible *and* that the sufferer can no longer help herself. It is this double-edged point, together with the specific constraints involved in a helping encounter, which will require some careful handling in the course of the interaction. It will be implicated in the moment-by-moment struggle for power between the participants.

Where Rehbein has provided an interpretative scheme for understanding how stories about suffering are structured and interpreted in conversations, especially in terms of the double point of this particular type of story, conversation analysis has focussed on the conversational activity itself: troubles-telling.

Troubles-telling in conversations

Considerable research has been done on how stories about troubles are told in the course of an ongoing conversation. Most notably, Jefferson (1980;1984a,b) and Jefferson and Lee (1981) have devoted considerable attention to the social activity of troubles-telling. As previously mentioned, troubles-telling can be a rather tricky undertaking, requiring some special *management*. Unlike other kinds of storytellings, it is subject to almost continual disruption. I shall now deal with these two aspects of troubles-telling in more detail.

Managing troubles-telling

In our society, it is not acceptable to complain.<19> If a person wants to do so anyway, she may have to put some effort into finding someone willing to lend a sympathetic ear to her troubles. In other words, she needs to organize a 'troubles-receptive hearing' (Jefferson 1980:166).

One way this may be accomplished is by orienting the other party to the presence of a trouble, while, at the same time, indicating that the teller of the trouble is not going to make an undo fuss about it. This entails a simultaneous display of the potential trouble as well as attention to 'business as usual.'

A device particularly suited to performing this double-duty is the *trouble-premonitory-response*. This is a downgraded, conventional response on the part of the person with the trouble to a query into her well-being, indicating that a report on a trouble might follow, but does not have to.

For example:

A: Hi, how are you?
B: Oh, pretty good, I guess.

Here we see A performing a rather standard greeting, which is followed by a response completely in tune with conversational conventions. At the same time, it foreshadows the presence of a potential trouble. That is, we (and presumably A) all suspect that things are not all that 'good' with B. B does not immediately unload her problems, however. A has been warned that there is a potential trouble and she has been reassured that B is not going to talk about it, unless A specifically asks her to. If the trouble is to be told, A will have to elicit it with something like:

A: So, what's the matter, then?

Not only does this query return the floor to B so that she may engage in a troubles-telling, but it puts her under a constraint to do so. After all, it is rude not to answer a question (Schegloff and Sacks 1974). Thus,

> ... the one with the trouble exhibits an observable reluctance to tell the trouble until her coparticipant produces an utterance that aligns her as a co-tracker of the trouble, at precisely which point the one with the trouble switches to an observable eagerness to tell the trouble (Jefferson 1980:176).

This handy interactional device has the following advantages over simply launching into a troubles-telling right off the bat.

- It prepares the recipient to 'track' the trouble.
- It indicates an awareness on the part of the party with the trouble of conventions and a willingness to respect them.
- The troubles-teller demonstrates that she is not going to make an undo fuss.
- The troubles-telling is displayed as a joint endeavor, which is open to negotiation by all parties involved.

Faced with the interactional difficulty of initiating a potentially unwelcome form of talk (Bremer 1980a),<20> the teller may improve her chances of a sympathetic hearing by managing its introduction into the ongoing talk. This is, however, only the beginning and in no way ensures that she will be able to tell the trouble.

Disruption of troubles-telling

Troubles-telling is subject to interruptions. This occurs with such regularity that Jefferson and Lee (1981) began to wonder whether it was a coherent conversational activity at all. Upon investigating the disruptions in more detail, however, they discovered that there was a certain 'method to the madness.' Troubles-tellings were being disordered in primarily two ways (Jefferson and Lee 1981:401-404):

1. It may not be clear that the troubles-teller is 'only' engaging in talk about her trouble. The recipient may 'hear' the troubles-telling as doing some *additional interactional activity* and respond accordingly. For example, a troubles-telling may seem like an excuse to avoid doing something. As is the case with all excuses, the recipient will, consequently, need to evaluate the grounds upon which the excuse is being proffered to see whether or not they are sufficient. Similarly, a trouble might seem to be an extenuating circumstance, an obstacle to the achievement of some goal. If the recipient hears it in this way, she will probably

engage in a discussion of the trouble and some negotiations around the plan itself, whereby the trouble may fade into the background.

In both instances, the trouble becomes secondary to some other concern. It becomes subject to dispute or debate rather than eliciting a sympathetic or affiliative hearing. This feature of being able to do 'double-duty' could account for troubles-telling having a disrupted or messy quality, when it occurs in conversations.

2. Another source of disorder is *interactional asynchrony*. In order for talk to proceed smoothly, co-participants need to be properly aligned in terms of membership categories and corresponding activities. A Troubles-teller and a Troubles-recipient, for example, constitute a fitted pair, as do a Service Supplier and a Service Recipient. In terms of general conversational categories, they are aligned. One is the speaker; the other the recipient.

   Put schematically:

conversational type	**conversational activity**	**speaker**	**recipient**
Troubles-talk	Troubles-telling	Troubles-teller	Troubles-recipient
Service Encounter	Advice Giving Provision of Help	Service Supplier	Service Recipient

As long as participants remain properly aligned, there is no problem in carrying off the interactional accomplishment of either activity. Jefferson and Lee (1981) discovered, however, that this is precisely what they do not do. Troubles-recipients apparently engage in Advice Giving in the course of a troubles-telling. This almost automatically leads to a dispute over the trouble.

The reason for this dispute is, among other things, that, by giving advice, the recipient moves from her recipient category to the category of Advice Giver. An Advice Giver is a Speaker category, whose fitted recipient-part is an Advice Recipient. As pair, Troubles-teller and Advice Giver are not fitted because two speaker categories are involved with no appropriate recipient.

Thus, the accepting of advice may bring with it removal

> from the category Troubles-teller and loss of whatever prerequisites that troubles-relevant category and its attendant conversational general category, Speaker, may entail. Correlatively, the delivering of advice may bring with it removal from the category Troubles-recipient and acquittal from whatever obligations that troubles-relevant category and its attendant conversation general category, Recipient, may entail (Jefferson and Lee 1981:410).

In conclusion, in everyday conversations, troubles-talk requires specific interactional work for its production and reception. Whereas the tasks facing participants differ, the accomplishment of a troubles-telling is a cooperative endeavor, presumably between interactional equals. Experiences involving suffering are organized in terms of characteristic interpretative schemes (Rehbein 1980). They are subject to special and specific interactional constraints when presented in conversations. Given the double-edged point of stories about suffering; i.e. 'this is terrible' and 'help me,' it is not surprising that the recipient will display a double response as well. She may find herself offering sympathy as well as advice. Offering advice, however, can lead to disruption of the ongoing troubles-telling. Thus, even in the context of ordinary talk, a troubles-telling can prove a problematic affair.

In a professional service setting, more is involved however. To begin with, participants are not interactional equals. This effects both how stories about troubles will be produced, as well as how they will be disrupted in the course of a conversation. Troubles-telling may prove even more difficult to manage in service encounters and, therefore, subject to more frequent disruption than is the case in ordinary conversations. Whereas storytelling may be the preferred form of talk for patients or clients, it is rarely preferred from the point of view of a professional service provider. Moreover, professionals have more powerful interactional resources at their disposal for disrupting stories than patients or clients have for initiating and sustaining them. And, finally, the business of providing help will presumably be even more constraining in a service encounter, where help is expected and oriented to by both participants. All of this lends particular significance to patients' attempts to tell stories, even in the face of opposition by a more powerful party. Storytelling in institutional encounters is clearly a good location for investigating power struggles, as well as how asymmetrical relations involving dependence and autonomy are being produced, reproduced, and undermined at the level of interaction.

On the basis of the current literature on storytelling and, more specifically, troubles-telling, I am now ready to formulate some theoretical guidelines for the analysis of these conversational activities as they may occur in general practice consultations.

STORY- AND TROUBLES-TELLING IN GENERAL PRACTICE CONSULTATIONS: SOME INFORMAL HYPOTHESES

Before formulating some descriptive or informal hypotheses as guidelines for the analysis to follow, a few words are in order concerning the difference between hypotheses, as they are generally formulated, and the descriptive guidelines I am offering here.<21>

Hypotheses are an inevitable ingredient of the hypothetical-deductive method, so common in the social sciences. They are meant to provide a theoretical statement that can be tested against the data and, thereby, verified or disconfirmed. My intention here is quite different.

First of all, my analysis is not theory-driven, but data-driven. It was my initial confrontation with my data, with its confusingly friendly physicians and Janus-faced power plays, which generated my search for both theory and analytic constructs, which would facilitate an understanding of this interaction. In other words, my data necessitated working out a framework for conceptualizing power (Chapter 4) as well as analytic concepts which could help me come to terms with it.

Second, since data, including my own, do not speak in an unambiguous voice, they need to be reconsidered in light of this theory and these analytic constructs. This means returning to the consultations to see whether story- or troubles-telling occurs as my preliminary analysis has suggested. Whether my theory and analytic categories can come to terms with the interaction in these consultations is, therefore, in the final analysis, an empirical question.

On the basis of the literature on story- and troubles-telling, I am assuming that the context of general practice will have an effect on how troubles-talk occurs within it. Moreover, I expect that this context will effect the production of troubles-talk in specific and systematic ways.

I shall now describe these expectations as a series of informal hypotheses, which will function as empirical questions for the analysis to follow. The first three are:

1. *If patients engage in any storytelling activity, then the story will be about some trouble.*

This implies two things. On the one hand, patients do not necessarily tell stories in the course of a medical interview. They may or may not chose to package their experiences in this particular form. On the other hand, if they do chose to engage in storytelling, their choice of story-type will be constrained by the situation in which they find themselves. In other words, they will be less likely to relate a heroic deed (Labov 1972) or a strange experience (Polanyi 1979) to the GP as they might to a friend or neighbor.

2. *If there is a story, there will always be some evaluative point concerning the unusual-ness of the event.*
   Stories may be differentiated from other forms of talk because they have a subjective or evaluative point. This point is what makes the teller's experiences worth telling. They must be demonstrated as unusual enough to warrant being told (Polanyi 1979). In the case of stories told by patients in consultations, this unusualness will involve experiences, which are (especially) painful, distressing, terrible, worrisome or whatever.
3. *If the story is about a trouble, the trouble will not only be marked as unusual, but it will be oriented toward some normal state of affairs or 'business-as-usual.'*
   This orientation towards normalcy may be expressed in both what the patient says as how she goes about saying it. We might expect it in the construction of suffering as a 'skandalon' (Rehbein), whereby the patient shows that she has become a victim of circumstances. Her present distress is not of her own making and she has, in fact, done everything in her power to avoid or alleviate it. She has finally been forced to admit defeat; she is powerless to do anything about it. By not accepting her present state of distress, she demonstrates that she is looking for a re-instatement of some other (more normal) state of affairs. She wishes to be helped.

   The patient will also be oriented to 'business as usual' in how she manages the troubles-telling within the ongoing conversation (Jefferson). She will be demonstrating that she is not going to make an undo fuss about the trouble and, ultimately, that she is still one of the fold.

These informal hypotheses are concerned with what might be expected of troubles-talk in general; i.e. troubles-talk in general practice consultation would simply be demonstrating the same features inherent in all troubles-talk. Storytelling, as previously noted, is almost identical with everyday talk. It is what friends and acquaintances are most likely to do when engaged in face-to-face interaction (Quasthoff 1980). Institutional talk, however, is not the same as everyday talk. Thus, we might expect stories about troubles to deviate from normal talk when produced in the context of a consultation. In particular, the following alterations might be expected:

4. *If storytelling is produced in a medical context, then patients will display an ongoing orientation to the consultation as medical encounter, even while engaging in storytelling.*
   I am assuming that, whereas patients may most routinely tell stories to acquaintances, they will not mistake the GP for a friend. Throughout the consultation, even as they engage in storytelling, they will demonstrate an awareness of the encounter as a medical one. In other words, storytelling efforts may be adding an additional frame to the situation, but not disregarding the institutional one.
5. *If story- or troubles-telling is produced in a medical context, it will not always be a welcome activity and as such will be subject to obstruction.*
   Storytelling produces and reproduces a friendship frame (Goffman 1974; Quasthoff 1980). At times, this will be facilitated or even utilized by the GP. At other times, it will be at odds with the institutional or service frame. We could, therefore, expect the GP, in his capacity as medical professional or service supplier, to - upon occasion - interrupt, disrupt or otherwise impede a patient's troubles-telling. In this sense, his talk would serve as an obstacle to the patient's talk about her troubles. Of course, from his point of view, these instances also mark the patient's talk as an obstacle to him or his medical agenda. Either way, troubles-talk in a general practice consultation could be investigated in terms of obstacles.
6. *If obstacles occur, once located, they will reveal struggles between participants over the definition of the situation.*
   Under 'normal' circumstances, troubles-talk is a collaborative affair requiring specific interactional tasks for its production and reception. Each troubles-telling will be examined in terms of the successful completion of these tasks. Obstacles may, but are not necessarily, located along these points:
   - getting and keeping the floor
   - recounting the trouble
   - making the point
   - returning to turn-by-turn talk

   I expect that obstacles to troubles-telling along any of these points would reveal an attempt to re-instate a medical or institutional frame. As such, these obstacles would be the locus of a power struggle between participants over who is going to define the ongoing situation and how they are going to do it.

7. *If obstacles are a locus of power struggles between the GP and the patient, then they should display the relational and multi-faceted features of power.*

   Viewed as interruption or disruption of troubles-talk, an obstacle marks an instance of the patient being obstructed in her activity by the GP. Her own activity may display resistance or acquiescence to this obstruction.

   If my theoretical assumptions are correct, however, this will only be one dimension of the power struggles occurring in the consultation. Obstacles will not only be displaying the constraining features of power, but the enabling ones as well.

   If the positive dimension of power is involved, then we would expect the GP to display an orientation to the patient's frame as well. He, too, would use it as a resource in his attempt to exercise control over the situation. In that case, the friendship frame, as constituted by troubles-telling, would be intimately connected with the mutual construction of relations of dominance and subordination between the GP and the patient.

Until now, the investigation has been proceeding slowly from a theoretical treatment of the research problem and a consideration of the various methodological implications, involved in its investigation, toward the empirical analysis of the material itself. It is to this task that I shall be addressing myself in the chapters to follow.

NOTES

1. Ehlich (1980:15) provides an interesting account of the recent interest in everyday storytelling in historical terms. He suggests that the ivory-tower approach to linguistic phenomena reflects the separation of intellectuals from the 'real world'. This applies to both those engaged in research on stories, as well as to those engaged in the production of stories. The result was an inability to come to terms with storytelling in general - because 'in general' means concrete, often institutional settings, with non-professional storytellers.

2. As Quasthoff (1980) has pointed out, if stories only rarely possess an ideal format in real-life, how useful is it to maintain such a format? How normal is the 'normal story format,' if it is almost never present in a 'normal' conversation? See, also,

Riessman (1988).

3. In German as well as Dutch, there is no separate term for narrative. 'Erzählung' and 'verhaal' can be used for storytelling or narrative. A distinction is made, however, between stories and reports.

4. Interestingly, Labov (1972:363), who does not, at first glance, distinguish between stories and narratives, does provide a clue for the difference between them. Among the elements of a narrative (or story) are a complicating action, followed by an evaluation and a result or resolution. Whereas result and resolution appear within the same segment, they are decidedly different. A *result* can refer to 'what happens next,' saying nothing about how the speaker or protagonist might feel about it. *Resolution*, on the other hand, suggests that the complicating event has brought about a state of tension, which is then altered in the course of the narrative. This change will be reflected in how both parties feel about the story. They will not just be 'informed,' but experience 'relief.' For a sensitive discussion of this difference, including examples convincingly demonstrating this point, see Ludwig (1984).

5. The following account rests heavily on Rehbein (1984), where a systematic discussion of the similarities and differences between the three narrative forms is provided. It was of particular interest to me as well, because he includes descriptions in his account as opposed to just reports and stories. This is useful for the present inquiry as patients often provide descriptions of the location of their symptoms, interspersed with reports on the progress of their illness, as well as stories about the experience of being ill.

6. The obvious exception is the literary description, which is often enjoyable for its own sake.

7. Within positivist social science, descriptions are the only narrative form considered useful as data. According to Rehbein (1984:85), this 'objectivity' is, in fact, nothing more than the way a description is constructed. In other words, it is assumed in advance that the person doing the describing is separate from what she is describing. Objectivity is one way of constructing reality, no more identical to that reality than another construction (report, story) would be.

8. See, for example, the extremely interesting work of Riessman (1988) on how people tell stories about why their marriages failed. She discovered that middle-class women are more likely to order events chronologically, showing what led up to the divorce, whereas working-class, non-Anglo women may order their experience in terms of a common theme. Similarly, men and women display special differences in storytelling with men focussing on one incident responsible for the breakup and women on an accumulated series of events.

   This is provocative in that it suggests that chronological order is but one way to organize a story and that there may be gender, class, and ethnic differences in the way stories are organized.

9. Interestingly, 'True Stories' are so designated, when the events in question are so

terrible (shocking, bizarre, etc.) that the recipient is inclined to doubt whether they actually could have happened.

10. Some authors maintain that the teller of the tale must also be the protagonist. Quasthoff (1979:104) rejects this distinction in favor of the tller having some personal involvement in the events in question. See the 'double-uterus'tape (Chapter 7) for an example of this.

11. Polanyi (1979:208) suggests that stories contain separate information structures: event, durative descriptive, and evaluative. A report might be compared to the event structure. Thus, reports and stories show some overlap. The evaluative structure, however, is found only in stories.

12. Stories about troubles are not the only kinds of stories patients tell. Aside from the occasional story as part of routine small talk, patients may also tell stories about their experiences with the medical profession. Webb and Stimson (1976) divide these accounts into 'atrocity stories,' 'critical stories,' and 'typical stories.' Whereas I did find some examples of this kind of storytelling (see the 'one-continuing-story' tape, where the patient describes her experiences with her therapist), the majority of the storytelling is about the patient's current distress and/or what led her to seek the services of the GP.

13. According to Rehbein (1980), these stories are frequently employed in courtroom hearings by defendants interested in building a case for their innocence or at least providing extenuating circumstances. In general, this type of story would be appropriate in any situation where the teller needs to provide an account for her subsequent actions.

14. See, also, Cicourel (1972:43-59) for a more detailed discussion of interpretative procedures.

15. In their analysis of storytelling attempts of patients in psychoanalytic interviews, Flader and Giesecke (1980) stress the psychological aspect of the stories. The storyteller attempts to gain emotional relief by 'working through' her experiences. Whereas this accounts for the repetitive quality of the stories, the authors seem to feel that it interferes with the actual story itself. It is my contention that they are confusing storytelling which is, in and of itself, an emotional form of narrative, with reports. In other words, by reiterating her feelings, the storyteller is simply doing what she needs to do to tell her story. Since these feelings are still present in a story about suffering, she will be obliged to repeat them.

16. In order to understand why this aspect of the point is necessary, it might be useful to consider what would happen if the storyteller was not a blameless victim. Self-inflicted suffering, for example, would need to be very carefully managed, if the recipient is to become aligned as sympathetic listener. Commiseration could only be obtained if the storyteller established herself as appropriately remorseful about her actions. In other words, additional accounting would be necessary like: she didn't

oversee her actions, she had a change of heart afterwards, etc. Otherwise, the troubles would stand to be re-formulated by the recipient, along the lines of 'So, what did you expect?' and the grounds for telling the story at all would become very shaky, indeed.

17. See Meigs (1961:861) and Siegel (1963), quoted in Zola (1972:678). In annual check-ups for even supposedly healthy people, there was scarcely an individual, who did not possess a symptom, which could be regarded as clinical entity worthy of treatment. General surveys indicate similar findings, warranting the conclusion that most of us are sick to some degree every day of our lives (Pearse and Crocker 1954; Commission on Chronic Illness 1957).

18. According to Goffman, this leads to somewhat paradoxical behavior on the part of people in need of help. For example:

> The interesting thing about medial symptoms is how utterly nice, how utterly plucky the patient can be in managing them... for each of these deviations from normal social appearances and functioning, the patient will be able to furnish a compensating mode of address.
>
> He gives accounts, belittles his discomfort, and presents an apologetic air, as if to say that in spite of appearances he is, deep in his social soul, someone to be counted on to know his place, someone who appreciates what he ought to be as a normal person and who is this person in spirit, regardless of what has happened to his flesh (1971:407).

19. See Bremer (1980a) for an interesting treatment of the phenomenon of complaining. In western society, it has been relegated to something, which deserves our dislike. All synonyms for complaining leave a negative after-taste (Bremer 1980a:13). The stiff-upper lip mentality is what receives respect. This negative view of an activity, which is, after all, a very human one, is due to the fact that we do not have social and/or collective rituals for complaining. Moreover, complaining is invariably linked with the business of providing help - something which complicates matters considerably and possibly accounts for the reticence with which complaining is received.

20. I do not mean to suggest that other forms of storytelling do not require special handling. For example, a story generally requires a kind of preface in order to be initiated in an ongoing conversation (Sacks, 1970).By providing a synopsis of what it will be about, the storyteller gives the recipient a chance to elicit it. In conversational settings where parties occupy asymmetrical positions, as in storytellings performed by children for adult recipients, the child will first give the floor away, only to have it returned because an answer 'must' be given to the adult's question:

> Child: You know what?
> Adult: What?
> Child: .... (story)
> (Sacks 1974:229)

These devices resemble the trouble-premonitory response in terms of how they function to get the floor for the storyteller and ensure a receptive hearing for the tale on the part of recipients. What makes the trouble-premonitory-response different is the attention paid to equivocality. Not only does the recipient have to elicit it, but she has to be reassured that there won't be endless complaining and that the teller is aware of the potentially unwelcome nature of what she has to say.

21. See, also, Schwartz and Jacobs (1979:322-327) and their account of 'conceptual laws,' which help the investigator find her way into the data. The rationale is that:

> There are patterns in the world which cannot be found unless one already knows (or believes) that they are there. The person who thinks he knows of such a pattern is in an unusual position. Ordinarily one has a hypothesis, knows pretty well what it means, and wants to know if it is true. Here one has a hypothesis, is pretty sure it is true, and wants to know what it means (wants to know *what* is true). In order to answer this question, he treats his hypothesis as a conceptual law and puts a bit of knowledge beyond the question for a while in order to 'see where it will lead' (Schwartz and Jacobs 1979:327).

7

SELECTING CONSULTATIONS FOR FURTHER ANALYSIS: A STRUCTURAL DESCRIPTION

Moving from a preliminary analysis of data to a theoretical framework and establishing analytic categories for investigating the research problem, I am now ready to embark on the actual analysis. Four audio-taped consultations have been selected from the 20 tapes, which were used in the preliminary analysis (Chapter 3). In this chapter, I shall discuss how these four tapes were selected and transcribed. This will be followed by a structural description of the consultations as a whole, which will serve as an orientation for the analysis to follow.

SELECTING MATERIAL FOR THE FINAL ANALYSIS

My first consideration was *theoretical*. Based on Glaser and Strauss-type notions on generating grounded theory by means of theoretical sampling, I asked myself whether the material at hand provided enough relevant material to adequately work out my research problem and do justice to theoretically important areas; i.e. the feminist critique of health care for women, medicine as an instance of social control, as well as the new general practice training programs and ideologies (Chapter 1). Was my material relevant for the issues mentioned earlier on in this investigation? Do the consultations, for example, enable me to show how women's initial troubles may be transformed in the course of a consultation? Are there good examples of potential obstacles she may encounter along the way? Does the material contain enough of the Janus-faced and ambiguous phenomena described in Chapter 3 to make a clarification of the subtle workings of power possible? Are there instances which allow a demonstration of how 'nice-ness' can also be oppressive to women? Obviously, such considerations are qualitative. This entails looking for the best possible examples for delineating these research questions.

My second consideration was *practical.* As anyone engaged in discourse or conversation analysis will readily attest, it is essential to limit the material to be analyzed. This is, quite simply, because the method itself involves viewing the material in terms of various levels of interactional organization, from utterance-to-utterance to the overall structure of the conversation and, ultimately, the way social structure is produced and reproduced in and through the ongoing talk. This yields not only a wealth of information, but also involves the painstaking and extremely time-consuming task of organizing all of this interactional detail into a structural analysis. With this in mind, one could say that the ideal would be one tape which provides all the information necessary for coming to terms with the research problem. Whereas this is sometimes possible,<1> in the present inquiry it was, unfortunately, *not.* This is presumably due to the fact that my research problem has several levels. For example, I needed tapes that would display how a patient's troubles were being transformed as well as how power was implicated in that process.

In making my choice of tapes, I selected those, which provided the best opportunity for analysis of both processes. Thus, throughout this selection procedure, my main concern was: can I do it with *less* than this? I wanted to select the smallest number of tapes, which would enable me to come to terms with the research problem. After several months of preliminary analysis, I ultimately reached a point of 'theoretical saturation' (Glaser and Strauss 1967:61-62). I was not finding any additional information in my tapes, but rather instances of the same phenomena over and over again. My final selection of tapes was based on it being just enough, but not more than, what I needed to develop a grounded theory of power in gender relations in medical encounters.

My third consideration was *methodological.* I was not simply approaching my tapes in terms of theoretical or practical considerations, but I needed to find tapes, which would enable me to embark on an analysis, using story- or troubles-telling as an analytic category. In particular, the consultations had to be sufficiently diverse to provide a good idea of some of the interactional tasks facing storytellers (and recipients) as well as how they can be obstructed in general practice consultations. Whereas it was not my intention to provide a typology of these obstacles, I would need enough material to be able to analyze the workings of power in consultations between male GPs and female patients.

With these considerations in mind, I returned to the 20 tapes used for the preliminary analysis. As was expected (Chapter 5), they all contained some instances of storytelling, although some were very fragmented.<2> Moreover, these stories were invariably about some sort of trouble, whether physical or otherwise (Chapter 6). The troubles generally concerned the patient, although in several cases, other people's troubles (friends, relatives) were mentioned as well. In one tape, atrocity stories were told about another professional helper (see Chapter 6, note 12).

Having established that stories about troubles occurred regularly in my material, the tapes were chosen to display a variety of storytelling patterns. In addition to selecting tapes, which enabled me to show how this activity may be accomplished, as well as some of the difficulties encountered by storytellers in the process, I chose two tapes, which might be regarded as extreme cases. In the first, the patient very rarely engages in storytelling activities; in the fourth, she has almost unlimited ability to do so. My ultimate selection included the following four tapes, which can be characterized in terms of their storytelling patterns as follows:<3>

1. *Story fragments*. In this consultation, the patient makes repeated attempts to get a story started and is cut off each time with routine medical questioning. It is a good illustration of a medical frame prevailing almost without interruption, as well as of the least successful effort at troubles talk.
2. *Two stories*. The patient initiates one story about a medical-type problem. When the GP appears at a loss to do anything about the problem, he elicits a second story, this time about a psychological difficulty. These stories are linked in a way, which can demonstrate how storytelling may be put to other interactional ends.
3. *An interrupted story*. The patient begins with a story and is interrupted by a combination of a 'joke' and a routine pill-check. As she is about to be ushered out the door, she manages to pick up the story again and proceed for another 2 minutes.
4. *Continuous stories*. In this consultation, the patient has more than ample opportunity to engage in troubles talk. She produces one story after the other and has an unusually affiliative recipient response from the GP. When the stories are, ultimately, obstructed, this is done in a very subtle, 'empathetic' way.

Having selected four tapes suitable for the analysis, the next step was to transcribe them.

TRANSCRIBING THE TAPES

As previously mentioned, conversation analysis has always stressed the importance of using naturally occurring conversations as data. As research material, painstaking and accurate transcriptions are made, based on repeated listenings of taped material. Whereas no transcription system is ever a literal representation of the conversation itself,<4> an attempt was made in the present inquiry to transcribe the consultations in a way, which provided enough detail for me to be able to uncover even the most subtle aspects of the sequential ordering of talk. Thus, interruptions, simultaneous speech, intonation patterns and the like were meticulously coded. Whereas some conversation analysts produce an orthography of the talk, I stuck to standard Dutch spelling in the original transcripts. The notational system used was a somewhat simplified version of Gail Jefferson's system (see Appendix). In the analysis itself, the original transcripts were used. Later on, the excerpts, which have been included in the text proper, were translated into English. This was done to facilitate reading, as well as to make the text accessible to a larger audience.

Whereas conversation analysts from countries where English is not the main language are often compelled to translate their transcripts for similar reasons, a few words are in order concerning some unavoidable limitations of this practice. If a transcript itself is a less-than-perfect representation of the actual conversation, then a translation of the transcript will be even less accurate. In my own translations, I was motivated primarily by two considerations. The first was that I wanted to capture the idiomatic quality of the talk, as much as possible. This proved a relatively easy undertaking, although the occasional snag in the translation has always been duly noted in a footnote. The second consideration was less simple, however, to implement. Since I had recorded interruptions, intonation, and overlapping speech with great care in the original transcript, I wanted to preserve the original word order as much as possible. This was often impossible, as it would have produced a translation which was virtually unintelligible. Therefore, I opted for a translation which did justice to what was said and was less precise about the exact point where an interruption occurred. This means that the translated excerpts should not be viewed as transcripts in the conversation analytic sense of the word, but rather as a more ethnographic rendition of the talk.

Having transcribed and translated the consultations, my next task was to provide a structural description of them in their entirety, before embarking on the analysis itself.

DESCRIBING THE CONSULTATIONS

The preliminary descriptions of the consultations as a whole provide a 'contextualized frame' (Cicourel 1982) from which specific segments will be selected and submitted to further analysis. Not only has relevant background information been provided, when available (for example, concerning the type of practice, age of the patient, etc.),<5> but an account of topics discussed and activities being performed has been given. In order to facilitate the investigation, each consultation has been organized in segments. A few words are now in order concerning the nature of this segmentation.

One of the truisms of conversational analysis is that talk is an orderly phenomenon, created ongoingly by members as a cooperative endeavor. Conversations, then, are organized events and it is possible, retrospectively, to delineate how this organizational work was achieved. This can be done at various levels: utterance-by-utterance, in terms of topics, or with respect to the overall structure of the conversation. In order to locate specific interactional activities (of which storytelling is but one possibility), it was necessary to describe the conversation as a whole. Whereas various 'structures' may be found within one and the same conversation,<6> I have organized the conversations around the activity of storytelling and potential obstacles to it. This produces a segmentation something like: opening-storytelling-other interactional activities or obstacles-storytelling-other activities ... etc. - closing.

I shall now turn to the four tapes.

A: Story Fragments (the 'classic case of gall-stones' tape)<7>

The physician is in his early forties. He has his own very busy practice in the suburbs of Amsterdam, which he describes as being 'very efficiently organized.' He has not received any vocational training above and beyond his medical examination.

The patient is fairly young, married, and has two children. This is not the first time she is consulting the GP. In this particular conversation, she introduces several topics:

- She wants to make an appointment for a *sterilization*.

- She is *tired* and *lethargic*.
- She has severe *abdominal pains*.
- She is *nervous*.

The consultation has been segmented in terms of how each of these topics is introduced and dealt with.

I *The Medical Entree: Sterilization*

Following a brief opening ritual, the patient indicates that she has come to make an appointment for a sterilization. She has been taking birth control pills for the last thirteen years and has finally decided to go ahead with the sterilization. After checking whether she or her husband want to do it and when, the GP proceeds to take her blood pressure and perform a pelvic exam as routine procedure for a pill check-up. He is starting to wind things up, when the patient introduces another complaint, thus establishing the pill check-up as a kind of entrée for other troubles.

II An Unsuccessful Story-Attempt: 'a little tired and lethargic'

The patient re-opens topic talk by requesting a blood test in a rather round-about way. Upon being questioned by the GP as to her reasons for this request, she indicates that she has been feeling tired and lethargic lately. She further justifies her wish for a blood test with the fact that a friend she knows had similar complaints and it turned out, that *she* had something wrong with her liver. The GP abruptly rejects her diagnosis, stating flatly that the woman 'must have had something else.' Her request for a blood test is met with a sarcastic remark concerning the absurdity of such a course of action. The patient concedes defeat and drops the subject.

This segment illustrates how a patient can display an orientation to the medical context in the way she introduces a new complaint. She requests a medical procedure (blood test), providing a lay-diagnosis of what might be causing the complaints.

Whereas it indicates the kinds of negotiating necessary for getting a story about troubles started, it is also an example of an unsuccessful strategy - unsuccessful, because it is immediately cut off by the GP and does not return as topic in the remainder of the conversation. It is as if there is a general rule in operation that patients are not supposed to talk like doctors (Fisher and Groce 1986).

III *Successful troubles-telling: 'a classic case of gall stones'*

In this segment, the patient initiates a third topic: she has terrible stomach pains. She introduces it, once again, in the form of a possible medical problem ('gall-bladder?'). It is impossible to say what the status of this particular complaint is for the patient, but, for the first time, a prolonged response on the part of the GP is forthcoming. He embarks on an extended question-answer sequence, whereby he appears to be working through the questions necessary to determine whether these symptoms are, indeed, indicative of a gall-bladder disturbance. At this point, he maintains tight control over the conversation by means of Q-A strategies (Merritt 1976) and the occasional formulation (Heritage and Watson 1979).

The patient is, on the surface, providing the information requested by the GP. Her responses, however, are not simply straightforward reports of her symptoms. She also provides a highly emotional account of her suffering, using complaining intonation, intensifiers, and direct speech. In short, her talk has the makings of what Rehbein (1980) has called a 'Leidens-geschichte,' albeit in fragmented form.

As a whole, this segment can be characterized as having two main features:

1. *Troubles talk as a successful and cooperative endeavor.*
   The segment is topically coherent. Both GP and patient are talking about the same trouble; i.e. stomach pains. They appear to be cooperatively engaged in developing it as topic within the ongoing consultation. From the *medical* point of view, it is a successful undertaking because the GP manages to reach a preliminary diagnosis and suggest further steps (x-rays, diet).<8> From the *patient's* point of view, she has been able to express her current distress to an attentive listener.

2. *Troubles-talk as an instance of crossed agendas.*
   Whereas this segment is clearly a harmonious achievement, it also displays the specific and different ways participants approach the trouble being told. This discrepancy becomes particularly apparent towards the end. The GP, having satisfied himself that the problem is the patient's gall bladder, begins to round off the conversation. The complaint has a name and, from his point of view, the business at hand has probably been accomplished. The patient, however, seems to

have other ideas. She launches into another round of troubles-telling, describing further difficulties she encounters with her stomach pains. This elicits an explanation of how the gall-bladder functions, as well as advice concerning her eating habits. The patient doesn't seem satisfied with this. She repeatedly relates her own experience of suffering, i.e. how it is for her, the sufferer. In other words, her talk does not center around a particular complaint, i.e. a gall-bladder problem, but rather around the activity of troubles-telling itself. Her focus is on *complaining*, not the *complaint* (Jefferson and Lee 1981). Thus, members, while talking ostensibly about the same topic, show markedly different perspectives in their way of approaching it.

At this point, the consultation has revolved around a - more or less - cooperative discussion of one particular problem: the patient's stomach pains. The conversation shifts, however, upon the introduction of a new trouble, i.e. the patient's nervousness. Whereas the topic of the 'gall-bladder colic' does not vanish from the conversational stage, it now has a fellow-trouble. How this new topic is initiated and dealt with will be the focus of the next segment of the consultation.

IV. *A Side Sequence<9>: 'a nervous question'*

This segment opens with the GP's remark that the patient seems to be, *in general*, a little nervous. This is linked to the prior talk, where the patient has explained that she always feels nervous when she eats out. She elaborates with instances of getting worked up when dining with other people and the fact that she is a rather high strung person.

This resembles the last segment in that the patient engages in troubles-telling - this time about her 'nervousness' - and proceeds to illustrate it with examples from her daily situation. The activity of the GP, however, has shifted. He is now engaged in rounding off the consultation, re-formulating the diagnosis, and making arrangements for x-rays. Although he has introduced the topic of the patient's 'nervousness,' to which she is duly responding with more troubles-talk, he is not treating it as a complaint to be dealt with like the previous one. He does not ask questions about it. Although he mentions this particular 'problem' several times, it is always incorporated in attempts to close the consultation, which is, apparently, for him already over. For example, he hastens to assure her that, despite her nervousness, this is *not* a 'matter of nerves' here, but rather a 'classic case' of gall-bladder problems, just what you'd find 'in the books.' Or, he makes an appointment for x-rays and indicates that she can always come and talk to him-some other time - about her nerves. In other words, the introduction of 'nervousness' is not so much an invitation for further troubles-talk, as it is a means for curtailing protracted stretches of complaining, which are - in terms of his agenda-no longer in order.

V *Closing*

In this final segment, the GP has tight control over the conversation. He summarizes the talk-thus-far, indicating what they have talked about as well as further steps to be taken. This functions to initiate a closing (Schegloff and Sacks 1974). The patient's abrupt attempt to suggest a homeopathic remedy for her insomnia is rejected as not being 'any solution at all.' Further talk is out of order at this point. He does add that she may come back and talk about it some other time and proceeds with the good-byes.

In summary, this tape contains four separate 'troubles' which are talked about in the course of the consultation. Of the four, the first is a routine 'entrée;' the second is rejected without being discussed at all; the third - the patient's stomach pains - is treated seriously; and the last one is not developed as a trouble, but rather used for rounding off the conversation. Only one of the four complaints - the 'classic case of gall stones' - got off the ground at all. This complaint alone goes through the stages described by Ten Have (1987) as being 'typical' for a general practice consultation.

Whereas the patient makes continued attempts at trouble-telling, they remain of a highly fragmented character. She rarely manages to align the GP as troubles-responsive-recipient (Jefferson and Lee 1981). Aside from engaging in troubles-talk, she displays considerable orientation to the medical setting. Diagnoses are offered on several occasions, answers to the doctor's questions are readily supplied, and suggestions are made concerning further courses of action. In short, she appears to be *helping* the doctor. Many of her attempts, however, elicit interruptions, abrupt changes in topic, or even challenges to her competence. It would appear that medical professionals do not greet such help with enthusiasm.

Throughout the consultation, the GP maintains a tight control over the topic talk (Hughes 1982). This is primarily accomplished by means of Q-A sequences. On occasion, he also employs formulations to sum up or round off parts of the talk-thus-far and, less frequently, overt challenges or criticism of what the patient has said.

This tape might be viewed as a routine medical encounter. The methods employed by the GP to get his work accomplished match standard descriptions found in other analyses of medical interaction. The complications and negotiations, which arise when patient's and the doctor's agendas conflict, resemble the 'routine complications' delineated by West (1984).

B: Two Stories (the 'a little bit nervous' tape)

The GP has his own practice in the suburbs of Arnhem. It is extremely busy. He is young and has followed a general practice training program. The patient is a middle-aged woman, married, and has two children. She has outside employment as well.

The consultation can be separated into two segments in terms of the problems the patient is talking about. They are:

- a *sore-throat* and *headache*.
- *being upset* (anxious, nervous, problems-of-living).

This is followed by a closing segment, similar to the last tape.

Whereas there is considerable overlap throughout the consultation, I shall, for analytic purposes, divide it into talk about a somatic trouble and talk about a psycho-social trouble. <10>

I. *Somatic Troubles: 'sore throat and headache, that's the only thing I have'*

Following the usual greeting ritual, the patient indicates her reason for coming as being her sore-throat and headache. She requests that the GP examine her. After a few questions concerning the location and duration of the pain, he proceeds to examine her throat. He is not able to find anything and he tells her so.

This elicits a new round of troubles-telling from the patient, during which she elaborates how terrible the pain is, that it keeps her awake at night, how she's done what she could to alleviate it (stopping with smoking), etc. She employs many of the elements used in a 'Leidensgeschichte'(Rehbein 1980):

complaining tone, intensifiers, contrasts, etc. She also indicates some reticence about complaining, displaying an orientation to 'business as usual'(Jefferson 1980).This is done by laughing frequently and playing down her illness. Thus, dramatization of the problem and unwillingness to complain-without-good-reason go hand in hand.

A dilemma has arisen, however, when the GP indicates that he hasn't found anything, which would account for her sore-throat. This elicits a strong reaction from the patient and a renewed round of troubles-talk with emphasis on the fact that the pain *is* very severe. The GP finds himself in a predicament. Faced with the fact that he cannot do anything about her present complaint, he can resign himself to 'just' lending a sympathetic ear or he can send her home empty-handed and, presumably, with considerable dissatisfaction.

Or, and this is what he does, he can look around for another problem to talk about, which does allow for some medical intervention. Given the mandates of differential diagnosis, the lack of organic evidence for a sore throat made its transformation into a psychosocial problem the order of the day.

II. *Social-psychological Troubles: 'whether or not I've been working myself up about that ...?'*

In this segment, the issue of the patient's sore-throat has been dropped altogether. Both participants seem to be working on the construction of a new topic. This construction process has two parts.

1. *Preliminaries to constructing the complaint*

   A potential new trouble is initiated by the GP. He asks her whether the sore-throat makes her frightened and, a bit later, worried that 'something' (cancer?) might 'be there.' The patient confirms this, indicating that her father had throat cancer and she's wondered about it herself. She augments this with further complaints (sore eyes, headaches, and frequent medical examinations, which don't reveal anything, despite the fact that it doesn't go away).

   Although the GP responds affiliatively to her anxiety, he also indicates that *he* hasn't been able to find anything. He supplies her with two alternatives: waiting awhile or taking something for her 'nerves.' This is the first time that the question of nervousness as a possible complaint, something to be treated with tranquillizers, has been mentioned.

2. *The Patient's Version*

The GP has managed to find something which may be diagnosed (anxiety) and treated (tranquillizer). The patient accepts the offer of a tranquillizer and proceeds to launch an extended stretch of troubles-talk as an account for her being upset. Her turn at talk resembles a series of incidents ('troubles') which might be unified under the same evaluative point; namely, that they are her reasons for being nervous. She underlines this point throughout her talk ('and if this is what I'm nervous about...').

Her problems are, indeed, numerous. She has terrible headaches, which make it impossible for her to work. Her mother-in-law has to be moved to a nursing home and this appears to be solely the patient's responsibility. Her disabled brother-in-law is constantly calling her up. She has two children who are ill with a pancreatic condition.<11> Her husband has his own worries and does not support her in this situation. He just tells her that she gets too worked up or that she should go to the doctor with her complaints.

The picture, which emerges, is that of a woman having to cope with a lot of extra demands placed on her by her family with little support. Whether her physical symptoms (sore-throat, headaches) merely aggravate an already-difficult situation or are the outward symptoms of psychological distress, as the GP seems to be suggesting, is unclear. What *is* clear is that as soon as she has the floor, she can find any number of problems to talk about based on the current realities of her life.

Throughout this extended stretch of troubles-talk, the GP provides affiliative responses in the form of listening noises (hmm-ing, yeah-ing), commiserates with her difficulties, and reassures her that she has good reason for feeling as she does. If she is nervous, it's because she 'really has a problem.'

III. *Closing*

The final segment involves an attempt by the GP to bring the troubles-talk to a close and conclude the consultation as well. As in the 'a classic case of gall stones' tape, the GP here also formulates the talk-thus-far, thereby providing it with a kind of signature (Heritage and Watson 1979). The trouble is not physical, but *psychological*. This is phrased as a reassurance, both for the patient and for her husband. There is nothing wrong with her health (only her head). The patient provides confirmation, indicating that she'll do her best, and they part company.

In this tape, there is considerably more troubles-telling going on than in the previous one. The troubles are not simply medical matters, but also problems-of-living, which, ultimately, are 'psychologized' and treated accordingly. The GP provides considerable affiliative response to the troubles-telling and, in fact, even invites the patient's account of her problems at home. Despite a seemingly endless supply of troubles, she is given ample conversational space to recount them.
There appears, however, to be some confusion concerning the nature of her troubles. 'Is all this just a matter of nerves?' she asks, again and again. The GP seems to be convinced that it is, indeed, just that. For him, the patient's troubles-telling comes as a welcome deliverance from his dilemma; namely, that he was not able to find out what was causing her original complaints and, consequently, could do nothing about them. With her nervousness, however, he can help. After listening sympathetically to her problems, relief can be provided. A prescription for tranquillizers will do the trick.

C: An Interrupted Story (the 'double-uterus' tape)

The GP is in his early thirties and has his own practice in a working-class district in Rotterdam. He has had vocational training. The patient is an elderly woman, who has obviously been to see him before for various complaints like arthritis in her knees, water retention, etc. In this consultation, the patient has one trouble, which she introduces at the beginning: her daughter has a double uterus. This trouble-telling is interrupted, however, by a rather lengthy medical side-sequence, and picked up again later on. Thus, the consultation may be divided into four segments: the double-uterus story; a medical side-sequence; return to the double uterus; and closing.

I. *Starting a Troubles-telling: 'she seems to have a double uterus'*

Following a ritual greeting segment, the patient indicates that she is 'a little nervous.' She orients the GP to the presence of a trouble without immediately embarking on a protracted turn at talking. It is only after the GP inquires into the nature of the problem that she proceeds to relate how her oldest daughter has a double uterus. This daughter has been married for six years and wants to have a child, something, which may now be impossible due to her double uterus. The patient also appears to be worried that her daughter might have to have an hysterectomy. The trouble-telling contains a request for information concerning what this anomaly means.

Throughout this first segment, the GP provides affiliative responses to the patient's tale. He elicits the trouble, makes listening noises, and even produces an occasional expletive ('Oh?'). At the same time, he interrupts the patient's talk twice, first with a joke and, later, with a medical question. The first interruption is parried by the patient; the second moves the conversation into another form of talk (Q-A sequence), as well as bringing a change in topic. Thus, the patient's attempted troubles-telling comes to a close for the time being.

II. *A Medical Side-Sequence: 'Palpitations and Pills*

This segment begins with a series of medical-type questions concerning the patient's heart palpitations, a topic mentioned at the beginning of the consultation. The GP controls the conversation, introducing two separate topics by means of his questions.

1. *Palpitations*

   In this Q-A sequence (Merritt 1976), the GP attempts to obtain information concerning the patient's complaint of heart palpitations; i.e. how fast they are, how often they occur, in which situations, etc. He also takes her blood pressure. His questions are often presented in the form of a choice between two alternatives, one of which is not particularly serious ('does your heart beat *real* fast or is it not so fast, but just a little irregular?'). The case, which is being built here, is that there is nothing to worry about. The patient provides answers to the questions, but does not concede that she has no reason to be worried. She does this in a mitigated way, by first agreeing with him and then producing a counter-example that she is, in fact, constantly worried - about her daughter.

2. *Pills*

This segment begins with the GP initiating a new topic; namely, which medications the patient is currently taking. As such, it resembles a kind of medical checking operation. Using her file as a reference, he tracks down yellow pills, blue pills, pills for water retention, and pills for muscular pain. In addition to requesting information on her pill consumption, he espouses his own philosophy concerning overly-extensive use of medication. Despite the fact that he was the one to bring up this particular topic himself and, in fact, offers the patient some more pills, he maintains the stance of someone well-aware of the problems inherent in taking too many drugs. The implication is that the patient is the one wanting the pills.

The patient is highly cooperative, answering the questions posed by the GP and expressing agreement with his remarks on pill consumption. Towards the end of the segment, the GP attempts to close the consultation. He does this by reformulating the results of the conversation (Schegloff and Sacks 1974).

III. *A Second Attempt at Telling the Trouble: '... my daughter?'*

In the preceding segment, the consultation has been rounded off, more or less, and the GP is helping the patient into her coat. At this juncture, she unexpectedly re-opens topic talk by returning to the unfinished business of the double uterus. She does this in the form of a request for information. The GP complies by providing an explanation - for the first time - of what a double uterus entails in terms of being able to get pregnant. This elicits a distinctly emotional reaction on the part of the patient. She indicates that she finds it very 'sad,' repeating what she has said to her daughter about it.

This sequence is cut off, for the second time, by a joke. As in the beginning segment, the joke has a reassuring dimension, as well as serving to prevent further troubles-talk. This is followed by a truism and the advice to try to look at the sunny side. Aside from his initial provision of a medical explanation for the double uterus, the GP has not allowed himself to become aligned to further troubles-telling.

The patient tries to maintain the floor by reiterating her reason for telling the tale but, after several attempts to interrupt the GP, concedes defeat and produces another medical request; namely, what should she do about her nervousness? What follows might be viewed as a cooperative attempt to bring the consultation to a successful and harmonious conclusion. The GP indicates that he does not want to give her tranquillizers, that she will just have to work out the problem with her daughter for herself, and that, of course, she can come back and talk about it any time.

The patient, her story having 'flopped' for a second time, confirms this reading of the talk-thus-far, even though she had not actually requested pills; nor can it be said that she was unwilling to talk about her problems. Her confirmation, however, enables them to part in a pleasant fashion.

IV. *Closing*

This final segment contains the same elements found in previous tapes. Following an opening-up closing (Schegloff and Sacks 1974), good-byes are produced, as well as a bit of friendly small talk.

This tape essentially involves one troubles-telling, which is interrupted by a medical time-out. The patient displays considerable facility in getting the story started and organizing the GP's attention. She even re-introduces the trouble at the very end of the consultation and manages to get the information she requested.
The GP, aside from minimal responses, is not well-aligned as story-recipient. In addition to interrupting her with medical questions, which were quite unrelated to what she herself had introduced as complaint, he produces several jokes, which tend to trivialize her trouble. Because these jokes are simultaneously well-intentioned and friendly, they serve as a particularly complicated obstacle to the ongoing troubles-telling.

D: Continuous Stories (the 'one-continuing-story' tape)

The GP works at a general practice training institute and has a part-time practice in Amsterdam with another colleague. The practice is in an urban renewal neighborhood with a lot of elderly and foreign patients.

The patient is a young woman, married, and currently on sick leave from her job. She has no children. She is seeing a therapist.

The consultation itself is 2½ times as long as the previous three. The patient engages in considerable troubles-telling concerning her present distress and she has the space to do this within the consultation. The GP is very empathetic and aligned to her troubles-telling, until the consultation has been underway for some time. He then finds himself faced with some dilemmas of his own. The rest of the consultation involves an interactional struggle between the two parties over control of the conversation.

Since the patient introduces a variety of troubles in the course of the talk, I have segmented this tape in terms of the global interactional activities being performed around the business of troubles-telling. Thus, the conversation may be divided into the following parts:

- Setting the Stage
- Series-of-Troubles
- Negotiating the Point
- The GP's Dilemmas
- Negotiating the Dilemmas
- An Unhappy Ending

I. *Setting the Stage: '... rather not have any more pills'*

In this opening segment, the patient produces two medical entrées: she wants to stop taking both birth control pills as well as the tranquillizer she is presently using. Whereas she is ostensibly requesting confirmation from a medical expert (similar to the 'double-uterus' tape), it soon becomes clear that the pills are connected to various other troubles in her current situation. These troubles are introduced in this first segment and return throughout the course of the consultation. They are:

- *difficulties with her husband*
- *wanting to have a child*
- *present 'state of affairs'*

These troubles are only hinted at here as potential topics for further talk. In addition to the problems explicitly mentioned by the patient, other potential difficulties are involved in the way she is presenting her problems in the consultation. The interactional dilemmas, which will emerge in the course of the consultation, are foreshadowed in this first segment:

- the patient is *angry* (at her husband, because she can't have a child now; at her therapist; and, later, at the GP)
- she feels *powerless* (she presents herself as a broken machine sent in for repair, as though she has no control over her life)

- she is trying to get the *GP involved in a coalition* (with her against her husband, against her therapist).

The GP displays affiliation with her troubles throughout this segment by means of expletives ('Oh?'), formulations of how she feels, questions, etc. He finds himself in the dilemma of being asked to provide information on contraception and, at the same time, being drawn into an ongoing dispute between the patient and her husband concerning whether or not she should be 'allowed' to have a child now. Faced with the possibilities of merely providing the information requested, rejecting the position of having to decide, or making the decision, he opts for the latter and agrees with her husband that now is not the time to have a child.<12> The patient complies and introduces another topic, her 'real' reason for coming.

II. *Series of Troubles: 'it's my life and I have to do it ... but how?'*

The following prolonged stretch of talk involves a series-of-stories (Ryave 1978), whereby the GP is aligned as story recipient. The patient presents her troubles as instances of the same underlying theme; i.e. her current distress. The difficulties are numerous: anxiety attacks, failure of therapy to help her, her cat dying, her husband's drinking problems, and an accident.

The turn shape of this segment alternates between *troubles-tellings*, with the patient employing devices commonly used in stories about suffering (Rehbein 1980) and the GP providing affiliative responses, and *negotiations around the point* of the story. As the consultation progresses, the troubles-tellings become not only instances of the patient's present distress, but an integral part of the negotiation process.

1. *Getting the Floor: 'anxiety'*

   The patient opens her troubles-telling by indicating that her 'real' reason for coming wasn't the issue of having a child or not, but rather her anxiety attacks. She does this in a self-deprecatory way (she's probably being too negative; talking with her therapist helps, but only temporarily; etc.). In other words, she indicates the presence of a trouble without launching into a recounting of it. It will be up to the GP to elicit it (Jefferson 1980), which is exactly what he does.

2. *The First Trouble: 'afraid to go outside'*
The first troubles-telling concerns the patient's anxieties, her inability to walk down the street without breaking out in a cold sweat. She provides vivid descriptions of her bodily symptoms, as well as the consequences her anxiety has for her job. Conversations with her therapist about how these fears should best be dealt with are re-enacted. The entire account is accompanied by evaluative statements about her despair and how she is going to have to do something about it, which seems to be the general point of her story (Polanyi 1979). Whereas the GP has provided minimal affiliative responses all along, he finally summarizes what he has heard her saying with a formulation, which, on the surface, indicates sympathy with her predicament: she is in a 'crisis' and is understandably frightened by it. Furthermore, he remarks that she doesn't seem to be getting much help. This initiates a round of negotiations, indicating that, apparently, this does not quite capture the point from the patient's point of view.

3. *Negotiating the Point: '...nobody can tell me what to do'*
The following segment is topically tied to the preceding one. They are still talking about the patient's anxiety. Both appear to be (re)formulating the point of the story and, whereas their utterances are similar in that they agree that the situation is distressful, there is an important difference. The GP, on two separate occasions, reiterates that she is being told by others that she'll have to make a decision to help herself. The implication is that this is part of the problem. The patient produces her own version of the point; namely, that she is, indeed, the person who will have to solve her problems. Thus, whereas they agree on her needing help, there appears to be some disagreement over the issue of who is to do the helping. This point is a recurrent one throughout the rest of the consultation.

4. *Another Trouble: Therapy*
The GP interrupts the patient with a question about how her therapy is going. His request not only underlines his reading of the patient's problem; i.e. that she needs (professional) help, but it serves to elicit a new round of troubles-telling. The patient embarks on an extensive story about her experiences in therapy, the gist of which appears to be that

it is not helping her at all and, in fact, the things she has to do (get up early, call the therapist if she gets up later, talk directly to her husband) are presented as ridiculous. Not only does the troubles-telling provide an answer to the GP's request for information about what is happening in her therapy and whether or not it is helping, but it also recapitulates the patient's point, which is that, although she needs help, she isn't being helped at present (and possibly can't be helped).

5. *More Negotiations around the Point: 'I'm afraid it's not going to help'*
   This segment begins, once again, with a formulation, in which the GP summarizes the emotional point of the patient's troubles-telling; i.e. she's upset. What ensues resembles a tug-of-war in which the GP produces formulations of the talk-thus-far (she's upset, she feels abandoned, she expected more from the therapist) and the patient counters with the fact that she feels angry, that she doesn't expect him to take her by the hand, and, at the same time, how bad the situation is and how powerless she feels to do anything about it. She appears to want to talk about how terrible everything is, without having to agree that therapy is what is going to help.

6. *Troubles-telling as Account: 'despite all the awful things'... 'able to cope after all'*
   In what appears, at first glance, to be a side-sequence (Jefferson 1972), the patient interrupts herself midstream and launches on a completely different topic. She cuts off her troubles with therapy and her therapist and begins to recount some of the worrisome things, which have been happening lately. Her much-beloved cat has died. Her husband drinks too much and he has just had a car accident while under the influence. Instead of telling her about it, she discovered it by finding a letter from the injured party. Both these incidents are presented as extremely upsetting. At the same time, she appears to be building a case that she's not working herself up without a good reason. This supplies support, retroactively, for her preceding account about her therapy. With such heavy-duty problems, getting up at 9 o'clock in the morning is clearly not going to be much of a solution.

   She closes with a remark that she can talk with the GP

better than with the therapist, once again drawing the GP into a position of being on her side, this time, against the therapist instead of against her husband.

From the vantage point of the patient, she has been able to engage in extensive troubles-telling, receiving, for the most part, a highly affiliative response. Whereas there has been some negotiation around the point of the stories, in particular, the rather thorny issue of being the recipient of professional help for one's problems, the GP has been aligned as story-recipient all along. Despite the cooperative nature of this segment, some of the difficulties, which the participants are going to have with one another have already made their entrance here.

With her expression of appreciation for his qualities as listener, the patient has brought the GP as professional helper in a dilemma. It is not surprising, then, that the consultation takes a turn at this point. It is no longer a primarily harmonious endeavor, where the patient's storytelling is encouraged. The GP now comes to the fore and begins to deal with some of his own troubles.

III. *The Doctor's Dilemma*

In this segment, the GP takes on a more active stance. The patient's troubles-telling is no longer the primary form of talk, although she does continue to introduce new troubles, as well recycle old ones. Her talk, however, no longer elicits a uniformly affiliative response from the GP. He appears to be grappling with some difficulties of his own.

The first concerns his own *role as professional helper*. The patient is drawing him into a coalition against her therapist. The therapist has failed to help her, but the GP hasn't (yet). She can still talk to him. This puts him in a difficult position. It is his task as primary care provider to refer patients to specialists for problems outside his domain. This has already been done. He will not want to involve himself in the work of another professional. Furthermore, professionals are leery of criticizing their colleagues in conversations with laypersons. This is an unspoken code. The patient, by drawing the GP into a coalition against her therapist, has (unwittingly) placed the GP in conflict with his duties as professional.

The second difficulty involves *terminating the consultation*. Until now the patient has engaged in, what for most consultations, is an unusually lengthy round of troubles-telling. Her point throughout has been that her current situation is terrible, but that there is nothing anyone can do about it. This could go on indefinitely. The GP will need to find a way of ending the consultation while, at the same time, maintaining it as a harmonious and, hopefully, successful endeavor.

In this segment, he deals with these dilemmas in the following way. He explains his own position as someone responsible for her bodily ills and an occasional chat. He clearly, albeit indirectly, differentiates this from what the therapist does. He also displays understanding for how she feels; i.e. her dissatisfaction with the therapist. Whereas he provides an emotionally reciprocal response, he also - for the first time - indicates some disagreement with her stance. His point seems to be that if she needs help, she should go back to the therapist; that's *his* job). At this point, the GP is steering a careful route between providing an empathetic response to the patient's troubles and paying attention to the constraints imposed upon him by the medical situation. If the patient gets the message, she will terminate her troubles-telling and leave. This is, however, not what she does.

IV. *The Struggle*

This segment takes up nearly one half of the consultation and might be characterized as a tug-of-war, conversationally speaking, between the GP and the patient. The patient continues to initiate troubles-tellings. Most have been mentioned previously: her agoraphobia, her dissatisfaction with the way her therapy is going, and problems with her husband. She continues to underline the emotional point of her storytelling: that it's terrible, that nothing is helping, and that she feels powerless to do anything. The entire segment has a repetitive quality, with the patient emerging as a 'victim of circumstances' (Rehbein 1980). She often resorts to dramatization to make her point: heavy intonation, intensifiers, contrasts. At one point, she even produces a veiled threat that she might be considering suicide. This is presented indirectly, as something the therapist was worried about, but the effect is a temporary mobilization of the GP's interest and sympathy.

The patient is clearly fighting to keep the floor. Instead of trying to re-align the GP as troubles-receptive recipient, she now rather randomly links her troubles-telling to whatever he says. This strategy is not particularly successful. Whereas she has the floor, she no longer has his affiliation. The GP is, at this point, engaged in dealing with the dilemma's mentioned in the preceding segment. Throughout this stretch of talk, he is doing two things:

First, he repeatedly states his position *vis-à-vis* the therapist, indicating that he does not want to come between the patient and her therapist and implying that she should talk to the therapist, rather than him, about her problems.

Second, terminating the consultation has become the major concern for the GP. Whereas he still provides affiliative responses to the patient's troubles-telling, he is also clearly engaged in bringing the conversation to a close. He does this primarily by recycling the topic of the tranquillizers, which had been the patient's entree (she wanted to stop taking them). He now returns to them and initiates a discussion about whether she wants to continue them or not. This ultimately leads to a prescription, albeit with several qualifications: she can fill the prescription and 'see' if she needs to take them or not or, if that doesn't work, she can talk to the therapist about them.

Just as the patient is attempting to maintain the floor for more storytelling, the GP is trying to regain control over the conversation so that he can bring it to a successful close. He is having trouble doing so and, in fact, the conversation threatens to go on indefinitely.

At this point, the GP resorts to a more challenging stance. His patience seems to have come to an end. Following a long stretch of troubles-telling and the patient's (rhetoric) remark: how long can a person go on like this, he remarks, drily, that she has managed for the last 10 years. This is anything but affiliative and, in fact, implicitly indicates that her complaints are now being taken with a grain of salt. Whereas he mitigates this somewhat later, the message is clear. The patient launches on a face-saving manoeuver and provides an account for her behavior: she has only just begun to realize how serious things are and, moreover, people with her 'high level of intelligence' always have more difficulties, according to her therapist.

Interactively speaking, the consultation has taken a definite turn for the worse. The 'nice' GP has had to resort to insults and the patient is defending herself to him. Gone is the quality of a friendly chat between friends. The task remains, however, to terminate the consultation with as little damage to the respective participants and their encounter as possible. This is accomplished in the next segment.

V. *Terminating the Consultation: An Unhappy Ending*

> The GP opens this segment with the beginning of a farewell sequence (I wish you all the best), thereby setting the stage for the final phase of the conversation. He sums up the talk-thus-far in a formulation, which expresses her difficulties in positive terms. The patient confirms this, but - perhaps noting the GP's reticent manner - does not immediately leave. She puts off ending the conversation by introducing a new trouble. She's afraid to leave because she might have an attack of hyperventilation. This performs several tasks at once. She prolongs the conversation. She renews her request that the GP help her with her present difficulties. She also indirectly criticizes him for sending her away in her present state.
>
> This conversational manoeuvre is reminiscent of the first segment of the consultation where the GP was also placed in a kind of conversational 'double-bind'. This time, however, he does problematize it. He immediately questions why she is saying this right now, thus forcing the patient to admit what she is 'really' getting at.
>
> What follows is a parody of a polite conversation. Normally, politeness conventions will restrict completely honest or direct responses (Brown and Levinson 1978). Here, the GP takes the patient's self-deprecatory remarks at face-value and the result is anything but polite. The consultation rolls to an end, which is probably unpleasant for both participants. 'Thanks a lot,' the patient says, in English, meaning, very probably, just the opposite.

This consultation differs from the other three in several ways. It is notably longer. The patient has lots of space to engage in troubles-telling. The GP's response - at least in the beginning of the conversation - displays considerable emotional reciprocity. He does not use protracted question-and-answer sequences, but is primarily engaged in formulating what the patient has said. In this respect, the consultation resembles a therapy conversation rather than a medical consultation (Davis 1984b,1986). It is clear, however, that he runs into some difficulties with this stance. He has trouble maintaining his own professional identity, as well as terminating the conversation. His attempts toward the end to regain control over the conversation lead to some not-so-nice moments and the interaction comes to an unpleasant end.

In this consultation, the conflict between a patient's troubles-telling and the constraints under which the GP as professional helper has to operate are thrown into relief. Moreover, it provides a good illustration of some of the pitfalls inherent in the modern approaches to general practice with their one-sided emphasis on social skills.

In conclusion, an attempt has been made to describe how material was selected for the final analysis, based on the findings of my preliminary analysis and various theoretical and methodological considerations. After transcribing the consultations, an overall description was provided of each consultation. In addition to providing information about the participants and the topics being discussed, the consultations were organized and described in a way, which would facilitate further analysis. These descriptions are meant to serve as a framework for locating and analyzing how stories about troubles are told and obstructed in the course of a general practice consultation.

NOTES

1. As it was, for example, in an earlier inquiry (Davis 1984b, 1986). In that investigation, I was able to restrict myself to one therapy session. This was possible because the entire research problem was a limited one; i.e. how the problems of a female client were transformed and individualized in the course of a therapy session. Having a good example of both processes, I needed to look no further.

2. It is important to note here that this fragmentation is not a feature of the patient's story or, as Mishler (1984) calls it, the 'voice of the lifeworld,' which is coherent enough to be reconstructed within an analysis. Fragmentation refers, on the contrary, to what happens to stories told in institutional settings and, as such, reflects asymmetrical relations of power between the person telling the story and the person(s) on the receiving end.

3. This is, by no means, a typology of the kinds of stories to be found in general practice consultations as a whole. It is sufficient, however, to illustrate the difficulties patients may encounter in telling stories about their troubles.

4. See, for example, Atkinson and Heritage, who admit that transcription systems are always selective, designed for accuracy, but also for

   > use in *close conjunction with* the tape-recorded materials that constitute the data base. The fact that publication conventions allow for the inclusion only of data extracts in written form should therefore not lead readers to the conclusion that the analyses are based exclusively on an inspection of

transcripts conducted independently of direct reference to the original recordings (Atkinson and Heritage 1984:12).

In my own case, I listened to each consultation at least 50 times, moving back and forth between the transcript and my analysis. Particularly in instances where I was in doubt about how to interpret an utterance or segment of the talk, I returned to the tape-recorded version to check.

5. The following information was recorded as part of the TRIPT-project: the age of the GP; year he finished medical school; where and when he completed his vocational training; the city in which his practice is located; the general characteristics of the practice, the patients, the neighborhood; and the number of years he has been in practice. The approximate age and sex of the patient were also recorded (Thomassen and Woltring 1982).

6. Another 'structure' is, of course, the one that has been found by many investigators of medical interaction and which I have described in Chapter 3: history-taking, examination, diagnosis, discussion of the problem, treatment. This structure could also have been 'discovered' in my data, just as I suspect my alternative structure could be 'discovered' in their material.

7. I have also given each of the tapes a name, which makes some reference to the patient's complaints or troubles. The idea behind this was to make an immediate and easy identification of the consultation possible. In this way, the reader is spared the bother of having to continually refer back to earlier chapters to find out which patient or what consultation is currently being analyzed.

8. The medical agenda in the case of a gall-bladder colic involves the actual location of the pain (over to the right slightly above the stomach), the patient's need to move ('rolling from the pain'), and a diet of fatty foods. The GP can be seen as orienting his questions to this scheme. Medical students will often receive the following memory aid for diagnosing gall-bladder problems: 'fat, forty, female.'

   Gall-bladder disturbances may be a particularly popular diagnosis for medical professionals by virtue of the availability of a fairly clear-cut diagnosis. The next step is equally straightforward: x-rays, a diet, and pain-killers. It is not surprising that a GP would prefer such a complaint to the more elusive 'shaky hands' or 'nervousness.'

9. 'Side Sequence'is a term coined by Gail Jefferson (1972), referring to a break in the ongoing activity of a conversation. It is not part of that activity, but is considered in some way relevant to it. Nor is it a termination, as the activity may be resumed at some later point in the conversation. It is often employed as a means for overcoming various interactional obstacles, which have emerged in the course of the talk.

10. This classification is for-all-practical-purposes and does not mean to reflect on the nature of the patient's problems. In other words, I do not wish to suggest that sore-throats are strictly somatic matters, nor that the patient's nervousness is unrelated to

her bodily well-being. The term 'nervousness' is one employed by both the GP and the patient throughout the consultation as a common denominator for a wide range of problems in her present situation. In effect, I have done the same.

11. The tape is unclear on this issue. Two readings are possible. The first is that the patient's children are suffering from a severe, hereditary ailment of the pancreas, which can lead to chronic bronchial complaints. It involves constant care. For example, they need to have their chests 'pounded' in order to loosen congestion. The second reading is that the patient's children are merely ill with something resembling bronchitis and that they saw a television program about this more severe illness and that frightened them. Either way, the message is clear. The patient has her hands full.

12. These options produce a double bind, where a person is confronted with two contradictory messages and the illusion of being able to choose. In this situation, there are three possibilities. The person can take one of the remarks at face value and ignore the other (the GP discusses the matter of stopping with the pill, but disregards his role in making that decision), *or* the interactants can continue for a limited time with this kind of conversation, knowing that it will eventually come to an end and they can revert to 'normal' talk (by, for example, changing the subject). The third solution, and according to Watzlawick *et al* (1967:234-235), also the best one, is to discuss the problem itself as a snag in the communication. This amounts to changing the rules of the game or ruling that the game is, in fact, over ('Am I supposed to be making this decision or what?' or 'Hey, whose side am I supposed to be on here?'). In the case of our GP, this might well have been the best alternative. As it is, he (implicitly) accepts the patient's relationship definition that he is the one to make the decision about whether she should have a child now or not.

III

DELINEATING POWER:

AN EMPIRICAL ANALYSIS

8

ANALYZING POWER STRUGGLES

Up until now, I have been engaged in the preparatory work necessary for an analysis of how female patients and male GPs talk about the troubles of the former and, more to the point, how asymmetrical power relations are constructed and sustained in and through that talk. This entailed not only finding a way of conceptualizing power at the level of interaction, but developing a method for analyzing it. The research problem had to be operationalized in a way that was in keeping with the theoretical and methodological considerations set out at the beginning of the inquiry. Taken together, theory and method provided the 'microscope' I needed in order to locate and delineate any power struggles, which might be occurring in my data.

In this chapter, a brief introduction will be provided for the analysis to follow (Chapters 9-12). I shall begin by describing how power struggles will be located and analyzed, using troubles-talk as a starting point. This will be the 'red thread' running through each of the chapters. An explanation will be provided for the organization of the analysis as a whole; i.e. the rationale for the ordering of the individual chapters. And, finally, a few words are in order concerning how material from the consultations described in Chapter 7 was used in the actual analysis.

FROM TROUBLES-TALK TO POWER: SOME SUBSTANTIVE CONSIDERATIONS

Troubles-talk in general practice consultations

To begin with, I shall be looking at how troubles-talk can be accomplished at all in the context of a general practice consultation. Story- or troubles-telling is, as previously mentioned, one possible way that patients can 'package' their experiences of suffering.

Whereas it most routinely occurs in a friendship frame, it can, and often does, happen in a professional service encounter as well (Chapter 6). Thus, I will be showing how patients and doctors engage in the activity of troubles-talk.

This particular activity was chosen out of all the other activities occurring in the context of a consultation for the simple reason that it ensures that the patient will, from the outset of the analysis, remain squarely in the forefront. In this way, it will be easier to view her as agent, actively engaged in constructing the encounter along with, but sometimes at odds with, the physician. As the reader will recall, this was a necessary pre-condition for the analysis of power (Chapter 4).

Like any other social activity, troubles-talk requires interactional work<1> for its accomplishment. The simple fact that two or more parties find themselves face-to-face does not automatically guarantee that a conversation will follow. On the contrary, participants produce conversations through various activities like initiating talk, keeping it going, and, ultimately, closing it down.

There is more to a conversation, however, than talking. Participants must also actively and ongoingly display their alignment to one another's talk.

> In a sense, every remark or turn at speaking should be seen as an *attempt* to interact. It may be an attempt to open or close a conversation. It may be a bid to continue interaction: to respond to what went before and elicit a further remark from one's interlocutor. Some attempts succeed; other fail. For an attempt to succeed, the other party must be willing to do further interactional work. That other person has the power to turn an attempt into a conversation or stop it dead.
> (Fishman 1978:399)

Viewed in this way, interaction becomes a potentially problematic endeavor, which can only be pulled off by the continuous, turn-by-turn efforts of the participants.

Thus, troubles-talk will be viewed, first of all, as a cooperative venture, a kind of interactive co-production involving specific tasks on the part of both the patient (as troubles-teller) and the GP (as troubles-recipient).

Obstacles to troubles-talk in general practice consultations

Whereas stories about troubles can proceed quite harmoniously for long stretches of the consultation, sooner or later they will run into obstacles. This is not surprising, of course. Particularly in an institutional encounter like a medical consultation, troubles-talk can prove an unwelcome occurrence, precisely because it conflicts with the constraints imposed by the service frame itself (Chapter 6). When troubles-talk is obstructed, the obstacles will require negotiation<2> on the part of both participants.

They mark moments in an ongoing story- or troubles-telling, which are problematic for *both* participants. For the patient as storyteller, an obstacle can mean that she is having difficulty getting the floor *or* that she cannot recount her trouble *or* that her point is not getting across. In short, she finds herself unable to organize a troubles-receptive hearing for her story. For the GP, on the other hand, these obstacles can indicate that he is running into some difficulties as well. In obstructing the patient's storytelling, the GP may simply be trying to re-instate a medical or service frame in order to meet the requirements of his own (professional) agenda.

Thus, obstacles are decidedly two-way affairs, marking points in the ongoing consultation where the patient and the GP find themselves working at cross-purposes.<3>

Obstacles to troubles-talk and the struggle for power

Although the negotiation of any obstacle to a story about trouble will be a reciprocal process, it will not be an equal one. Patients and GPs have differential and unequal resources at their disposal for influencing the course and outcome of the negotiations. Whereas it is reasonable to assume that, ultimately, the structured asymmetry in the distribution and quality of the resources available to each of them will account for how the consultation progresses, including the way that the patient's troubles come to be defined, diagnosed and treated, the empirical question under consideration here is how this asymmetry is accomplished in the ongoing interaction. By locating obstacles to troubles-talk within the medical encounter, we can see where power comes to the surface in the form of a struggle (Chapter 6). Whereas these obstacles do not account for all the ways that power may be exercised in a consultation, they do provide a convenient locus for its investigation. This is because they represent moments where a relatively overt struggle to gain control over the interaction is taking place. Thus, negotiations around the obstacles to troubles-talk are a good place to begin an analysis of power.

In order to delineate how power is exercised, I shall be returning to the theoretical framework set out in Chapter 4. This will be used as a heuristic device for analyzing how power works in the negotiation of obstacles to the patient's troubles-telling. Obstacles emerging in the course of a troubles-telling will be examined in terms of the following dimensions of power:

1. If power is integral to social interaction, then attention should be paid to the specific *skills* and *resources* drawn upon by both the GP and the patient as they negotiate interactional outcomes. These resources may be involved in the production of meaning (interpretative schemes), as well as in ensuring compliance or resistance to moral claims (sanctions).
2. If the *agency* is *central* to any analysis of power, then the activities of both the GP and the patient should be delineated. This entails showing how the GP works to curtail the ongoing troubles-telling and how the patient attempts to proceed with it. Since both activities are 'bounded' (by unacknowledged conditions and unintended consequences), attention should also be devoted to what participants did not do (but might have done).
3. If power involves a *dialectic of control*; i.e. a relationship of autonomy and dependence in both directions, a careful sorting out of how both GPs and patients attempt to exercise control as they negotiate the obstacles to troubles-talk is necessary. This requires showing how the dialectic of control reflects asymmetrical power relations, involving structured forms of domination and subordination, as well as how they can be systematically undermined.
4. If power often has a Janus-face, each obstacle should be investigated for its *enabling* as well as its *constraining* features. In particular, attention must be paid to how these features are intertwined and, ultimately, implicated in the outcome of the negotiations.
5. If the exercise of power is always a *process*, then each obstacle must be viewed as an ongoing series of negotiations, whereby asymmetrical power relations are being produced, maintained, but also altered from moment to moment throughout the interaction.

In short, each obstacle needs to be treated as a negotiated interactional *practice*; i.e. something which is 'done' by the participants as they go along. Whereas the outcomes or products of these power practices will be considered as well, the focus will be on the process itself.

Ultimately, by uncovering the step-by-step construction of relations involving autonomy and dependence, it will become clear how domination and subordination can be produced, reproduced, but also undermined and transformed in the context of a medical encounter between a male GP and a female patient. More specifically, this will enable us to see *how* power relations work to construct women's complaints in negative and/or repressive ways (Chapter 13).

FROM GETTING THE FLOOR TO RELINQUISHING IT: SOME ORGANIZATIONAL CONSIDERATIONS

Since both the achievement of a troubles-telling, as well as its obstruction, are being viewed as interactional accomplishments, ongoingly negotiated by the GP and the patient in the course of a consultation, I have chosen to organize the analysis in terms of the kinds of tasks required for a story about trouble to be told or, as the case may be, curtailed. As the reader will recall, these tasks are:

- getting the floor
- recounting the trouble
- making the point
- relinquishing the floor

Each chapter will have one of these tasks as its topic. First, I will show how participants perform the task in question, both from the position of storyteller, as well as story recipient. After demonstrating how this can be a cooperative endeavor, a look will be taken at what happens when the troubles-telling is not so cooperative; i.e. when the necessary tasks are not performed or even actively obstructed. When this occurs, the ensuing negotiations will be examined as a moment-by-moment struggle for power. Ultimately, these power practices will be analyzed in terms of how they produce, reproduce, but also undermine relations involving dependence and autonomy between the GP and the patient.

Excerpts have been selected from the consultations described in Chapter 7 for the analysis. At the beginning of each chapter, examples have been chosen to illustrate how a story about trouble can be told in the context of a general practice consultation. This is, by no means, an exhaustive selection, since similar examples can be found throughout most of the consultations. My aim here is to highlight the various interactional tasks involved in telling a story about trouble.

Excerpts were selected because they were especially good illustrations of the task in question. They provide the take-off points for 'finding one's way into the phenomenon' (Schwartz and Jacobs 1979:292) of troubles-talk.

When it came to showing how stories about trouble could be obstructed in the course of a general practice consultation, the selection of suitable excerpts for the analysis became more complicated. Obstacles are never one-turn affairs, but are negotiated over long stretches of talk. This makes it necessary to delineate them *in progress*. For this reason, the selection of excerpts for delineating the obstacles to troubles-talk are limited to one consultation. In this way, they could be traced, step-by-step, as they proceeded toward some specific outcome.

Obstacles were also analyzed in terms of the task to be performed. The choice was based on the overall relevance of the obstacles to the consultation in question. In some cases, the troubles-talk was obstructed in the same way, again and again. For example, in the 'classic case of gall stones' tape, the patient's stories rarely got past the most fragmentary beginning. Unsurprisingly, obstacles tended to occur around the activity of getting the floor rather than making the point. In other tapes, obstacles could be found at all stages of the ongoing troubles-telling. The patient in the 'double uterus' tape, for example, not only has difficulties getting the floor, but she cannot recount her trouble or make her point. In that case, obstacles were selected for analysis because they were essential for understanding the quality of the interaction, as well as how power was being exercised in that particular consultation. Thus, considerations of a more substantative nature also constrained the selection of excerpts.

These introductory remarks having been made, I am now ready to embark on the analysis of power practices in general practice consultations.

NOTES

1. The term 'interactional work' refers to the myriad temporary, tacit, but also explicit or even expert procedures and practices, which are employed in order to accomplish any conversation, whether routine or specialist. Interactional work is, for example, required to chair a meeting, take a medical history or prevent an embarrassing lull at a family gathering. It is often, as Goffman (1967) has convincingly demonstrated, primarily a matter of 'face work,' undertaken by members to display approval and/or deference for one another's self-image. Or it may be 'interactional shitwork' (Fishman 1978), that gendered specialism performed by women to make sure that any conversation proceed in a smooth and well-oiled fashion.

2. The term 'negotiation' is a familiar one in the annals of conversation analysis. It refers to the bargaining quality of interaction, whereby participants provide each other with a series of offers, which may be followed by acceptance or rejection. It is not at all unlike the way buyers and sellers haggle over a used car, eventually arriving at a price satisfactory to both. See Berger and Luckmann (1966); Strauss (1963); and Scheff (1968).

3. I wish to avoid the impression that obstacles are something that the GP, willfully and with malicious intent, lays in the path of the hapless patient. Whereas I am examining them in terms of what they mean for the storytelling (and, consequently, the storyteller), they could just as easily be viewed as moments when the GP is floundering in the conversation and the patient is making things difficult for him.

9

POWER PRACTICES: GETTING THE FLOOR

GETTING THE FLOOR

Conversations proceed 'normally' in turn-by-turn fashion (Sacks, Schegloff, Jefferson 1978). Story- or troubles-telling constitutes a disruption of this turn-taking pattern for the simple reason that it takes more than one utterance to tell a story, making it an exceptionally long turn at talk. Thus, the first interactional task facing the patient as troubles-teller is to orient the GP to the fact that a story is coming. It will be his task as recipient to recognize that a story is on its way and to relinquish his conversational turn long enough so that it can be told.

Several patterns for going about this emerged from my data.

Teller-initiated troubles-tellings

The person with the trouble can simply announce it.

EXCERPT 1
[the 'classic case of gall stones' tape]

```
73 P: Well something else I had another question
74    O.K.? Nowadays I have uhhh every time that
      I've been out to eat, then I'm lying there
75    rolling from the pain in my stomach during
76    night until the next day.
77    And it's here on this side.
78    And then I ask myself, could it b- be like
      maybe my gall-bladder?
79    And 'm I'm uh looking around for every-
      thing that I can do about it.
```

In this excerpt, the patient launches into a description of her current stomach pains without waiting for any affiliative response from the GP. However, she goes to considerable lengths to package her trouble in a medically-appropriate manner; i.e. she wonders tentatively whether it might be a problem with her gall-bladder. In this way, she displays an orientation to the medical frame, where troubles are not being told for their own sake, but for the purpose of being diagnosed. Moreover, she presents herself as the layperson, who is asking questions, whereas the GP as medical expert will be the one to provide an answer.

Recipient-initiated troubles-tellings

Another way of embarking on a troubles-telling is for the recipient to elicit it. In the next excerpt, the GP inquires into the patient's 'nervousness' and the patient provides a troubles-telling as elaborated response.

EXCERPT 2
[the 'classic case of gall stones' tape]

151 D: *You're ALTOGE:THER a little bit nervous*
152 *aren't you? Or not?*
153 P: *Yeah that's me, I do have that, yeah.*
154 D: *You do impress me as being kind - kind of*
155 *nervous(now(or something-(*
156 P: *(Yeah (yeah (Yeah.*
Well like- ((complaining, rapid)) that's
157 *what you get from - like when I go out*
158 *with people, you know? -and you start-*
and you start talking or something-
159 *then I get excited about something and*
then uh-
160 *I get nervous about it I don't know what*
that is.

The patient has been engaging in a rather protracted round of troubles-telling concerning her stomach complaints. In particular, she has difficulties after eating out. The GP introduces a new topic at this point by remarking that she seems pretty nervous in general. Not only is this a 'candidate reading' of her talk-thus-far (Heritage and Watson 1979), but it generates an additional trouble - the patient's nervousness. The patient may confirm or disconfirm the reading. In this excerpt, she confirms the GP's reading of her problem and proceeds to talk about situations in which she feels nervous and how this comes about. Thus, troubles-telling may be employed as response to another person's inquiry.

As Jefferson (1980) has pointed out, getting the floor for troubles-telling is not always such a straightforward affair as the above instances might suggest. In some cases, potential troubles-recipients are not willing to engage in troubles-telling without preliminary assurances that the troubles-teller is not going to engage in endless rounds of complaining. She will have to display her orientation to 'business-as-usual' first.

Trouble-premonitory response

One way of doing this is to provide what Jefferson (1980) calls a trouble-premonitory response. Following a routine inquiry into one's health or well-being, a response is made, which is both conventional and marks the presence of a trouble without going into it. It will then be up to the other party to elicit the trouble.

EXCERPT 3
[the 'double uterus' tape]

4 D: Have a seat. ((7.0)) How are you?
5 P: Well: ((complaining)) all right, a little
6 nervous, you know? I don't know where it
7 comes from, nervous and now I've been
getting heart palpitations ((sighs, 3.0))
8 maybe I'm having trouble because of my
blood pressure.
9 D: Then we'll measure that. ((4.0)) You're
10 nervous but you don't know how come.
11 P: WELL: kind of I do:
12 D: Yeah?
13 P: Yeah(hh).
14 My oldest daughter - she's been married
15 for six years and she'd love to have a
baby now, after all, you know? But
(Mmhmm
16 now: there's been an examination and well:
17 she's - she seems to have a double uterus.
18 D: ((surprised)) OH!

In this excerpt, the patient provides a trouble-premonitory response to the GP's conventional 'how are you?' (line 4). Note that she does not elaborate, but merely marks the presence of a trouble ('a little nervous'). In addition to displaying an orientation to the problematic nature of troubles-telling in everyday talk, she also demonstrates an awareness of the medical context by providing medical options for further topic talk; i.e. heart palpitations and blood pressure. Thus, the GP has the choice of tracking the trouble by asking the reasons for her being 'a little nervous' or he can look into her palpitations or take her blood pressure.

He does, in fact, inquire about her nervousness (lines 9-10), but it is only after an additional trouble-premonitory response (line 11), that the patient finally recounts her reasons for being nervous (lines 14-17): her oldest daughter has just discovered that she has a double uterus and, therefore, might not be able to have a child.

The patient in this example may be viewed as working in a competent and methodical fashion towards achieving a 'troubles-receptive hearing' (Jefferson 1980). The GP displays his alignment by picking up her cue and eliciting a troubles-telling, thereby relinquishing the conversational floor to the patient as troubles-teller.

The above examples show how troubles may be introduced into the ongoing talk by either the troubles-teller or the recipient. Another possibility is to return to previously-mentioned troubles and 're-cycle' them.

Recycling troubles

Members have the option of re-introducing 'old materials' at some later time in the ongoing conversation. This is generally done by having them arise 'naturally' from whatever is being talked about at that particular moment (Schegloff and Sacks 1974). In the following excerpt, the same patient takes advantage of this possibility to return to her initial trouble.

EXCERPT 4
[the 'double uterus' tape]

123 D: O.K.?
124 P: Yes.
125 D: Otherwise your life will start to be one continuous string of pills-
126 P: ((loudly)) No, no thank you. Because uhh-

127 ((laughs)) I'd rather not. ((laughs))
((3.0))
128 D: I'll just help you into your coat, just
129 like at the hair dresser's.
130 P: Yeah. ((laughs))
Would- yeah- would it uhhh- well yeah-
131 would that really have an effect on - with
my daughter?
132 D: With your daughter?
133 Uhh- it has- uhhh in fact it does probably
134 effect the possibility of getting
pregnant. But that doesn't mean that if
135 she has a double uterus like that that she
136 for sure won't be able to get pregnant.
137 P: Pecu:liar.

The patient in the above example is about to leave. Nevertheless, she pauses to recycle some 'old material,' as legitimate entry, even at this late stage of the consultation.

To begin with, she does not reiterate the entire trouble, but suffices with indexical references like 'it' and 'that' along with 'my daughter' (lines 130-131). The GP is placed in the position of having to fill in the rest. He can only do this, of course, because the trouble has already been talked about.

Furthermore, she packages the trouble as a request for information. On the one hand, this links it to the medical encounter, where such requests may be reasonably made of medical experts. Not only is such a request in order, but the GP will be under some constraint to provide the information. The patient's utterance is not only linked to the ongoing talk by virtue of its format as request for medical-type information, however. The request is itself an instance of 'old material.' The patient's initial query concerning her daughter's trouble has still not been answered by the GP. Thus, the patient is completely within her rights - conversationally speaking- to return to an unanswered question. Her recycle also provides him the opportunity to collect his thoughts and respond affiliatively, thus displaying an orientation to those conversational rights. Although he gives the patient the opportunity to talk about her trouble for another two minutes, this is not done in an unproblematic way. Busy physicians are not particularly fond of the 'by-the-way-syndrome;' something Heath (1981) has called the 'nightmare of everyday consultative practice.' The GP also displays some astonishment at this untimely recycle in his echoed 'with your *daughter*?'

(line 132). Counter to his expectations, the consultation is not over.

In summary, the above examples indicate some of the interactional devices, which may be employed in order to get the conversational floor for troubles-telling. It can require careful organization. Not only do patients tend to display some reluctance to engage in troubles-talk without a demonstration of interest on the part of the GP, but they show an orientation to the medical context itself by packaging the trouble as a medically-relevant candidate diagnosis or as a request for medical information. In doing so, they also show deference to the physician's position as the expert who, in the final analysis, will be the one to decide what the current problem is. Thus, troubles-talk is constructed as part and parcel of the medical encounter and not just something to be engaged in for its own sake.

The GPs above displayed alignment as troubles-recipients by giving up their turn at talk and allowing the patient to introduce a troubles-telling. By abandoning their own positions as speaker, they enabled patients to embark on a story about trouble.

OBSTACLES TO GETTING THE FLOOR

[the 'classic case of gall stones' tape]

Having shown how patients and GPs cooperatively embark on a troubles-telling, I shall now take a look at what happens when the patient has difficulties in getting the floor. The 'classic case of gall stones' tape has been chosen as example for analyzing what happens when the floor is contested for a troubles-telling. The patient makes repeated attempts to launch a troubles-telling without ever getting past a most fragmentary beginning. The obstacles to her troubles-telling are centered around the business of getting (or re-gaining) the floor long enough to engage in troubles-talk. In the course of the consultation, she and the GP talk about three complaints: her being tired and listless, her stomach pains, and her being nervous. Only one of these - her stomach pains - is dealt with at any length, possibly because the GP sees them as the only complaint he can straightforwardly do something about - a 'classic case' of gall bladder colic (see Chapter 7, p.176). Although this complaint receives the most attention, the patient is still consistently impeded in her attempts to contribute to the medical encounter by means of troubles-talk. In this way, an asymmetrical relationship is ongoingly being established, in which the GP sets himself up as undisputed authority over how the patient's complaints are to be diagnosed and treated.

I shall now take a closer look at some of the ways the patient's attempts to get started with a troubles-telling are opposed.

EXCERPT 5

52 D: But why did- did you want to have
that examined?
53 P: (-but yeah be-
54 because you do get curious about it some-
55 times. Look, (because I actually do often
((starter:))
56 feel pretty much uhh- a little
57 tired and lethargic, (actually,
Hmm
you know (
Yeah-
58 and because I - another lady -
59 D: (what are you thinking of then ...
60 P: I did hear that she did have that back
61 then- And as a result she turned out to
(Yeah-
62 actually have something with her
liver. (
63 ((surprised)) Oh?
64 And uhh- (
65 D: ((brusque))-But then she had that
66 because she had other complaints-
(because with a routine pill check-up
67 P: th- that's what I thought
68 D: we don't do that (((sarcastic))
69 No she had it- ((un-
clear))
70 because then there's no end to it <1>
come now-
71 P: -s-she must have like had something else.
72 D: Yeah.

In this excerpt, the patient introduces a potential trouble as topic: she often feels tired and listless. It is not initiated directly, but requires considerable organizational work on her part. To begin with, she asks whether blood tests routinely accompany a pill check-up (which she has just had). This is followed by an account of a woman she knows with the same complaints, who discovered, presumably by means of a blood test, that she had something wrong with her liver.

By framing her account in diagnostic terms as a potential liver ailment, she competently organizes her troubles as suitable topic for a medical encounter. At the same time, she displays herself as knowledgeable in matters concerning her health. She lets the GP know that she is not only informed about which complaints belong in a medical consultation, but also that she has some understanding of their possible etiologies and the medical routines necessary for finding out more about them. In this way, she establishes herself as a serious conversational partner, as someone interested in and capable of helping the GP with the business at hand.

Whereas the patient's interactional work could serve to establish her competency as co-partner in the medical encounter, as an attempt to launch a troubles-telling, it proves singularly unsuccessful. After reacting with a surprised 'Oh?,' the GP interrupts the patient with a rather brusque rebuttal (lines 65-70), which takes the floor away from her. The information she has provided is contested on no uncertain terms. *If* the patient's acquaintance had a blood test, *then* she must have had different complaints than what the patient has. In other words, *either* the woman was tired and listless (but didn't have a blood test) *or* she had a blood test (but for some other complaints). Either way, the patient's grounds for suggesting that she might need a blood test are sanctioned as insufficient <2> and she is being reminded in a not-so-gentle fashion that physicians, *not* patients, decide under what conditions tests will be performed. Thus, the GP's remark (lines 65-68) does more than correct a possible bit of misinformation on the part of the patient. By informing her that blood tests do not belong to routine pill check-ups, he sets her right in terms about how things are done in the medical world and who is in charge of doing them. Her suggestion is further *disqualified* as not only irrelevant to a pill check-up, but a waste of the busy doctor's time (line 70). Rapid speech tempo and a sarcastic tone serve to underline the fact that the patient's request was definitively out of order.<3>

Despite several unsuccessful attempts to re-gain the floor, the patient must finally concede that her acquaintance 'must have had something else' and in so doing she bows to the GP and his undisputed authority in medical matters.

The (unintended) consequence of this particular method of getting the floor for further troubles-talk is that the troubles themselves are banished from the conversational arena. Paradoxically, displays of knowledgeability on the part of the patient in medical matters do not ensure that these troubles will receive conversational space within a consultation. On the contrary, it can make it necessary for the GP to re-establish his own position of authority -something, which apparently takes precedence over the patient's present health problems.

The above obstacle does not, by any means, signal the end of the troubles-talk in this consultation. Shortly after this incident, the patient produces another candidate for further troubles-talk: her stomach pains.

Unlike the last trouble, this one does become topic for a protracted stretch of talk (see Chapter 7,pp.174-176). Moreover, it is the trouble, which is, ultimately, graced with a diagnosis, thereby establishing it as something meriting further examination and treatment. Despite its success, the patient's attempts to get the floor for a troubles-telling about her experience with stomach pains do not get off the ground any more easily than the prior attempt. Let us examine some of these obstacles more closely.

EXCERPT 6

78 P: *And then I ask myself, could it b-be like*
my gall-bladder maybe?
79 *And 'm I'm uh looking around for every-*
thing that I can do about it.

Once again, the patient packages her trouble in medical terms, indicating not only that she has pain, but naming the offending organ as well. 'I ask myself,' 'could it be,' 'maybe' and 'like' all display interactional uncertainty and perform the necessary work of displaying deference to the GP as medical authority. <4> The final confirmation of her 'diagnosis' is being left to him. The patient goes on to display herself as competent agent in matters concerning her own health: she is doing everything she can to help herself (Rehbein 1980). Thus, in the talk which follows, she not only describes her pain (lines 100-102), but also her own trial-and-error attempts to find out what is causing it (lines 105-113).

EXCERPT 7

100 P: *((complaining)) WELL- t-the last time it*
101 *was a tiny bit less but before that, then*
I was lying there really rolling back and
102 *forth and I didn't know where to tu:rn-*
103 D: *-And you aren't able to just lie still?*
[No.

((2.0))

104 Yeah. [((starter))
105 P: [And uhh the last time then I got up
106 and took a hot bath in the middle of the
night and- ((1.0))
107 and- [then I stick my finger down
108 [Listen, uhh- where-
my throat ((complaining)) thinking maybe
109 that will help and then only after my
sto:mach is like empty, yeah I did want to
110 look around a little I thought yeah maybe
111 it's because of the wine at the meal But
112 at home I also drink a glass of wine some-
113 times and- [there I don't have any
114 D: [Look well just listen-
trouble with it.
115 That pain. So it's located up towards the
116 right. Does it spread out anywhere else?

This segment marks the beginning of a long stretch of talk during which the patient repeatedly initiates a troubles-telling, only to be interrupted by medical-type questions. In this example, her description of her pain is intercepted by questions concerning its location. Topically speaking, the GP's questions are linked to the prior talk. Both he and the patient are referring to the latter's pain. His utterance, however, does not constitute an affiliative response to a troubles-telling, enabling the troubles-teller to continue. On the contrary, he places himself in the position of the one asking the questions, constraining the patient to provide answers. This constraint is all the stronger, since the questions are so manifestly drawn from the GP's professional schemes of relevance. Having ousted the patient from her position as troubles-teller, and more generally, speaker (Jefferson and Lee 1981), the GP reinstates his own conversational control over the topicality.<5>

This kind of control is fairly straightforward and easily attributable to the 'misbegotten hybrid' of troubles-talk and service encounters (Jefferson and Lee 1981). Medical question-answer sequences invariably shift membership categories between the speaker and the recipient, thereby obstructing an ongoing troubles-telling. Whereas medical Q-A sequences constitute the most straightforward obstacle to the patient's being able to get the floor for troubles-talk, they are not the only type of obstacle in the 'classic case of gall stones' tape. Let us examine the following excerpt, taken from the end of the segment. The trouble is still the patient's stomach pain:

EXCERPT 8

124 D: *Yeah, well, that really makes me think this story ((of the gall-bladder))*
125 P: *Yeahh: that- same feeling I had too-*
126 D: *((rapid)) Yeah, I know almost for sure that it's not supposed to be like this.*
((2.0))
127 *Well: we would have to find that out by taking some pictures.*
((3.0))
Pictures of your gall-bladder-

Having gathered enough information, the GP is making a tentative diagnosis of the patient's complaint. Interestingly, it corresponds exactly to how the patient 'diagnosed' the pain herself earlier on (Excerpt 6). At this point, she displays agreement with his diagnosis, confirming it as something she felt 'too.' At the same time, she undermines his exclusive authority over the activity of diagnosing her problems. Perhaps this accounts for why he does not honor her utterance as an agreement, but rather interrupts her with a second assessment, which serves to contrast *his diagnosis with hers*. They may both agree on the pain being a gall-bladder problem, but that is where the agreement ends. The contrast being established is not between *what* they know, but rather the status of their respective versions. The GP emphasizes that he knows *almost* for *sure* what the patient (only) has a *feeling* about. In this way, a difference is constructed between professional knowledge based on scientific fact and lay experience based on unscientific feelings.<6>

This is further bolstered by the reference to x-rays as the prerequisite for making a final diagnosis. It is stated as a matter of medical protocol; i.e. *first* x-rays, *then* diagnosis. X-rays are unquestionably the prerogative of trained medical professionals, presumably constituting the editorial 'we' in line 127. Ultimately, professionals will read the x-ray and make the final diagnosis. Whereas the GP is part of this 'we,' the patient is definitely not.

She is debarred from having anything more to say about the diagnosis. Thus, not only is the fact obfuscated that the patient had the correct diagnosis for her complaint, but her right to proffer a diagnosis at all has become an issue of contention. By mentioning the 'normal' order, an implicit criticism is made of the patient's attempting to diagnose herself prior to x-rays. Once again, she has committed an interactional misdemeanor by encroaching upon medical terrain, demonstrating that she lacks sufficient knowledge of the 'way things are done.'<7> The outcome of this struggle for power concerning the diagnosis of the patient's stomach pain is that she forfeits the floor for further talk about this particular trouble.

I shall now examine one last obstacle to the patient's troubles-talk, which occurs at the very end of the consultation. It displays many of the same interactional practices found in earlier examples. The GP is rounding off the conversation by listing the patient's complaints and what is to be done about them:

EXCERPT 9

214 D: So you'll come talk about the steriliza-
215 tion, together with your husband. (We'll
(Yeah.
216 take pictures of your gall-bladder.
217 And then we'll just have to take a little
uhh- loo- ((soft)) <u>And</u> you should just
218 stop by sometime maybe about the nervous-
219 ness what we can do about that,
maybe ((you have-))
220 P: (Yeah, I don't- don't want to be taking
221 pills or (or anything like that.
(Mm
222 I am willing once and awhile uhh- t- well
223 I have a bottle of <u>dormisal</u> from Doctor
Vogel for my son then when I (go to
(Hmm
bed-
((rapid)) what I do sometimes have is that
224 I (like can't get to sleep- (or
(Hmm (Yeah.
225 so(mething like that
226 D: (((rapid)) That - it doesn't offer any
solution at all.
227 P: No.
228 D: You know?
229 P: ((softly)) No:
230 D: ((softly)) You have to tackle the problem
231 at its roots, don't you?
Where it came from.
232 P: Yeah, that's - I would- that would- (
233 D: (<u>Well</u>,

234 *just come- do come* <8> *and talk about it sometime, O.K.?*
235 P: *Yeah. O.K.*

Although the final phase of a consultation is obviously not the best time for patient-initiatives, the patient, nevertheless, interrupts the GP, expressing disagreement with his suggestion that she might want to *do* something about her *nervousness.* Prefacing it with an agreement ('yeah'), she provides an account which can be heard as a suggestion running counter to what the GP has said.<9> After asserting that she does not want to take any pills, she re-formulates the GP's assessment 'nervousness' (lines 223-225) as a common, everyday event: not being able to get to sleep sometimes. Whereas the remedy for 'nervousness' might well be tranquillizers, for the patient's assessment of the problem, a mild, herbal remedy will do the job just as well.<10> The fact that she uses this for her son underlines both the ordinary quality of the complaint, as well as the harmlessness of the remedy. Although the patient undermines both the GP's definition of her trouble, as well as the implied solution to it, she does so in an 'off the record' fashion. <11> The remark can *either* be interpreted as a criticism of the GP's suggestion and yet another attempt on the part of the patient at self-diagnosis *or* as the simple provision of information. The patient can, therefore, not be held accountable for challenging his diagnosis.

Whereas members ordinarily observe politeness conventions and allow one another a bit of leeway for the sake of interactional harmony, the GP does not. Instead he sanctions her remark as illegitimate inroad on his territory. Far from being treated as a possible course of action, the patient's suggestion is characterized as 'not being any solution at all' (line 226) as it does not 'tackle the problem at its roots' (lines 230-231).

Thus, the GP reinstates his authority in the face of the patient's challenge. The consultation rolls to a close with the GP establishing *his* solution as the preferred one: the patient should 'come and talk about it sometime' (line 234). Whereas this implies that there *would* be an opportunity for the patient to talk about her troubles at some future time, the present interview provides little reason for optimism.

STRUGGLE FOR POWER

[the 'classic case of gall stones' tape]

The obstacles in the 'classic case of gall stones' tape will now be examined for how negotiations around the activity of getting the floor for troubles-talk reflect an ongoing struggle for power between the GP and the patient. In order to do this, I shall return to the dimensions of power as set out in earlier chapters (Chapter 4, Chapter 8).

Power in interaction

Both the GP and the patient drew upon skills and resources, which were structurally available to them in order to negotiate the obstacles around getting the floor for troubles-talk. These resources included systems of signification (interpretative schemes), as well as sanctions to be mobilized.

The patient's most characteristic strategy for getting the floor for troubles-talk was to present her troubles in medical terms. For example, she pre-packaged her current distress as a candidate for a medical diagnosis ('something with my liver?' or 'maybe my gall bladder?'). In this way, she could display them, right off the bat, as suitable topics for a medical consultation. Moreover, she drew upon shared understandings of medical procedures for dealing with illness. After trying to find out herself what the problem might be (Excerpts 1,6,8), she sought the services of a professional who had access to procedures unavailable to her (blood tests). While displaying familiarity with medical routines, she continually demonstrated deference to the GP's position as expert, who would, ultimately, be the one to authorize any diagnosis or treatment decisions.

In addition to drawing on shared understandings about what constitutes proper material for medical troubles-talk, the patient made use of her personal experiences with her illness. She supplied relevant information to elaborate the GP's diagnosis (Excerpt 7) or to indicate that she had her own ideas about how to deal with her complaints (Excerpt 9). Information taken from her biography provided counter-examples for what the GP was proposing, thereby tactfully challenging his assessment of her problems.

Like the patient, the GP also drew upon skills and resources taken from the medical frame. Unlike the patient, however, he employed these resources in order to prevent troubles-talk from being initiated. In most cases, this was accomplished by means of interruptions, usually in the form of a medical question. Questions are, generally speaking, a useful way of controlling topicality because they place the next speaker under some obligation to answer back (Schegloff and Sacks 1974). In a medical context, however, this constraint is even more pronounced. Questions, which are clearly derived from professional schemes of relevance, not only compel the patient to respond (in the interests of her own health and well-being), but they do not have to be fitted into prior talk. Whereas interruptions are normally disruptive to the ongoing talk, in a medical setting the physician's interruptions are likely to be chalked up by *both* participants as necessary for the business at hand (West 1984).Thus, the GP was able to initiate Q-A sequences, thereby unproblematically taking the floor away from the patient (Excerpt 7).

In addition to controlling access to the conversational floor by means of interruptions and medical Q-A sequences, the GP cut off the patient's attempts to embark on a troubles-telling by challenging her competence. This was done when she formulated her troubles as a potential diagnosis (Excerpt 5,8) or when she made a suggestion for treatment, which ran counter to his own (Excerpt 9). Her contributions were promptly disqualified as incorrect ('We don't do that') or misinformed ('She must have had something else then') or simply out of line ('There'd be no end to it. Come now!'). These conversational sanctions could be mobilized by virtue of the GP's privileged knowledge concerning the etiology of complaints, as well as the appropriate procedures for diagnosing and treating them. By sanctioning the patient's contributions to this process, his assessments could be established as the only authorized version, thereby cementing their respective positions as expert and layperson.

Power and agency

The exercise of power cannot be separated from agency. It involves a subtle mix of what actors actually do, whether intentionally or not, and what they might have done, but didn't. In other words, their activities are bounded by conditions of which they may not be aware, as well as by consequences, which may not have been intended. The 'classic case of gall stones' tape is no exception.

In this consultation, the patient was easily discernible as competent and knowledgeable agent. Not only did she indicate that she had already put considerable effort at home into trying to find out what was causing or contributing to her illness (Excerpts 6,7,9), but she continued this active stance within the consultation as well. In fact, she went out of her way to 'help' the GP perform the task at hand; namely, finding a suitable medical diagnosis and treatment for her current troubles. In view of recent developments in medical thinking about the desirability of patients' 'responsibility' or 'autonomy,' as well as the trend toward a more 'cooperative' model of health care (Chapter 1), she would seem, in fact, to have all the makings of a particularly *good patient*.

Medical *ideology*, however, does not necessarily correspond with medical *practice*. In this particular consultation, most of the patient's attempts to shoulder some of the responsibility for her own health problems within the medical encounter backfired, indicating that whereas autonomous or competent patients may be a desirable item in current medical literature, they may be singularly 'incompetent', interactionally speaking.<12> Rather than paving the way for further troubles-talk, the patient's displays of knowledgeability in medical matters led, paradoxically, to a forfeit of her right to talk about the troubles at all (Excerpt 5). Similarly, the fact that her own assessment of her current symptoms as 'something with my gall-bladder' (Excerpt 1) exactly matched the GP's assessment (Excerpt 8) did not evoke an admiring or appreciate response on the part of the GP or entail her to some share of the credit for having found a diagnosis for the problem at hand. On the contrary, these efforts elicited a brusque and even insulting reception on the part of the GP (Excerpts 5,8,9).

Judging by the 'classic case of gall stones' tape, the much-heralded 'coming of age' of patients as 'subjects' rather than just the 'objects' of health care may not be nearly as welcome a phenomenon for practitioners as one is led to believe (Brinkgreve *et al* 1979; Aghina 1982; Willems 1983).<13> In fact, the patient, who, in the wake of her newly-found subjecthood, attempts to take some responsibility over how her problems are being defined, diagnosed or treated in the context of the medical encounter, may find herself being rudely returned to her former position as uninformed and passive object of the GP's ministrations. Thus, an (unintended) consequence of increasing patient autonomy in health matters may be a paradoxical loss of actual control over the interaction in medical encounters with physicians.

Power and the dialectic of control

Power will presumably be relational, involving a dialectic of control. This means that relations of autonomy and dependence are being constructed, sustained, and undermined in *both* directions. In the 'classic case of gall stones' tape, the power dialectic centered primarily around the production and reproduction of the GP's *professional* authority over medical matters like diagnosing and treating the patient's problems. Far from being a stable structural parameter of the medical encounter, this authority had to be ongoingly re-negotiated against the patient's attempts to contribute the diagnostic process and, thereby, undermine the GP's undisputed control over this particular feature of the interaction.

In another problem-solving context, displays of knowledgeability concerning the task to be performed might be expected to legitimate the person exercising more (rather than less) control over the interaction. When participants are not interactional peers - as is the case in a professional-layperson relationship - the activities, which they may appropriately perform, become hierarchically ordered as well. In fact, the problem to be solved can even become secondary to the business of establishing the relationship as an asymmetrical one. A case in point is the patient's initial presentation of her complaints; i.e. that she was 'tired' and 'lethargic' (Excerpt 5). Despite the fact that these complaints were presented prior to the problem of 'stomach pains,' they did not receive any medical attention. There is no reason to assume that these particular complaints were any less serious than her subsequent troubles and, in fact, their position as first topics in the patient's presentation would even suggest that they may have been more problematic for her than the subsequently-mentioned 'stomach pains.' However, by introducing her initial troubles as a potential diagnosis ('something with my liver'), she made the mistake of encroaching upon the GP's territory, thereby jeopardizing her right to the conversational floor. The GP, drawing upon his privileged access to medical and technical knowledge as a resource, was able to overrule the patient and prevent these troubles from becoming topics for further talk. Although this decision was not necessarily warranted on medical or bio-technical grounds, it was displayed *as though* this were the case. In this way, the GP could establish his authority over decisions concerning which complaints would be suitable candidates for further talk within the consultation - and which would not be.

Thus, the patient's organization of her talk around the business at hand; the diagnosis and treatment of her problems, did not ensure her a piece of the interactional pie. However competent in terms of the task to be performed, her activities, without the GP's authorization, were, ultimately, doomed to failure. Ironically, within this particular dialectic of control, a less 'competent' stance might have enabled the patient to exercise more power over how her problems were being defined, diagnosed, and treated.

Although the GP in the 'classic case of gall stones' tape was able to keep the conversational reins well in hand, his control was, by no means, absolute. Like any other medical practitioner, he had no direct access to information taken from the patient's biography, her experiences outside the consultation, prior medical treatments, etc. Since this information, however, was necessary in order for him to perform his own specialist activities; namely, diagnosing and treating complaints, he was dependent on the patient to supply the relevant facts. In view of the much-lamented problem of time shortage, he presumably preferred to elicit the necessary information himself, rather than waiting around for the patient to produce it on her own. This entailed keeping the direction of the medical history-taking firmly under *his* control, thereby enabling him to gather the desired information, while keeping extraneous talk at a minimum. Thus, for the GP in the 'classic case of gall stones' tape, the patient's attempts to direct the course of the medical history-taking were a potentially problematic and, therefore, unwelcome activity (Bliesener and Siegrist 1981).

Seen in this light, it becomes clear why he allowed the patient to initiate a troubles-telling based on her experiences ('I was lying there, *rolling* from the pain') unimpeded, while abruptly cutting off talk about the location of the pain (Excerpt 8). By the same token, her suggestions for treatment, particularly when couched in argumentative terms ('I don't want to be taking any pills...') could hardly be allowed to stand uncontested. In order to have the final word concerning the diagnosis and treatment of the patient's complaints, the GP not only had to assert his authority, but had to have it affirmed ongoingly by the patient.

In the 'classic case of gall stones' tape, the struggle for power seems to have been engendered by the patient's having, unwittingly perhaps, encroached upon the physician's turf. This led to a kind of 'territorial dispute,'<14> whereby the stake became *who* would be allowed to use *which* resources in the process of defining, diagnosing, and treating her current troubles. Whereas the GP, ultimately, did win the battle, it required considerable mobilization of troops to do so. The fact that a patient, simply by engaging in a bit of diagnostic activity, could elicit such a reaction from a professional with all the resources and legitimation of the medical establishment at his disposal indicates that even the most powerful manifest the occasional 'chink in the armor' (Giddens 1984).

Power as enabling and constraining

In this consultation, the outcome of the struggle around the conversational floor did not fall in favor of the patient. As we have seen, she only rarely succeeded in getting the floor and, when she did, it was primarily because she was 'allowed' to do so by the GP (Excerpt 7). Not only were her contributions to the process of diagnosing and treating her present complaints curtailed, but her efforts to display herself as a worthy helpmeet to the GP were not honored as such. This could lead to the conclusion that the most distinguishing feature of the present consultation was control and repression. If power, however, rarely involves just constraint, then the 'classic case of gall stones' tape should show the enabling features of power as well.

By exercising control over the conversational floor as well as the unfolding topic talk, the GP not only placed limitations on the patient's opportunities for introducing new complaints or elaborating on those already discussed, but through the same activities he also facilitated - to quote Hughes:

> an organization of talk that permits the production of ordered sequences of topically relevant talk offers obvious mutual advantages (1982:374).

By ensuring that topics were discussed which fit into his specialist schemes of relevance, the GP enabled the medical history-taking to proceed along the appropriate paths, culminating in a diagnosis of (at least) one of the patient's current complaints.

The construction of the consultation as a medical event required more, however, than the deployment of specialist knowledge and procedures. It also involved the constitution of a social relationship in which one of the participants was a specialist. To this end, the GP had to establish himself as the medical expert with the patient in the concomitant position of layperson seeking his help. In the 'classic case of gall stones' tape, the GP's institutional authority was not enough, in and of itself, to ensure that the interaction could be channeled along unproblematic medical paths. On the contrary, the asymmetry itself had to be interactionally accomplished and, in fact, required continual shoring up. Even in the short space of a 7-minute consultation, ongoing surveillance was required against the ever-present problem of 'territorial fuzziness.'

Thus, the control exercised in the 'classic case of gall stones' tape might be viewed as both enabling as well as constraining. It helped to constitute the consultation as a medical event as well as an institutionalised relationship of authority, while at the same time curtailing the patient's power to contribute to the process of defining, diagnosing and treating her current problems.

Power as process

Power practices at the level of face-to-face interaction produce and sustain social structure - a structure routinely involving relations of domination and subordination. This duality of structure should be observable in how power is exercised in the course of any medical consultation.

In the 'classic case of gall stones' tape, the fact that the GP had sole authority over medical knowledge and procedures already constituted a power imbalance between himself and the patient. Even when she employed the same resources 'correctly,' this did not effect the balance, but only led to a re-drawing of the boundaries between their respective competencies. The consultation involved a moment-by-moment struggle to establish an institutionalized relationship of authority between a professional and a layperson. As such, it represents an instance of the power of the medical profession being produced and reproduced at the micro-level of medical talk.

Along with establishing the authority of the medical professional over myriad and sundry health problems, a corresponding recipient of his ministrations needs to be created as well. She will have to receive instruction concerning how she should behave within the professional encounter and otherwise 'learn the ropes' of what being a patient entails (Bringreve et al, 1979). Laypersons display an increasing tendency towards proto-professionalization; i.e. turning to professional sources of help and attempting to package their ailments in accordance with the routines and jargon of the profession.

In the 'classic case of gall stones' tape, both the process of establishing professional authority, as well as the complementary process of the patient trying to 'learn the ropes'<15> were observable in the interaction. The GP, drawing upon his professional schemes of relevance, ongoingly worked to establish his authority within the medical province of diagnosing and treating problems. When this authority was challenged, it had to be re-established by means of sanctioning processes, serving to prevent the patient from usurping this control and placing her squarely in the position of layperson.

The patient, on the other hand, could be viewed as a suitably proto-professionalised partner for the medical expert. Whether her complaint was 'lethargy' or 'stomach pains,' she invariably formulated it as an appropriately medical problem and, subsequently, proceeded to deal with it in accordance with the medical procedures of diagnosis and treatment.

Whereas the process of creating professional authority and the process of establishing a group of laypersons, who are in a position to seek the services of these professions, are two sides of the same coin, the former necessitating the latter, the 'classic case of gall stones' tape indicated that their relationship may not be an altogether happy one. Instead of two neatly interlocking social processes, this consultation demonstrated that the production and reproduction of the relationship between a professional caretaker and a layperson is not without its conflicts, albeit unintentional ones, and that, ultimately, the issue of who has the upper hand can become a matter taking precedence over the quality of the health care being provided.

NOTES

1. Literally, the expression is: you can keep on and on. This translation was less awkward.

2. Interestingly, the patient's request is, medically speaking, not entirely out of line. GPs are, of course, warned not to start performing unnecessary tests when patients present vague symptoms like 'feeling tired and lethargic.' In the majority of the cases, the symptoms will go away on their own, with or without a pill, and with or without any counselling (de Haan and Bleys 1980:70). The tests themselves can be tiresome, frightening or even risky and should, therefore, be avoided for potentially 'innocent' symptoms. On the other hand, 'tiredness' is often the first warning that the patient has a serious infectious disease - one of which is hepatitis, a disease of the liver. Thus, whereas the GP may be following standard medical procedure, it goes too far to say that the patient's request is ridiculous or merited the treatment it got.

3. Sarcasm is the production of ironically-worded utterances with the intent to hurt the recipient. It entails saying the opposite of what one actually means, but, because of prosodic (nasality) or kinesic (smirk) or contextual cues, the recipient can be expected to get the underlying meaning (Brown and Levinson 1978:226). In this remark, the GP obviously doesn't mean that 'we' can 'keep on going' indefinitely with blood tests. Since everybody knows how busy doctors are, this is ridiculous. By implying that the patient *has* wanted such a thing, he can be heard as being critical, albeit in a strictly 'off the record' way.

4. Mitigation as basic feature of interaction has been discussed at length by Labov and Fanshel (1977:84). It is the opposite pole of aggravation and refers to members' tendency to soften or modify an expression in order to avoid giving offense. If mitigation is displayed in most interaction, this is especially the case in professional encounters. In the above example, the patient formulates her candidate diagnosis (a gall bladder problem) as a request for information. An assertion ('It's a gall bladder colic') or just a tag ('It could be my gall bladder, don't you think?') would have been

less mitigating (Labov and Fanshel 1977:85). In this way, she shows deference for the GP's position as medical expert.

5. West provides an excellent analysis of Q-A sequences in medical interviews. Questions are asymmetrically distributed, with physicians asking 98% of them. Physicians display a marked preference for questions, which elicit a 'yes' or 'no' answer as opposed to conditional queries ('What would you do if ...'), which would entail further negotiation (West 1984:103).

6. Both lay 'feelings' and professional 'facts' are socially constructed. The latter, however, seems to possess more validity; it appears to be rooted in empirical reality, whereas unscientific intuition or emotions are regarded as inevitably equivocal. See Latour and Woolgar (1979).

7. It is debatable, of course, whether the patient has simply displayed lack of knowledgeability in medical matters; i.e. that a conclusive medical diagnosis of a gall bladder disorder requires taking x-rays first *or* that her incompetency resides in her continual tendency to step-out-of-line.

8. Literally: just come and calmly talk it over sometime. This is a bit ambiguous, referring possibly to having a 'calm conversation' in the future. I am inclined, however, to think that this is not what the GP meant, but rather that she should not have any apprehensions about coming back any time at some later date to consult him.

9. According to Pomerantz (1975; 1984), an initial assessment invites a second assessment, whereby agreements are, by far, the preferred next action since they involve less interactional work. One elegant way of expressing disagreement is to begin with a token agreement ('yeah') and follow it up with a stretch of talk, which conveys just the opposite. The patient begins with a 'yeah', followed by a disagreement to the effect that she does not want to 'do' something about her nervousness, if that means 'having to swallow things.'

10. *Dr. Vogel* refers to a Swiss botanist, who devoted himself to natural medicine and developed a series of herbal and homeopathic remedies for various common ailments. In The Netherlands, these remedies can be purchased in any drug store and *Dr. Vogel* is a household name. See Vogel (1979).

11. 'Off the record' refers to performing a communicative act in such a way that it is not possible to attribute a discrete and unambivalent intention to the act.
The actor leaves himself an 'out' by providing himself with a number of defensible interpretations; he cannot be held to have committed himself to just one particular interpretation of his act (Brown and Levinson 1978:216). The advantages of such behavior, particularly when the speaker is engaging in potentially face-threatening activities, are clear.

12. By this, I do not mean to suggest that the patient was actually familiar with the literature on patient responsibility and autonomy to be found in many a treatise on

modern medical practice. This is rather unlikely and, in any case, it would only have added confusion to her other troubles, interactional and otherwise.

13. For a good discussion of the contradictions surrounding the discovery of the responsible or autonomous patient, the reader is referred to Heydelberg (1984).

14. Lacoste (1981), drawing her inspiration from Labov and Fanshel (1977), divides medical interaction into 'doctor's events' (bodily indications, examination results) and 'patient's events' (biographical data, experienced symptoms) as well as 'disputable events.' The latter refer to events, where the boundaries have become 'fuzzy' and open to negotiation. This includes determination of vocabulary. For example, patients may say that they 'have a headache,' but not that they have a 'hypertrophic left ventricle.' Boundaries may also become unclear concerning the status of the event in question. Is a patient really feeling better or merely saying so because she doesn't want to have an operation? Is the doctor being honest about her prognosis or does he want her to worry because he thinks she should have an operation? Such 'territorial disputes' are part and parcel of the construction of any relationship between a professional and a layperson (Lacoste 1981:172-173).

15. Brinkgreve et al (1979) refer to these processes as *professionalization* and *proto-professionalization*, showing how they go hand in hand in the development of the care professions. I shall be returning to a more structural account of professional power in the last chapter of this inquiry.

10

POWER PRACTICES: RECOUNTING THE TROUBLE

RECOUNTING THE TROUBLE

Recounting the trouble

Once the patient/troubles-teller has gotten the floor, her next task will be to relate what the trouble is. She will need to provide enough information for the GP/recipient to follow her troubles-telling. He will need to know what happened, who it happened to, and *that* what happened constitutes something troublesome. There are various ways of going about this.

Sometimes a simple statement will do the job:

EXCERPT 1
[the 'a little bit nervous' tape]

```
 9 P: Sore-throat and headache, that's the only
10    thing that ((laughs)) I h-have-(
                                      (Oh yeah
```

Sometimes, however, troubles do not lend themselves to a single utterance, but involve so many characters and events that an extended stretch of talk is needed in order to convey all the information necessary to understand what the present distress is all about. The following excerpt shows how the same patient manages to recount the necessary information for understanding her present trouble:

EXCERPT 2

[the 'a little bit nervous' tape]

```
106 P: Yeah(hh). And then my brother-in-law calls
       me up. My mother-in-law has to be put in a
107    nursing home. And of course my husband is
108    already, too ((laughs)), you know?
       Where should she go and everything and
109    uhhh- she's in an old people's home now,
       you know, but she can't do anything any
       more so uhhh-
110 D: Who's in the old people's home?
111 P: My husband his mother, you know?
112 D: Oh dear.
113 P: And uhhh- now she also has to go to ((un-
114    clear)) but we'll have to see, huh
       ((laughs))[
                  [Yeah. Yeah.
115    Yeah. And then my brother-in-law is always
       calling up about this and that and, well
116    whether I've been working myself up about
       that I don't know. <1>
117    ((soft)) I have two sick children too, you
118    know? I don't know what it's called, it
       was on television a while back. that they-
119    have to be pounded, you know what I mean?
120 D: Ohh:   yeah.
```

This excerpt is taken from a protracted stretch of troubles-talk (see Chapter 7, pp.178-179). In contrast to her initial statement (Excerpt 1) that 'all' she had was a sore-throat and headache, it now becomes clear that the patient's troubles comprise a whole series of problematic events, involving a veritable cast of characters - her mother-in-law, her brother-in-law, her husband, and her children. Not only are various persons implicated in her present distress, but the troubles themselves are numerous and varied.

In this recounting, the patient has the task of providing the GP with a picture of what her current troubles are about. She will need to supply the relevant details for understanding the trouble, while avoiding too many extraneous items, which could lead to boredom or confusion on the part of the recipient. It will be up to the GP as recipient to display that he is paying attention and following her talk.

For the most part, this proceeds in a fairly straightforward fashion. The patient enumerates her present troubles:

- her mother-in-law, who is presently in a home for the elderly, has to be moved to a nursing home
- her brother-in-law, an invalid, is constantly calling her on the phone
- her husband doesn't seem to be much help
- her children may be seriously ill.

The patient not only provides the relevant details concerning the particular event and the parties involved, but she links them together by means of connecting devices like 'and' or 'and then.' In this way, a seriality is constructed, reminiscent of the chronological ordering found in other types of storytellings. These particular events, however, are not related temporally to one another. It is, for example, irrelevant for the troubles-telling, whether the patient's brother-in-law called her before or after the children got sick. The seriality being constructed is much more one of a continuing tale of woe (Rehbein 1980). As separate daily cares, each trouble might not be considered to be particularly bothersome. Taken together, however, they constitute reasonable grounds for being upset.<2>

In addition to providing information concerning the troubles themselves, as well as for their being heard as a collection, the patient establishes her own relationship to the events in question. This is necessary, since they involve other people; i.e. she is not directly a protagonist (Quasthoff 1980). They are presented as potential reasons for her working herself up, as an account for her nervousness (line 116).

In this segment, the GP has aligned himself as troubles-recipient. Not only does he give her space to talk (by not interrupting or asking questions, for example), but he produces minimal responses to indicate that he is paying attention. He even displays some emotional reciprocity with her troubles as in line 112 with his 'oh dear'. At one point in the patient's recounting, the GP seems to be confused about who exactly has to be put in a nursing home. In asking for clarification concerning this detail, he not only contributes to the unfolding troubles-telling, but displays himself as attentive recipient, bent on following and understanding what the troubles are about.

Just as getting the floor requires interactional work on the part of both participants, recounting the trouble is also an interactional co-production. The troubles-teller has had to delineate the events, as well as her own position in relation to them. In this way, she accomplishes the groundwork for understanding her experience of suffering. It is not enough that the GP as recipient simply listens passively to what the troubles-teller says. He must ongoingly indicate that he has heard and understood the talk as well. If something is unclear, he has to do a quick check and request information to fill in the gaps in his understanding.

In summary, recounting the trouble, like getting the floor is an interactional accomplishment, requiring specific tasks from both the patient as troubles-teller and from the GP as troubles-recipient. In the 'a little bit nervous' tape, the patient not only obtains the floor easily, but receives affiliative responses throughout her troubles-telling from the GP. Beginning with the complaints 'sore-throat' and 'headache,' she is eventually able to embark on a full-fledged story about her numerous daily cares, as well as the problems she has with various people in her life. Unlike the previous consultation (the 'classic case of gall stones' tape), she has ample opportunity to engage in troubles-talk and is even encouraged to produce more troubles. However, this encouragement is not without its shadow-side. It is to this dimension of the troubles-talk that I shall now direct my attention.

OBSTACLES TO RECOUNTING THE TROUBLE

[the 'a little bit nervous' tape]

In this consultation, the patient and the GP appear to have rather different ideas about what constitutes the trouble or, more to the point, what belongs to its recounting and what does not. This gives rise to a series of negotiations concerning what the trouble is and, subsequently, what should be done about it. Although these negotiations occur in a highly cooperative fashion and result in a mutual understanding of what the patient's complaint is, a closer look will reveal that the patient has encountered subtle, but, nevertheless, systematic obstacles in the course of her attempts to relate her troubles.

The patient has already announced that 'the only thing she has' are a sore-throat and headache (Excerpt 1). The GP, however, upon examining her, has not been able to find anything:

EXCERPT 3

```
21 D: ((unclear)) -and also in your throat I
      don't- don't find - find any-[
                                   [No:
22    Actually it does look normal[       there.
                                  [Yeah?
23    Yes.
24 P: How can that be- I here when it feels so
```

25 so strange. ((complaining)) I wake up
during the night from it that there's such
26 pain here ((groans)) on the left on the
left side?
27 D: On the left side?

In this utterance, the GP contrasts what he has found (nothing) with the patient's experience of pain. His findings, based on the way her throat 'actually' looks when examined, are labelled 'normal.' In this way, he draws upon his institutional authority as the one who, ultimately, will decide whether subjectively-experienced symptoms can be classified as a sign of medical pathology. Despite his assessment of the patient's pain as 'normal,' he clearly has difficulties making this statement, as evidenced by the numerous self-repairs in his first utterance (line 21).

The patient doesn't make it any easier on him. By expressing surprised disbelief and dramatically recounting her troubles once again, she not only undermines his assessment of her sore-throat as normal, but contests his authority to define her trouble in a way, which runs counter to her experience as well. She feels the pain and it is simply too serious to be 'normal.' Thus, an interactional power struggle has arisen concerning whether the patient's sore-throat constitutes a suitable topic for troubles-talk in the context of a medical consultation.

The outcome of this struggle is postponed, however. Another solution has presented itself a bit later on in the consultation.

EXCERPT 4

38 D: So you stopped (smoking too?
(Yeah:
39 P: Yeahh: I don't do that any more either. So
yeah, once and a while just one, you know.
40 ((distinct)) Bu- but last week I was I
41 didn't do it all week but it ((soft))
42 doesn't help not even a little bit.
(No
43 D: Yeah. Does it make you anxious that
feeling?
44 P: Yes. Then I start thinking what is this.
45 (((a kind of uhh-))
46 ((laughs)) (And if that's why I have
(Yeah

such a headache. I don't know, see. ((com-
47 *plaining)) <u>Such</u> <u>headaches</u>. Well and*
48 *whether it's because of that, look I don't*
know, or because of my dreams or
49 *something [And <u>sore</u> <u>eyes</u>.*
[Mmhmm. Yeah
Terrible.
50 D: *Yeah. But that you have the feeling uhh-*
51 *that uhhh- god, is there*
something- [there or some-
52 P: *[Yeah, I do sometimes. Because*
53 *my father ((laughs)) had it in his throat,*
too. I think well, maybe I've got it, too.
54 *Do you understand, he had throat cancer*
55 *too, so I think, God, maybe that's what I*
have too, you know what I mean ((laughs)).
56 D: *Well: after all it's a uhh- an under-*
standable [thought.
[Yeah(hh)

The GP's announcement that he can't find anything has not elicited a compliant response from the patient, but rather more troubles-telling in which she emphasizes how terrible her complaint is. In this segment, a possibility for re-directing the talk towards a new trouble has emerged. Whereas this is done in a harmonious fashion, the participants not only have distinctive ideas about how the old trouble should be recounted, but about the new one as well.

In this segment, the GP, having checked up on possible aggravating factors for the patient's sore-throat, asks whether 'that feeling' makes her 'anxious.' Although this utterance is tied to prior talk by means of 'that,' it works to shift the talk to another dimension. The trouble is no longer the patient's actual physical pain, but rather how she *feels* about it. By labelling the feeling as anxiety, the GP works to make it a potential trouble of its own.<3> This will have real consequences for the outcome of the struggle around the definition of her problems, as we shall see.

The patient begins by confirming the GP's suggestion that she might be anxious. However, she immediately returns to one of the previously-mentioned *physical* complaints: i.e. her headache. She accomplishes this in a way, which displays a certain amount of deference to his reading of her problem, as well as a shared understanding that physical complaints like headaches can have psychological origins (line 44-46). On the other hand, she expresses a lack of conviction that anxiety can sufficiently explain her present distress. This not only works to stave off the potentially face-threatening dimension, which an admission of anxiety could entail, but it re-opens troubles-talk about her headache. It is the physical aspects of her complaints, which she repeatedly emphasizes (lines 47-49), thereby attempting to establish them as the most important topic.

Despite her valiant attempts to recycle her preferred complaint, she has not managed to get the GP to relinquish his topic. Setting his next utterance off from the preceding talk ('but'), he proceeds to elaborate on the anxiety, suggesting that it might be a fear of 'something, there', localizable in the patient's body. This is dramatized by means of 'god' and direct speech. By not expressing directly what that 'something' might be, the implication is that it is cancer; i.e. something too serious to mention in-so-many-words.<4>

The GP's intervention is successful, working to elicit an affirmative response from the patient and a round of troubles-telling about her father who had throat cancer. Thus, by initiating a new trouble, the GP is able to eliminate the patient's original complaint from the conversation, at least for the time being. Despite her attempts to recycle her headache for further troubles-talk, she is forced to abandon it in favor of the new one; i.e. anxiety. In this way, the GP can begin to work on a complaint, which is more amenable to diagnosis, thereby enabling him to come to the fore, once again, as medical expert.

The new trouble, once introduced, now becomes the topic of further troubles-talk. Although both the GP and the patient cooperatively engage in talk about the patient's 'anxiety,' they display rather different orientations toward how it should be recounted. The patient continually links these troubles to her current physical symptoms, as well as the concrete events of her daily life. She displays a marked reticence of having her distress labelled as anxiety. In her version, troubles are presented inclusively, as part and parcel of her overall situation. His concern appears to be her experience of suffering, rather than the trouble *per se* (Jefferson and Lee 1981).

The GP, on the other hand, appears to be working towards finding a label for what the patient has. His utterances proceed from her 'anxious feeling' to 'something there?' This is but a step away from a potential diagnostic category 'fear of cancer' (cancerophobia). In attempting to re-formulate the patient's troubles as 'anxiety,' the patient's recounting is disrupted. Instead of treating them as a unified whole, he isolates one aspect and provides it with a label. By exercising his institutional authority to define and diagnose the patient's problems, he transforms her ongoing story about her troubles from an experience to an 'object to be serviced' (Jefferson and Lee 1981). Ultimately, he has the power of definition; not the patient.

This is only the beginning. Throughout the rest of the consultation, the patient's recounting of her troubles is obstructed in much the same way.

EXCERPT 5

```
61 P: Yeah. And then again the whole ((month))
62    then you think, it'll blow over. But it
63    doesn't blow over at all. It's a very
      awful feeling. ((3.0)) So yeah. ((laughs))
64 D: What do you think? Yeah. So I just can't
65    find anything in the examination I don't
      find any uhh- (    anything unusual
66 P:              (No. You can't see anything-
67 D: I can't see anything. (   Do you want to
                            (No.
      just wait a little while?
68 P: Yeah?
69 D: Or uhh-
70    Yeah. Do you want a pill, yeah uhh- if
71    you're very nervous (    about something
                          (Yeah
      uhhh- something uhh something for the
      nervousness
72 P: Yeah. Just do that. I guess
73    that that (      I do-  Yeah: And I am
                 (Yeah?
      ((laughs)) a little nervous. (
                                   (Yeah.
74    We have so many uhhh- strange things at
      the moment.
```

This segment, occurring a bit later on in the consultation, illuminates the GP's dilemma even more clearly than the last. Although he has gotten a confirmation from the patient that her difficulties may be related to her anxiety, he is no closer than before to providing a solution to the problem of her sore-throat. Whereas he formulates this dilemma briefly (lines 64-67), he immediately switches back to the second trouble; her 'nervousness.' It is for this trouble that he can provide some advice. His first suggestion is that the patient 'wait.' Since this is probably an unattractive option for someone, who has already been suffering for a long period of time, he softens it with 'just' and 'a little while.' The second possibility is that she take a pill for her nervousness. By placing pill in the diminutive <5> and stressing that they should only be taken if the patient is very nervous (line 71), the GP displays himself as a progressive proponent of an integral approach to medicine. Not only does he indicate an appropriate reluctance towards prescribing psychotropic drugs without good cause, but, at the same time, he places the responsibility for the decision squarely at the patient's door. By allowing the patient to decide how her problems are to be treated, he can also obfuscate the fact that the options between which she must choose have been, in fact, set by him: her 'choice' is to either do nothing or to do something and that 'something' is the familiar standby for women's problems: a tranquillizer.

Taken as a whole, the patient's response in Excerpt 5 confirms the GP's assessment of her trouble as a matter of anxiety as well as his proposed solution for it. If we examine how she recounts the trouble, however, it appears to run counter to the GP's version in the same way as the examples above. Once again, she downgrades the label of nervousness by means of a mitigating 'a little' and laughter (Jefferson 1984a). Furthermore, she embarks on a protracted round of troubles-talk, prefaced by the announcement that a lot of 'strange things' have been happening lately. These strange things, as previously mentioned, concern the fact that she is currently overloaded with responsibilities for sick children, demanding relatives, job, and illness, without getting very much support. Thus, whereas she may agree with the GP *that* she is nervous and may even more thankfully accept any help he might have to offer, she works to display her distress as firmly linked to the contingencies of her daily life. Her 'being a little nervous' is nothing more than a legitimate response to a difficult situation. In this way, she undermines the GP's assessment of her troubles as strictly psychological.

The interactional struggle around the way the patient's trouble is to be defined comes to a head in the next segment, occurring towards the end of the consultation:

EXCERPT 6

127 D: Can you talk to your husband a
little [about it?
128 P: [Oh: sure, you know, yeah sure.
((3.0))
129 But my husband says, just go to the doctor
130 ((laughs)) with that headache. [
[Yeah(hh)

131 D: Well. You can tell your husband that it
really isn't uhh- [
132 P: [not uhh
133 D: isn't physical-
134 P: Ohhh:
((2.0))
135 D: ((clears throat)) No. I really thought
136 that uhh- the
problems [((clears throat)) - were
137 P: [Yeah, that's what I thought too
138 D: just getting caught in your [throat
[would that-
139 P: Could that be it? Did you think.
140 D: Yes.
141 P: Because I got those seresta pills from you
142 the last time and ((animated)) when I did-
143 it's really crazy - when I take one every
once and awhile, then I do feel uhhh- do
feel fitter.
144 D: Yeah.
((2.0))
145 P: Well, I guess that I have gotten myself a
little bit ((laughs)) too upset-
146 D: Just ((use)) this.

Up until now, the patient and GP have been cooperatively working toward a definition of the patient's present troubles. Whereas they seem to be in agreement about the problem and what should be done about it, they have distinctive ways of recounting it. The GP's reduction of the patient's complaint to a psychological label has been ongoingly countered by the patient, who insists that it be treated as part and parcel of a whole series of experiences.

In this last segment, the consultation is coming to a close. At this point, the GP abandons his position as troubles-recipient to ask a question. Whereas this question is tied to prior talk in terms of topic; i.e. whether the patient can talk to her husband about these problems, it also serves to cut off further troubles-talk and establish his reading of her problem in a more definitive way.

The patient has continually presented herself as currently having to bear too much responsibility. She has mentioned her husband on two occasions in the course of her talk. The first time (see Chapter 7, p.179) involved his telling her not to let things get on her nerves so much. This is hardly a reaction, which invites further talk about problems. The second time he is mentioned is to describe him as being upset about his mother having to be moved to a nursing home. This is not presented as a sign that he is supporting the patient with this particular problem, but rather that he himself is just one more problem for her to have to cope with (Excerpt 2). Judging by these remarks, there is little reason to believe that the patient receives much help from her husband and, in fact, he may be something of an additional burden to her.

What, then, can be made of the GP's query about whether the patient can talk to her husband?

To begin with, it displays a certain inattentiveness on the part of the GP to portions of the patient's troubles-talk. If he has, indeed, heard what she has been saying about her husband, then the question seems, at best, a bit rhetorical or patronizing. Requests for information rely on the person doing the requesting not knowing the answer beforehand. Otherwise the question can be heard as a test or baiting, as any school child can readily attest (Labov and Fanshel 1977:82).

The question, however, also works as an indirect suggestion that the patient *should* talk to her husband.<6> Given the fact that many professional helpers place a high premium on the curative effects of talk with significant others, the GP here may well be providing what he considers some helpful advice. Implicit in this advice is also the suggestion that what the patient has, up until now, been doing with the GP (talking about problems), she might just as well be doing with her husband. Thus, the GP's request does double-duty: it curtails further troubles-telling and constitutes his authority as medical expert; i.e. someone who can provide advice or help for the patient's problems.

Although the GP's question cuts off further troubles-telling, it does not do so unproblematically. Far from taking him up on his suggestion, the patient tosses the ball back to him. She states on no uncertain terms that she *does* talk to her husband, but he just tells her to talk to her doctor. This utterance accomplishes several tasks, simultaneously. To begin with, by claiming that she already talks to her husband, she refuses the GP's suggestion that talking to her husband could be a solution for her present difficulties.

Her utterance does more than that, however. It also implicitly challenges the GP, thereby undermining his authority as medical expert. By recycling her original complaint - her headache - she draws attention to the fact that this is what she came to the doctor for in the first place. She has, up until now, received no advice for it, even though advice is what he is supposed to provide. All of this is accomplished in a mitigated fashion. By attributing her potentially offensive remark to her husband, the patient is able to remain conveniently 'off the record' herself (Brown and Levinson 1978:216).

A somewhat ticklish situation has emerged here. Not only has the patient called the GP's assessment of her current distress and its possible alleviation into question, but she has returned to her original trouble which, as we know, was something the GP could do nothing about. His response is predictable. No matter how much negotiation there has been concerning the exact nature of the patient's problems, it will be the GP, who has the final say <7> concerning how they are diagnosed and treated. By mobilizing the resources available to him as professional, he provides the definitive diagnosis <8> of the patient's complaints - a diagnosis, which relegates them squarely to the realm of the psychosomatic. The problem is not physical at all, but a matter of problems, which have 'gotten caught in her throat.' With this neat bit of body imagery, he not only defines her physical symptoms as psychological in origin, but he indicates the direction to be taken, should she wish to alleviate them: talking about her problems, presumably with her husband. He overrides her version of her problems in favor of his own, but does so in a way which allows him to remain 'off the record.' Using the same ploy adapted earlier by the patient, the GP mobilizes the patient's husband as topic, suggesting that he can rest at ease that her headache is not the real problem. In this way, he can have the last word regarding the nature of the patient's troubles, as well as establishing his authority over hers in the business of defining her problems.

In the closing segment, the patient capitulates, agreeing that tranquillizers may well be the answer to her problems and that she is, indeed, 'a little too nervous.'

STRUGGLE FOR POWER

[the 'a little bit nervous' tape]

It has been demonstrated how the activity of recounting troubles in the 'a little bit nervous' tape could proceed in a cooperative fashion, as well as be subject to impediment. Unlike the 'classic case of gall stones' tape (Chapter 11), the patient's rights to the conversational floor were not being contested here. The power struggle centered instead around the way that her troubles were being recounted and, more to the point, which version of these troubles would ultimately be authorized as the correct one.

Power in interaction

In the previous consultation (the 'classic case of gall stones' tape), both the GP and the patient drew upon resources available to them within the medical frame in order to exercise control over the interaction. In the 'a little bit nervous' tape, however, the interaction was not restricted to a medical frame. On occasion, a friendship frame (Chapter 5) was also employed to negotiate obstacles, which emerged in the course of recounting the patient's troubles. Similarly, normative sanctions were mobilized from both frames.

As the patient proceeded to relate her current distress, the GP made use of professional schemes of relevance available to him as medical expert in order to fit these troubles into something resembling a medical diagnosis. These interpretative schemes were not of a bio-medical nature, but based on more recent understandings of the psycho-social dimension inherent in many somatic complaints (Menges 1986). Unlike the GP in the 'classic case of gall stones' tape, he did not impede the patient's account of her troubles by interrupting her. In fact, he even actively elicited new troubles (Excerpt 4) by initiating another round of troubles-telling, after the previous trouble had proven unamenable to further medical ministrations. A more promising candidate having been discovered, he proceeded to shape it up as a problem in its own right and, later on (Excerpt 6), as the basis for much of the patient's current distress. This was accomplished by means of formulating and re-formulating the patient's nervousness as problem (Davis 1986). Thus, the GP not only mobilized the patient's activity of troubles-telling itself, but drew upon (shared) understandings of the psychosomatic nature of her illness in order to establish his version of her troubles as a matter of 'nervousness.'

When this diagnosis was not immediately acceptable to the patient, as evidenced by her continually returning to previously-mentioned troubles, the GP organized some outside help from a fellow member of his sex. This time he did not mobilize support from his professional community of colleagues, but rather looked to the patient's own husband for a helping hand (Excerpt 6). By constructing a coalition <9> between himself and her husband, the decision over what was wrong with her could be taken out of her hands and presented as a matter to be agreed upon between her doctor and her husband.

Although the interaction between the GP and the patient in the 'a little bit nervous' tape was extremely cooperative throughout the consultation, they displayed different and, to some extent, conflicting understandings of the patient's current distress. The patient clearly saw her troubles as *either* somatic and, therefore, 'something for the doctor,' *or*, in the case of her other troubles, as the 'normal' response to a particulary trying living situation. Whereas she was quite willing to talk about these troubles and even to accept a prescription for tranquillizers in the hope that they would make her feel better (Excerpt 6), she continued to resist having her problems defined as a straightforward case of 'nervousness.'

This could be seen in her refusal to relinquish her original complaints ('sore-throat and headache, that's the only thing that I have'), thereby rejecting the GP's assessment that everything was 'normal.' She continued to come back to these complaints, somewhat to the GP's chagrin. Thus, despite her cooperative stance in confirming the GP's assessment that she might *also* be a bit nervous, she did not go along with having her physical complaints discarded without further ado.

Not only did she hold on to her initial complaints as her primary reason for seeking the services of a physician, but she recounted additional troubles in such a way that they were clearly just a part of her present circumstances. This was accomplished by means of storytelling devices, which enabled her to link her feelings (anxiety, nervousness) to the contingencies of her daily life. In this way, she could show that her 'nervousness' was nothing more, but also nothing less, than the legitimate response to a difficult situation. She might be in need of tranquillizers, but that was only because she was 'a little nervous' due to all the 'strange things happening lately.' Thus, she resisted any overly-simple or face-threatening psychiatric categorizations of her current distress.

When this did not work (Excerpt 6), she resorted to stronger measures. In order to counter the GP's tendency to psychologize her present troubles, she was, ultimately, compelled to switch over to a more traditional medical frame, mobilizing sanctions based on shared understandings of what constitutes appropriate behavior for a physician. By directing the GP's attention to his responsibility for diagnosing and treating somatic complaints, the patient indirectly challenged him in his competence as medical expert. Somatic complaints are, after all, what doctors are supposed to be treating. In this way, she also undermined his authority to define her troubles in a way, which ran counter to her own experiences or understanding of them.

Thus, both the GP and the patient in the 'a little bit nervous' tape mobilized skills and resources available to them in order to influence the course of the consultation as well as the way the patient's complaints were being defined, diagnosed, and treated.

Power and agency

Power will be exercised by agents who, despite asymmetrical access to resources, knowledgeably attempt to influence the course of the interaction. In the 'a little bit nervous' tape, the patient not only cooperatively contributed to the process of finding a suitable diagnosis and treatment for her present distress, but she attempted to influence the way her problems were being defined, even when this was at odds with the GP's assessment. Whereas the patient in the previous consultation (the 'classic case of gall stones' tape) tried to influence the interaction by providing her own diagnosis of her complaints, displaying her knowledgeability of medical routines and procedures, the patient in the 'a little bit nervous' tape left the business of supplying a diagnosis to the GP. Rather than intruding upon his 'territory,' she engaged in the indisputedly lay activity of troubles-telling. By engaging in troubles-telling, she could influence how her problems were defined; i.e. as a normal and understandable reaction to a difficult living situation.

Ultimately, she did accept a definition of her current distress as an individual, psychological problem, serious enough to merit a prescription of tranquillizers. This could be interpreted as a sign that she had finally, after considerable interactional struggle around how her problems were to be defined, come to agree with the GP's assessment of 'nervousness.' However, her ongoing assertion that she was 'only a little bit nervous,' as well as her repeated emphasis on the contextual reasons for this nervousness make another interpretation more plausible. Her confirmation of the GP's assessment may not have been so much the result of her agreeing with his assessment, but rather of a practical kind of compliance based on a rational weighing of the alternatives (Giddens 1984). In view of her probable interest in getting some sort of help for her problems as well as her previous, positive experiences with tranquillizers (Excerpt 6), she may well have settled for the diagnosis and treatment as a package deal.

Accepting this kind of help had some additional consequences, which the patient may not have intended. In order to be treated, she obviously had to have an appropriately treatable ailment. Since such an ailment was not part of her initial presentation of complaints, another candidate had to be discovered in the course of the consultation. Whereas the success of this endeavor led to the provision of help in the form of a prescription, the price was another problem to add to her already-lengthy list of troubles.

The GP in the 'a little bit nervous' tape was also clearly viewable as a competent actor, skillfully attempting to come to grips with that familiar bane of any physician's life; namely, the complaint with no discernible organic cause. Having unsuccessfully attempted to reassure the patient (Excerpt 3, 5), he was faced with the unsavory option of letting her leave his office empty-handed or of finding another complaint, which would lend itself more easily to treatment. Whereas the former course of action would obviously have been more straightforward, its potential interactional consequences left something to be desired. Even in a case like the present one, where nothing can be done for the patient's complaints, both she and the GP will expect that some sort of help be provided. This shared orientation - the unacknowledged condition of any professional service encounter - structured the subsequent interaction along predictable paths. Upon discovering that he could do nothing about the patient's initial complaints, the GP devoted his attention to finding something else. Generally speaking, a physician will look for etiologies within the individual patient, whether somatic or psychological, rather than attempting to grapple with her social context and its bothersome array of insoluables like sexual inequality, unsatisfactory work, poor housing, and the like. Whereas this explains our GP seeking solace in the realm of the psychological, his diagnosis of 'nervousness' served, at the same time, to mask the social dimension of the patient's current troubles. Thus, an unintended consequence of the struggle to find a diagnosis at any cost can be the psychologization <10> of the patient's problems.

Power and the dialectic of control

In medical encounters, diagnoses are interactional accomplishments, requiring ongoing negotiation on the part of both participants. Whereas the GP as medical expert will generally have the final say on how the patient's problems are diagnosed, the patient will have a vested interest in the process as well. Particularly when their respective understandings of her problems diverge, we might expect an interactional power struggle to emerge. In the 'a little bit nervous' tape, the dialectic of control centered precisely around this issue; namely, *which* complaint was to be selected for diagnosis and, having been selected, *how* it would be defined.

The first problems presented by the patient were physical symptoms - a sore-throat and headache. On the surface, such complaints seem just the thing for a medical consultation. By presenting them first, the patient endowed them with a privileged status: her reason for seeking the services of a medical expert. This is not surprising. Patients tend to package any trouble in medical-terms, thereby showing that they know what kinds of problems are suitable material for the context at hand (Heath 1981).

Patients are not the only ones to give physical symptoms a preferred position within their troubles-talk in medical consultations; physicians prefer them as well. Bodily aches and pains are what physicians are trained to treat. Their professional power is most securely rooted in this medical expertise. Unfortunately, GPs are not only confronted with straightforwardly broken fingers, needing to be set or infections, which will respond nicely to a round of penicillin. In fact, such problems may even have taken a backseat to the ubiquitous psycho-social complaint (Chapter 1). Since the GP's job remains the provision of help and this help is still contingent on his being able to diagnose the patient's problems and provide some form of treatment, he will also have to find a way to help these problems. Unlike broken fingers, however, problems arising from stressful working conditions, family relations, or, for that matter, social inequality cannot be mended quite so easily. In fact, the solution, if there is one, is rarely within the reach of an individual GP. Faced with this dilemma, a GP may opt for the solution of medicalizing these social problems, thereby enabling them to be treated within a frame of reference where he has at least some semblance of control.

The power struggle in the 'a little bit nervous' tape reflected both the powerlessness of the medical professional to alleviate the problems presented by patients as well as its insistence on providing help anyway, even when this help serves to medicalize the patient's problems. When it became clear that he could do nothing for the patient's initial complaints, but that her troubles were, by no means, limited to these symptoms, another possible course of action presented itself for the GP. He could either proceed, in accordance with the medical frame, and find another complaint, which he *could* treat, or he could switch frames and engage in an activity potentially available to anyone; i.e. lending a sympathetic ear. Whereas this latter activity would have been appropriate in terms of the kinds of troubles recounted by the patient (the burden of caring for sick children, having to find a nursing home for her mother-in-law, her brother-in-law's overly-frequent phone calls, etc.), *just* lending a sympathetic ear would, at the same time, undermine the GP's position as medical expert. Medical expertise is grounded in the fact that the practitioner does something different than what anyone else (the patient's mother, her next-door neighbor, etc.) could do. Thus, labelling the patient's problems as a case of 'nervousness' was felicitous in that it worked as a diagnosis of sorts, calling for a clear-cut course of action; namely, a prescription for tranquillizers.

The patient did not sit quietly by the wayside, passively awaiting the doctor's verdict. She made it quite clear that she expected something be done for her complaints, as demonstrated by her disbelief that the GP couldn't find anything causing her sore-throat (Excerpt 3). The help she seemed primarily interested in receiving, however, was for her physical complaints, something she continued to assert, up until the end of the consultation. Once it became clear, however, that no help would be forthcoming for these complaints, she quite readily took up the GP's invitation to produce some more 'troubles.' As it turned out, she had plenty to spare, her present situation being a veritable treasure trove of minor and not-so-minor difficulties.

By engaging in troubles-telling, she could make use of the opportunity offered in order to enlist the GP's sympathy, however temporarily, for her present predicament. In this way, she could let off steam and otherwise enjoy some of the benefits inherent in telling a story about her experiences (Chapter 5). Whereas this did not promise a remedy for her sore-throat and headache, it could, at least potentially, win an emotionally reciprocative response from the GP as sympathetic troubles-recipient. Even if her complaints did not become objects for medical treatment, at least she could have a friendly chat with her physicians about her daily cares.

The expression of emotional reciprocity can take various forms in social interaction, but a psychiatric diagnosis is not one of them. As Jefferson and Lee (1981) have noted, *any* attempt to do something about the trouble in the course of its being told is risky business. At best, it means taking the floor away from the troubles-teller and interrupting her ongoing tale of woe. At worst, it turns her experience of suffering into an object to be treated and she herself becomes a body to be serviced (Goffman 1961). Whereas the GP was sympathetic, he did not provide an affiliative response to her troubles-telling. Instead he offered her a diagnosis.

In part, both the GP and the patient, in their ongoing struggle to define her problems, were caught in a mutual dilemma centering around expectations that the GP provide help, but that there was no help available for those problems presented by the patient.<11> The GP attempted to solve the problem by doing something anyway; i.e. fixing up the problem so that it seemed as though it could be helped. The patient alternated between insisting that her initial complaints be remedied and, when that failed to produce results, engaging in some therapeutic troubles-telling about her other difficulties. The outcome of this shift in interactional frames was mixed: on the one hand, the patient, did, in fact, receive some help, although not for the complaints for which she had requested it. She was less successful, however, in establishing her version of her problem as being 'just a little bit nervous' due to 'all the strange things' in her present living situation. Accepting this version would have meant that the GP enter, at least temporarily, the interactive frame set by the patient through her troubles-telling: a chat between equals. Despite modern medical ideology about 'teamwork' (Chapter 1), doctors and patients apparently are only on the same team when the doctor decides what game they are playing.

Power as enabling and constraining

Defining (other) people's problems has been cited by various authors as an activity which has far-reaching political consequences. Having the mandate to relegate specific kinds of events or behaviors to the realm of the problematic, to designate what kind of problem they are, and to indicate what should be done about them opens up myriad avenues for social control (Schneider and Conrad 1980). The process of defining problems involves constraint and limitation; it can, however, be enabling as well. In fact, it may well be this intertwining of enablement and constraint, which produces a more effective, intensive form of control than would have been possible by more repressive means.

In the 'a little bit nervous' tape, the GP was, ultimately, able to draw upon his authority in the course of the interaction to legitimate and enforce his definition of the patient's problems - a definition, which, on the surface, seemed no more 'scientific' than the patient's ('the problems were just getting stuck in your throat').<12> It did not, however, entail the imposition of a straightforward constraint upon the patient's version of her difficulties. If it placed limits, the definition of 'nervousness' opened up possibilities as well.

To begin with, the definition itself provided the rationale for a concrete form of help. In view of the patient's reaction, this was, by no means, unwelcome and, in fact, she appeared to embrace a prescription for tranquillizers with open arms (Excerpt 6). If one subscribes to Parsonian notions concerning the advantages inherent in the sick-role (Parsons 1951), then defining the patient's problems of living in medical terms, to be treated with pills just like any other illness, might also have had the positive effect of making the patient feel less responsible for the fact that she is having trouble coping with her present situation. If she couldn't cope, it was, at least, not her fault. Her illness was to blame.

Furthermore, by not limiting himself to the patient's physical symptoms, the GP paved the way for a potentially holistic approach to the patient's health care needs. The patient's headache did not have to be rejected as irrelevant just because it lacked an identifiable organic etiology, but could be considered within the context of her everyday life, a bodily response to a situation characterized by heavy responsibilities and little support. Such a perspective might have opened up the possibility for a more adequate understanding and, perhaps, even alleviation for the present problem.<13>

Despite the obvious attractions this kind of definition can have for the individual, who is just trying to get through her day without having a splitting headache, it has its shadow side as well. The reduction of her complaints to an individual, psychological problem did not simply 'let her off the hook' in the sense that she could resort to the relatively simple solution of a tranquillizer, but it served to shift attention away from the structural roots of her distress. In this way, possibilities potentially open to her should she take the view that her difficulties were part of her situation rather than her psychological make-up, were closed off as the interaction progressed.

Defining the patient's current distress in such a way that it became her problem rather than, for example, a marital problem, involving her husband as well, or, at a more structural level, a problem stemming from the sexual division of labor, which puts the entire burden of caring for family and relatives on the shoulders of the wife, could (and frequently does) neutralize dissatisfaction (Barrett and Roberts 1978). As Waitzkin (1983) puts it, psychologizing problems tends to 'cool the patient out.'

It is in this context of conflicting definitions that the struggle for power takes place. In the 'a little bit nervous' tape, the patient continually presented her problems as a normal response to her present situation. She resisted the GP's assessment, placing her feelings in context and reformulating the problem as 'maybe just a little bit nervous.' In this way, she undermined his authority to define her current troubles as a strictly psychological problem.

The patient's troubles-telling did not only work as a limitation on how her problems were being assessed, but, ironically, it also facilitated the very process by which these problems could be defined in a way which ran counter to her own version of them. By packaging her experiences in story-form, just as she might have done <14> in a chat with her husband, a friend or relative (Chapter 5), she introduced an alternative interactive frame to the medical one in progress. As part of a friendship frame, storytelling can undermine or transform the dominant medical frame, transforming relations between the physician and the patient to an encounter between equals, at least for the moment. It can, however, also feed into and sustain the medical frame, as well as asymmetrical relations between the participants, by facilitating the process whereby the patient's problems are diagnosed as individual, psychological problems. Despite her protests that she was 'only a little bit' nervous, just a normal, 'member of the fold' (Jefferson and Lee 1981), the patient also provided grist for the GP's definition of her distress as something abnormal enough to warrant a prescription for tranquillizers.

Thus, the interactional struggle around the way the patient's troubles were being recounted demonstrated that power can be both enabling as well as constraining. This Janus-faced quality permeated the way *both* participants exercised power and, ultimately, resulted in a psychological definition of the patient's current difficulties with her living situation, warranting a prescription for tranquillizers. Similarly, the presentation of her current distress in story-form allowed the patient to present her experiences as she saw them, as well as put herself on more equal footing with the GP in the process. At the same time, it facilitated the very definition of her distress as an individual, psychological disturbance, something other than just a normal reaction to a difficult situation. Whereas expanding a somatic approach to illness to include psychological dimensions can open up productive avenues for health care, it inevitably limits the way these problems are viewed as well; namely, as problems for the doctor, rather than problems of living.

Power as process

Structural relations of domination and subordination are produced, reproduced, and undermined by interactional power practices. In the 'a little bit nervous' tape, this duality of structure was visible in the interactive power struggles around the definition of the patient's problem.

Although the patient's initial complaints were, at first glance, suitably medical, they proved problematic later on, as it became clear that the GP couldn't find any organic reason for them. This constituted a dilemma for both participants - the patient because she wanted relief for her suffering and the physician because he was the one expected to provide that relief. Confirming the patient's version of her troubles as, indeed, 'the only thing' she 'has' (Excerpt 1) would have entailed, at the same time, shouldering responsibility for the fact that he could do nothing about them, thereby undercutting his authority as medical expert. By finding another complaint, which could be treated, this unhappy state of affairs was prevented and the GP's professional competence restored. The general credibility of medical science as undisputed instance for dealing with patient's problems, whatever they may be, was thus re-instated.

By providing a psychological label for the patient's current troubles and legitimating it with a prescription for tranquillizers, ideological sustenance was given to cultural assumptions that women are especially susceptible to psychiatric illness (Chapter 1). If the problem is one of faulty (female) psychology, then it becomes self-evident that women's living and working situations are not what need to be altered. By means of psychologizing the patient's troubles, the structural roots of her personal distress could be mystified and depoliticized. Endurance rather than change became the order of the day (Smith and David 1975).

In his discussion of nature of medical control, Waitzkin notes:

> From a position of relative dominance, doctors can make ideologic statements that convey the symbolic trappings of science. These messages reinforce the hegemonic ideology that emanates from other institutions - the family, educational system, mass media, and so forth - and that pervade a society. The same messages tend to direct clients' behaviors into safe, acceptable, and nondisruptive channels; this is the essence of social control in medicine (1983:142-143).

In conclusion, the interactional process of defining the patient's problems of living as an individual, psychological problem served to establish the GP's wavering medical authority, enabling him to provide ostensibly expert help. This was accomplished, despite the fact that the help, upon closer inspection, would hardly pass muster. It is difficult to provide a scientific rationale for how tranquillizers might have helped the patient's sore-throat and headache or, for that matter, alleviated the overly-heavy responsibilities for the care of various family members. And, if a prescription for psychotropic drugs was not, in and of itself, a dubious enough solution for the patient's current difficulties (Cooperstock 1978; Nellis 1980), it had the additional disadvantage that, in order to emerge as solution, the patient's authority in defining her problems had to be undermined. Her initial complaints were greeted with skepticism. Since they couldn't be observed, they were apparently not 'really' there. Similarly, her version of her feelings as normal response to a difficult situation was discounted. The power imbalance around the definition of the patient's troubles reflected and sustained GP's authority to normatively authorize what would count as objective reality - whether this reality involved a bodily complaint or problems of living.<15>

NOTES

1. In Dutch, this is literally: whether I make myself nervous about that, I don't know. This is a bit awkward in English where a person tends to *be* nervous or, more actively, to work herself up. A middle-of-the-road translation would be: I get upset. Here, I have opted for 'working myself up,' because the patient seems so intent on taking responsibility for her own feelings and reticent about blaming anyone else for them (see, also, Chapter 7).

2. The patient's account might be viewed as a layperson's version of the coping model in stress theories (Kuiper 1978:187). Medical experts assume that people do not get sick due to any one event, but rather to the unprecipitous combination of factors. It is the overload, which, ultimately, leads to mental or physical breakdown.

3. For a description of a similar process, see Schwartz (1979) and Davis (1986). By means of meta-linguistic listening, the GP is able to summarize prior talk and, at the same time, introduce new topics for consideration.

4. See Sontag (1978) for a discussion of how people tend to conceal cancer, not referring to it directly by name. It is only when the GP questions her further that we learn that the patient's father had throat cancer. This was clear to the GP, however, even with the ellipsis, because 'it' in conjunction with anxiety will be heard by most members as cancer.

5. Throughout my material, examples can be found of GPs using the diminutive form for pills ('a little pill'). It has a mitigating effect on the GP's prescribing tranquillizers, something, which is not an entirely unproblematic activity. See Gabe and Lipshitz-Phillips (1984) for a sensitive discussion of how physicians can be both concerned about preventing their patients from becoming dependent on medical solutions to their problems, while, at the same time, prescribing tranquillizers *en masse*.

6. Interaction requires that demands for action be made indirectly. The imperative 'Do X!' is made very rarely. One way to request an action is to express interest in it. Such expressions can then be interpreted as a request that X be performed, provided that various pre-conditions be met (need for action, need for request, ability, willingness, or obligation). In the above example, the GP could hardly say 'Talk to you husband!' whereas his inquiry as to whether she does might be heard as saying

something to this effect (Labov and Fanshel 1977:82).

7. An interesting exception to this tendency may be found in the doctoral thesis *Huisarts & Patiënt in Gesprek* (1985) by Kaag *et al.* They analyzed the interaction between a (feminist) female GP and an elderly female patient. Although the physician clearly did not agree with the patient concerning the nature of her present complaints and said so, she did not insist on having her version accepted as the correct one at the end of the consultation.

8. 'Diagnosis' is used here in a praxiological sense, rather than as a standard category for classification. At some point in each of the consultations, the GP sums up what is wrong with the patient, sometimes as a medical-type classification (for example, gall stones), but, more often than not, as an everyday designation ('nervousness').

9. For a similar case of a GP forming a coalition with a patient's husband in order to bolster a contested definition of her problem, see the 'one-continuing-story' tape (Chapter 7). Both lend support to the thesis that the therapeutic relationship between a male professional and female patient displays structural similarities to a marital relationship (Chesler 1972). Arguing along the same lines, Hearn (1987) locates male power in the control or management of women's emotions. In his view, the professions, most notably medicine, have simply taken over tasks which were formerly performed within the family. The domination of men over women which was formerly exercised by husbands and fathers in the private sector of the home and family has been transferred to the public sector where it has been taken on by physicians and other professions. I shall be dealing with this in more detail in Chapter 14.

10. Psychologization and medicalization are processes often mentioned in the same breath and are generally used in the context of criticizing the expanding jurisdiction of the various care professions over 'problems of living.' The assumption is that in defining and diagnosing the problems of clients or patients, attention is shifted from structural (social, economic, ideological) conditions, which contribute to the distress. By mystifying the structural roots of personal problems, the problems become depoliticized and their solution comes to reside in the client/patient rather than oppressive social arrangements.

    In either case, the point of contention is not so much whether the problem is defined as a *medical* or a *psychological* entity, but whether such states can be assigned and treated on an *individualistic* basis

    > A view of human beings as socially constituted, as having emotions, beliefs, abilities, and so on only insofar as they are embedded in a social web of interpretation that serves to give meaning to the bare data of inner experience and behavior, would in fact seem to be incompatible with a social and political theory that sees social groups as built on the independently existing characteristics of individuals ... only if psychological states can be seen as attaching to individuals in abstraction from their social

> setting can we expect to appeal to them to justify forms of social organization; otherwise, we are in a position of attempting to evaluate a system that is constructing the data on which the evaluation is based (Scheman 1983:232).

11. It is interesting to note that whenever the discussion turns to the issue of patient autonomy or responsibility, that participants, whether they are medical professionals or not, are quick to point out that patients 'want' the physician to diagnose their problems; that they are the ones, who are so insistent about getting pills or tests or referrals to specialists and so on; that it is extremely difficult for the physician *not* to comply with the patients' expectations, etc. Whereas I do not doubt that this is an adequate description of the way much of the interaction between patients and physicians proceeds, it tends to shift the responsibility for any of the ensuing inequities onto the patients' shoulders, leaving the medical professional (suddenly) quite without power to do anything about it. In view of the amount of time and effort that goes into training GPs in social skills, it seems that one thing they might be able to practice is sticking to their guns, even if it means having to cope with a dissatisfied patient. This will be discussed in more detail in Chapter 13.

12. Freidson describes this 'moral entrepreneur' function of medicine as follows:

> The profession does treat the illnesses laymen take to it, but it also seeks to discover illness of which the laymen may not even be aware. One of the greatest ambitions of the physician is to discover and describe a 'new' disease or syndrome... (1970:252).

13. This is, unfortunately, not what happened. Taking the patient's social context into account would certainly have entailed discussing the fact that she was overburdened as the sole caretaker in the family. A suggestion for treatment for this particular difficulty might have involved her husband shouldering some of the care for the sick children, as well as coping with his own mother and brother instead of leaving it all to the patient.

14. But, in fact, does not. The impression is inescapable that the patient does more listening at home to the problems of others than that she is listened to.

15. For a more detailed treatment of this practice as structural feature of psychiatry, the reader is referred to Chesler (1972), Smith and David (1975), Scheman (1983), and Davis (1984b, 1986).

11

POWER PRACTICES: ESTABLISHING THE POINT

ESTABLISHING THE POINT

Storytelling always involves more than the simple recounting of 'what happened.' Stories, including stories about trouble, will have a point; i.e. a reason for being told (Polanyi 1979). Compared to the previously-mentioned activities of getting the floor or recounting the trouble, establishing the point of a story about trouble is a particularly complicated one. This is because, unlike other kinds of stories, a story about trouble has a *double point* (Chapter 6). Getting this (double) point across will, upon occasion, demand somewhat contradictory interactional activities from both the troubles-teller as well as the recipient.

As in any story, the first part of the point of a story about trouble will entail *evaluating* the events in question as unusual enough to warrant being told (Chapter 6). This means establishing them *as particularly distressful/painful/terrible*. In order for the evaluation to work, the recipient must recognize that the events in question constitute a deviation from what the troubles-teller might be expected to endure under 'normal circumstances.' Making the point will involve mobilizing the recipient's background assumptions about the events, what makes them especially terrible, and why they go beyond what anyone should, normally speaking, have to bear. Whereas background assumptions are sometimes shared, quite often they are not, as if the case, for example, when participants' life-worlds are separated by gender, class, ethnic or cultural differences.

The second part of the point of a story about trouble will involve *establishing* the troubles-teller's powerlessness to alleviate her present distress. She will need to show that she is a 'victim of circumstances,' struck down by events or conditions beyond her control, and, therefore, in *need of help*. Getting the point of a story about troubles across will, therefore, entail not only showing how terrible the trouble is, but also showing that a 'skandalon,' as

Rehbein (1980) calls it, has occurred.

Establishing herself as powerlessness and in need of help will also require that the troubles-teller orient the recipient to 'normal circumstances.' The issue of needing help is a potentially problematic affair, bringing with it the possibility of a loss of face to the troubles-teller, who runs the risk of relinquishing her status as a competent adult, able to take care of herself and manage her own affairs (Labov and Fanshel 1977). Thus, whereas she will need to demonstrate her present powerlessness as part of the story's point, she will, at the same time, go to some length to establish that this powerlessness is temporary and that she is normally someone, who can cope with her difficulties on her own. Establishing one's competency under normal circumstances can, however, conflict with the special constraints imposed by a service encounter, where the person with the trouble has, by the very act of seeking the services of a professional, forfeited her claim to normalcy.

I shall now turn to my material and show some of the tasks involved in the interactive accomplishment of the point of a story about trouble. The two dimensions of the point - evaluating the trouble as terrible and establishing the trouble teller's need of help - will be dealt with separately. It will be demonstrated how this (double) point is interactionally achieved by both participants and how its accomplishment entails an ongoing and somewhat paradoxical orientation to both the *unusualness* of the patient's experience of distress and her powerlessness in the face of it as well as to *normalcy*; i.e. how things should be under normal circumstances.

Evaluating the trouble

The first task is to establish the event being recounted as something troublesome enough to warrant being told. This is done primarily by means of constructing a relation of contrast (Quasthoff 1980). Specific segments of the troubles-telling are set off from the rest, thereby marking them as particularly relevant to the trouble at hand. The recipient will be required to display his alignment by acknowledging just how terrible the trouble is. In my data, I discovered primarily three ways in which evaluation could be accomplished: <1>

Complaining

One way patients have of marking off the point of the story from the other elements is to complain. By complaining, I do not mean the general activity of talking about painful experiences (Bremer 1980b), but rather the noticeable alteration in the way such talk is done. Early on in the present inquiry, I discovered a sudden shift in the rhythm of patient's trouble talk.<2> It took on a sing-song quality with certain words being heavily stressed and uttered more slowly and distinctly or louder than the preceding talk. Intonation patterns changed whereby selected words were pronounced with elongated vowels. This shift was often accompanied by sound effects like groaning, sighing, gasping or gagging.

In my material, this characteristic way of talking about troubles was not uniformly present throughout the troubles-telling, but rather interspersed at regular intervals. Interestingly, it was present in every tape I listened to. Upon closer examination, it became clear that these paralinguistic cues were performing an important interactive function in terms of getting the point of the story telling across.<3>

Whereas it is difficult to capture this shift in pitch and rhythm in my transcript, the following excerpt illustrates how such segments are located within an ongoing troubles-telling. I have marked intonational changes with arrows:

EXCERPT 1

[the 'classic case of gall stones' tape]

```
   85 D: Do you only get that when you've been
                                  Yeah
         eating out?
   86 P: Yeah.
   87 D: Not when you cook yourself?
   88 P: No.
         At home I don't have- one time I did have
   89    trouble but that was different.
→  90    ((complaining)) Those were real stomach
→        cramps. Then I was- was also sick the
   91    whole day, just one time and that's-
→  92    Then it goes away and ((groans))
→        ooh-h-h-h-h-     ((that's so-))
                     Mmhmm.
   93    Then you double over.
   94    But this pain is something different, it's
         constant as if - yeah just as if you had
→  95    gotten a kick in the stomach that kind of
```

→ 96 feeling and it's here.
→ like- ((groans)) ooh-h-h-h-h-h- or some-
97 thing, you know what I mean, like-
98 D: -But you said that you lie there rolling
99 with pain. Do you have to move, with that
pain?
→ 100 P: ((complaining)) WELL- t-the last time it
101 was a tiny bit less but before that, then
I was lying there really rolling back and
→ 102 forth and I didn't know where to tu:rn-
103 D: -And you aren't able to just lie [still?
[No.

The above segment has a Q-A structure (Merritt 1976), with the GP trying to determine whether or not this is, indeed, a 'classic case of gall stones' and with the patient recounting an experience of having had terrible stomach cramps after a home meal. Whereas the patient in this consultation rarely gets beyond a description of her bodily symptoms (Chapter 9), she does relate them in such a way that her experience of suffering is set off and especially marked. This becomes the point of her talk in contrast, perhaps, to the surface description required by the GP in order to reach a diagnosis (Rehbein 1984). Complaining, with its sing-song intonation, uneven stress, and intermittent placement, is one way of doing this.

Repetition

Another frequently used device for evaluating a particular event as troublesome enough to warrant being told is repetition. Aspects of the ongoing troubles-telling deemed significant by the troubles-teller can be underlined and marked for special attention. In the 'a little bit nervous' tape (see Chapter 8 and 10), the patient engages in a protracted round of troubles-telling about a series of difficulties in her current situation. Her recounting of the events is punctuated from time to time with utterances like: 'yeah, I am a little nervous. We've got so many strange things at the moment; 'I'm saddled with my mother-in-law and my brother-in-law, everything, and whether it's also getting on my nerves, I don't know-;' 'my brother-in-law, he can't well walk any more and whether that's been making me so-o-o-o-o nervous, I don't know, huh?;' 'my husband says, don't let it get on your nerves so much. I say, yeah- ((laughs));' 'And then my brother-in-law is always calling up about this and that and, well whether I've been working myself up about that, I don't know.' And, finally, in response to the GP's offer of a prescription for tranquillizers: 'Well, I guess that I've been getting a little ((laughs)) too nervous-.'

In each of the above examples, the troubles-teller adds a significance statement (Ryave 1978) to her recounting of some troublesome event. By doing this, the recipient can be alerted to hear the trouble as an instance of or an account for her being 'a little nervous.' His attention is focused on the effects the various events have on the troubles-teller and this effect becomes the point of the telling as a whole. Thus, repetition serves not only as a reflexive means of relating clusters of stories, but also to highlight the essential aspect of these stories for the recipient. He is being helped to evaluate the events as not only problematic in and of themselves, but especially so by virtue of the fact that they 'get on her nerves.' In this way, he can be helped to get the point.

Dramatization

A final procedure for evaluating the point of a troubles-telling is dramatization. This means delineating the dramatic potential of the events being recounted and, thus, focussing the recipient's attention on them. It is the way storytellers catapult everyday occurrences into the realm of the unusual.<4> Dramatization can occur at all levels of discourse: within a single utterance, as a device for organizing topicality, and as a feature of the troubles-telling as a whole.

In the following troubles-telling, various devices are being employed, which make it a highly dramatic and emotional account of one of the patient's troubles: her husband has recently had a car accident. A factual construction of the same incident (for example, a police report) would look quite different. By constructing a contrast between the events themselves and the evaluation, the recipient is being helped to build a model of relevant information so that he can understand what the patient's point is - not the accident itself, but what it means to her in terms of her marriage, her present distress, etc.

EXCERPT 2

[the 'one-continuing-story' tape]

200 P: Plus that my husband uhh- under- uhhh a
201 really- uhh sometimes drinks too
much- And uhh- still does, you
(Yeah(hh))
know. Th- that's just the way it is some-
202 how, so uhh- because it's not going well
203 with him either And uhh- well last
(Mmhmm)
week I got - he called up at ten ((was
204 sitting ... at the station)) something
uhhh had been drinking and had run some-
205 one over and kept it from me.
206 So I find out about it a week later from
207 a registered letter from the parents of
208 the girl. WELL YEAH(HH). ((2.0))
(Yeah(hh))
209 You think, God, what am I - NOW WHAT ARE
WE DOING?
((complaining)) What's happening now. My
210 whole marriage is like falling
apart and I'm falling apart and- and
(Mmhmm)
211 that man who says well, you've got to get
up at 9 o'clock.
212 Yeah.
213 Listen, I'm sorry I'm-

In this segment, the troubles-teller is not simply recounting what happened: that her husband has a drinking problem and has recently had an accident without telling her about it so that she found out when a special delivery letter arrived from the victim's parents. The entire recounting is full of intensifiers ('really' and 'still, you know'), heavy intonation ('drinking,' 'run over,' 'kept it from me') and slow, distinct speech ('NOW WHAT ARE WE DOING?').

In describing the events, she alternates between mitigating what happened (her husband drinks 'sometimes,' it's 'just' part of the present situation, which is bad for him, too) and aggravating it (he drinks a lot, he still drinks a lot, her whole marriage is falling apart, etc.). Her narrative account of what happened is interrupted by a sudden shift to direct, emotional speech ('god,' 'now what are we doing?'). And, finally, the entire troubles-telling is contrasted with an earlier mentioned trouble: her dissatisfaction with her therapist, who makes unhelpful suggestions along the lines that she ought to get up earlier (see Chapter 7). Not only does the contrast serve to confirm a previous point retrospectively (she's not getting much help), but it further highlights her troubles, and, more generally, the dramatic quality of her life. All of these devices enable the troubles-recipient to evaluate the events in question as highly problematic and difficult for the troubles-teller.

Up until now, attention has been directed at the devices employed by the troubles-teller for evaluating the point of her troubles-telling. Considerable effort goes into constructing the experience as particularly painful, distressful, and even dramatic for the person(s) involved. Since the point is an interactional accomplishment, just like the other tasks involved in a storytelling, the recipient will be required to *acknowledge* it, thereby displaying his alignment with the patient's troubles-telling. He can do this in various ways.

In most of the examples above, this task was accomplished by means of minimal responses like 'yeah,' 'mmhmm' and 'oh,' which were sufficient to demonstrate that the recipient had been listening.
The recipient can, of course, do more than simply acknowledge that he has heard the point. He can also explicitly express *agreement* with it.

EXCERPT 3
[the 'a little bit nervous' tape]

```
82 P: My brother-in-law, he ((unclear)) and he
83    can't well walk any more ((unclear)) and
      whether that's been making me so: nervous,
84    I don't know, you know?(
                              (Yeah.
85 D: That could very well be having an
      influence(       you know,(on those ehhh-
86 P:          (       Yeah-    (         it could be?
87 D: Yeah.
```

In this segment, acknowledgement of the evaluative point goes beyond the obligatory listening noise. The patient's reading of her present distress is confirmed by means of the formulation in line 85 (Heritage and Watson 1979). One of the patient's many problems, her invalid brother-in-law, is established as sufficiently troublesome to warrant her being very ('so:') nervous at present. In this way, the

GP as recipient of the troubles-telling conveys not only that he has heard her, but he agrees with her assessment as well.

Agreement can go beyond a mere confirmation of the trouble-teller's point. It may also entail an upgrading of the trouble. In the next excerpt, the recipient not only agrees, but finds the event in question even more problematic than the teller herself does:

EXCERPT 4
[the 'one-continuing-story' tape]

```
278 P: So, I've got to keep swallowing the pill
       because otherwise it would- ((laughs)) we
279    just couldn't use that now, even though I
280    really would like to and(       all these
                               (Yeah-
281    years would have wanted to.(
                                  (Yeah.
       ((2.0))
282    Yeah.
       ((2.0))
283 D: Yeah. You've been taking the pill all
284    these years even though you would really
       just love to have( a baby.     Yes.
                        (Uhh- YES!
285    This is also ehh- a situation which is
       enough to drive you completely crazy, I'd
286    imagine, isn't it? <5>
287 P: All things considered,(it is, yeah.
       (((starter))          (Yeah. Yeah.
```

A previously mentioned trouble is being recycled here.<6> The patient would like to have a baby, but her husband refuses. Not only does the GP as troubles-recipient support this assessment, but he strengthens it by suggesting that the whole situation is enough to make anyone crazy (line 285). This is done by using stronger evaluative terms than the troubles-teller herself did along with intensifiers (Pomerantz 1984).

Up until now, some of the interactional devices have been discussed, which are available to troubles-tellers and recipients for evaluating events. In order to establish the trouble as problematic, it is necessary to mobilize shared background assumptions concerning what makes an event more troublesome than normal.

Establishing shared background assumptions

Shared meanings about what constitutes a potential trouble are essential for getting the point across in the course of a troubles-telling. Since participants do not necessarily share the same background assumptions about what constitutes a trouble, they may have to help each other along by providing additional information.

EXCERPT 5
[the 'double-uterus' tape]

14 P: My oldest daughter - she's been married
15 for six years and she'd love to have a
baby now, after all, you know? (But
(Mmhmm
16 now: there's been an examination and well:
17 she's - she seems to have a double <u>uterus</u>.
18 D: ((surprised)) OH!
19 P: Have you ever heard of that?
20 D: It happens sometimes yeah. Sure enough.
21 P: And well: -
(Yeah -
22 D: (All rabbits have them, I believe, but-
23 P: - and then you work (yourself (up there
(Mm (Mmm
again (about it, you know?
(Yes
24 D: I can imagine.

This segment illustrates how a troubles-teller can make use of background assumptions to enable the recipient to appreciate the seriousness of the trouble. In this case, the trouble is that the patient's daughter has just discovered that she has a double uterus and is now worried about whether or not she will be able to have children. Not only does the patient/storyteller provide the recipient with the facts, but she supplies information, which, although extraneous to the trouble at hand, can help him realize how problematic it is for her and her daughter. She indicates that her daughter is married, has been married for 6 years, and apparently does not have children yet. Moreover, she doesn't *just want* a child now; she would *love* to have one. All of this information, presented economically in the first utterance, helps to establish the double uterus as a problem of some urgency. Whether or not the GP hears the matter as urgent will depend, in part, on whether he shares certain background assumptions with the patient. For example, he will, at the very least, need to sympathize with what it might be like for a woman, who wants to get pregnant, not to be able to do so. Moreover, if he believes that marriage makes having children compulsory, then six years of childlessness will seem a reasonable account for the present situation being a matter of considerable concern.

The GP as recipient may acknowledge the information, using devices described in the preceding segment (minimal responses, upgrading, agreement), thereby displaying his affiliation and cooperation in producing the troubles-telling as a mutual endeavor. Or, he may make use of some background assumptions of his own to help establish the point. In the excerpt above, the remark that rabbits have double uteri, too (line 22) is a case in point. If the patient remembers how prolific rabbits are, she might find such an assertion eminently reassuring. In this way, the point of the troubles-telling can be cooperatively shaped up into a matter of greater or lesser seriousness for the parties in question.<7> Establishing the point of a troubles-telling as especially painful/distressful, etc. depends, then, on what the troubles-teller and the recipient view as a deviation from what the teller should, under normal circumstances, be expected to bear without complaining. Working this out is not the only task facing those engaging in a troubles-telling, however. The point is not simply that the events are terrible, but also that the person with the trouble is in need of help.

Establishing the need of help (the 'skandalon')

Powerlessness

The troubles-teller not only has terrible troubles, but she is powerless to do anything about them. She is a victim of circumstances and part of the task of establishing her point will be to show that her current suffering is, indeed, a 'skandalon' (Rehbein 1980). Not only will her troubles-telling involve displaying her powerlessness as protagonist, but it will be oriented toward her need of outside help for the same.

In a medical context, needing and getting help is the order of the day. This can be accomplished in a straightforward fashion; i.e. the patient simply asks for help. For example, in the 'classic case of gall stones' tape, the patient requests information on something she cannot diagnose alone (Chapter 9, Excerpt 6). In the 'a little bit nervous' tape, an admission of nervousness elicits a prescription for tranquillizers (Chapter 10, Excerpt 5).

A troubles-teller's powerlessness and her need of help can also be displayed by embedding them within the ongoing story. Notice how this patient accomplishes her point:

EXCERPT 6
[the 'one-continuing-story' tape]

```
55 P: You can't do it on your own, yeah sure you
56    can think like well: I'll just walk down
      the street ((rapid)) and what do those
      people matter to you, right. Just feel bad
57    again. Just go ahead. ((complaining)) BUT
58    I DON'T WAN- TO FEEL LIKE THAT AT ALL so
59    I don't go down the street, do you under-
      stand? I-I shut myse(lf     I'd rather stay
                          (Yeah-
60    in bed the whole day or something. And uh-
      yeah, it's up to you, the choice whether
61    uhh- ((clears throat)) to break through
      it. That's something I can't do
      EITHER(     The only thing I can tell you
            (Mm
62    is that you are frittering away your life
      at the moment and uhh-(          ((1.0))
                            (Yeah(hh)
63    Yeah. ((complaining)) So, I have to make
64    the decision myself whether I will now or
65    in twenty years, I have to see myself how
      I have to get through it.
      ((3.0))
```

This segment shows how the patient's anxiety not only severely limits her activities (fear of walking down the street, staying in bed, etc.), but is ruining her life as well (lines 61-62). In addition to expressing her helplessness in the face of her anxiety attacks, she re-enacts conversations with her therapist, who has informed her that it is her choice to 'break through' or to go on like this for another 20 years. Whereas she seems to accept this injunction, she is also unable to follow through. Without asking the GP explicitly for help, she has dramatically and clearly demonstrated that she not only needs help, but hasn't, up until now, gotten it - at least not from her therapist.

We can now look at how the troubles-recipients can react, when confronted with such displays of powerlessness. There are, basically, two ways recipients may acknowledge the troubles-teller's powerlessness, thereby displaying alignment to this dimension of the story's point.<8> The first way is to show *sympathy* or *commiseration* for her predicament, as in the following example:

EXCERPT 7
[the 'one-continuing-story' tape]

77 D: Well, small wonder. It would seem really
78 crazy to me, too, if you didn't find that
scary. Uhh- yeah? When you start seeing
79 your world getting smaller and smaller
and- and you're being notified from the
80 other side that, well, listen here, it's
your choice and you have to do it uhh-
uhh-

Sympathizing with the teller's helplessness is not the only option open to the recipient, however. He can also provide some *advice*. This entails treating a display of helplessness as an indirect request for help and, subsequently, providing it:

EXCERPT 8
[the 'one-continuing-story' tape]

423 D: Listen, I've prescribed another thirty
424 tablets here <9> uhh- I would really like
(*Hmm*
425 it if uhh- uhh- if you like especially
used these pills too and like, that you do
426 that as much as possible in consultation
with the psychologist. ((slow)) Uh- I do
427 want us to - let's say- also also once and
428 awhile to talk to you. But that TOO I
429 would r-r- rather do in consultation as
much as possible.

Whereas there is also a connection between needing help and getting it in everyday troubles-talk, it is considerably more tenuous than in a medical encounter. A person with troubles may appreciate good advice, but she cannot always expect the recipient of her talk to provide it. Similarly, the recipient may provide some form of help, but is not constrained by virtue of his or her position to do so. Thus, whereas members do engage in advice-giving and receiving in the course of everyday troubles-talk (Jefferson and Lee 1981), the two activities are not inextricably connected as they are in medical troubles-talk, where both participants will expect help or, at least, some sort of advice to be provided.

Business-as-usual

Establishing that one is powerless and in need of help goes hand in hand with displaying an orientation to business-as-usual (Jefferson and Lee 1981). Just as dramatic renditions of the 'terrible-ness' of the troubles are followed up by reassurances that the troubles-teller will not make an undo fuss, a similar tendency can be found when the participants attempt to establish the second part of the point: the troubles-teller's powerlessness and her need of help. In a culture where a high premium is placed on taking care of one's self, it is not easy to admit that the present trouble has made the person unable to handle her own affairs. When confronted with this potential threat to face (Goffman 1967), members may attempt to show that, whereas they need help, they still want to be considered as responsible, competent adults, 'members of the fold' entitled to the same conventional treatment meted out to everyone else.

The following excerpt demonstrates how both the patient as troubles-teller and the GP as recipient take special care to ensure that the former has not strayed too far beyond the boundaries of normalcy:

EXCERPT 9
[the 'a little bit nervous' tape]

```
 96 D: Yeah, well yeah, you have gotten yourself
       uhhh-[
 97 P:      [M-my husband says, don't let it get
 98    on your nerves so much. I say, yeah(hh)-
       ((laughs))
 99 D: Yeah. You're a little ashamed about
       it[  I've got the impression.
100 P:   [Yeah: I think well uhh-
101 D: Yes, but there's a difference between-
102    between nervousness[     and if you really
                          [Yeah-
       have something something something[
                                         [Yeah.
```

```
103      are sitting with a problem,
         huh?      like      you  that uh- that
104 P:   sitting with a problem, yeah.
105 D:   member of your family who is really
                        Yeah
         seriously ill.
```

As we have already seen (Chapter 10), the patient recounted the various distressful events in her life, indicating, on the one hand, that they are terrible, but, on the other hand, that she is 'only a little nervous.' This same ambiguity is expressed in the above excerpt as well. First, she confirms her husband's assessment that she shouldn't get so upset, but follows it up with a laugh, letting her 'yeah-' trail off as if to say: easier said than done! At this point, the GP picks up on her reluctance to have her troubles given the overly-heavy designation of 'nervousness,' suggesting that she is ashamed and hastily amending that she shouldn't see the problem as 'nervousness,' but rather as something 'really' being wrong. In this way, he places her difficulties, albeit temporarily, in the realm of normal everyday affairs, where people 'just' have problems. Thus, the downplaying of the troubles-teller's helplessness in the face of her current distress by presenting them as ordinary concerns is a co-production in this particular stretch of talk. Both participants work to establish the present trouble, as well as the teller's feelings of powerlessness, as not exceeding some normal limit.

In conclusion, the point of a story about troubles is a double one. On the one hand, participants need to establish how terrible the trouble is. On the other hand, the 'skandalon'; i.e. the troubles-teller as powerless victim-of-circumstances in need of outside help, must be constituted as part of the story. This requires displaying how unusual the trouble is. It is not only worse than what one might expect ordinarily, but storyteller's need of help is especially great as well. Even while stressing the unusualness of her suffering, care will be given that it does not extend beyond the boundaries of normalcy or that her general competency comes into question.

This double point may be achieved as a cooperatively co-production. Like the other interactive tasks required for troubles-talk, however, the point may also be subject to impediment. It is to these obstacles that I shall now direct my attention. Since the point is a double one, the two dimensions of the point will be dealt with separately. Obstacles to the first dimension ('this is terrible') will be analyzed using the 'double-uterus' tape; obstacles to the 'skandalon' using the 'one-continuing-story' tape. Each analysis of the negotiations around the obstacles to the point will be followed by a discussion of the ongoing struggle for power inherent in the respective consultations.

Whereas the obstacles and the analysis of power struggles concerning the evaluative point have been separated into two dimensions, this distinction is strictly an analytic one. In practice, both dimensions are an integral part of the ongoing story about trouble, as well as power struggles concerning establishing its evaluative point. At the end of this chapter, a synthesis will be made of both dimensions, showing how they produce a specific kind of power. Thus, the remainder of this chapter will proceed as follows:

- Obstacles to the point I: 'this is terrible' (the 'double-uterus' tape)
- Struggle for power I (the 'double-uterus' tape)
- Obstacles to the point II: the 'skandalon' (the 'one-continuing-story' tape)
- Struggle for power II (the 'one-continuing-story' tape)
- A synthesis: the Janus-Face of Power.

OBSTACLES TO THE POINT I: 'THIS IS TERRIBLE'

[the 'double-uterus' tape]

In this consultation, the patient makes two attempts to engage in troubles-talk. Her first attempt is abruptly interrupted by a lengthy medical Q-A sequence. Just as she is about to leave, having been helped into her coat by the GP, she returns to the original trouble.

Whereas obstacles to her troubles-talk appear throughout the consultation, I have chosen to pay attention especially to those encountered in the business of establishing the point of the story. The patient is having a particularly difficult time in displaying her trouble as serious enough to even warrant being told. It is this issue of reportability; i.e. that it is *really* terrible, which is being negotiated here and is intimately involved in the ensuing power struggle.

I shall now examine two attempts to talk about her trouble (her daughter's double uterus), showing how, in each case, obstacles can arise around this first dimension of the point:

EXCERPT 10

```
14 P: My oldest daughter - she's been married
15    for six years and she'd love to have a
      baby now, after all, you know?(       But
                                    (Mmhmm
```

16 now: there's been an examination and well:
17 she's - she seems to have a double uterus.
18 D: ((surprised)) OH!
19 P: Have you ever heard of that?
20 D: It happens sometimes yeah. Sure enough.
21 P: And well: -
Yeah -
22 D: (All rabbits have them, I believe, but-
23 P: - and then you work yourself up there
(Mm (Mmm
again about it, you know?
(Yes
24 D: I can imagine.
25 P: -got to go in for a uhhh- Thursday uhh
26 Friday she's got to go in for a kidney
examination they'll be turning her
(Mmm
27 inside out again for sure.
28 D: How come again? Is she already has she
29 already had that examination
in connection with the uhh-
30 P: (No: ((laughs)) not that but yeah:
31 No. Not that. No. Not that.
(Mmhmm.
32 But yeah, that uh- then you start think-
ing about yourself ((laughs)). Back then
33 with my kidneys they and took took all
(Mm
34 mine too uhhh ((2.0)) So:
35 D: And you get yourself worked up like it'll
36 all be coming out right-
37 P: (Yeah: then uh
38 then uh you get terribly upset why,
(Mm (Mmhmm
you would be simply delighted too if
(Yeah
39 if she uhhh well, not especially for me
40 you know because uh that's not the point
now to-
41 D: But those palpitations. Do you feel them
42 now especially when you think about that
43 or uhhh- and then you notice really
(Yes.
when you-

44 P: Yeah: [then I get-
             [when you

In lines 14-17, the patient initiates the trouble, providing the GP with sufficient information that he can appreciate the potential seriousness of the matter. After indicating surprise (line 18) and establishing himself as someone who is, indeed, familiar with the problem (line 20), he makes a remark, which obstructs treating the trouble as something serious enough to warrant further talk: rabbits have double uteri, too.

At first glance, this remark constitutes an interruption. The patient has the floor and is about to continue her troubles-telling ('And well: ...'), when the GP, rather abruptly, takes it away from her. Interruptions are, of course, not necessarily simple violations of the 'only-one-party-may-speak-at-a-time' rule (Sacks *et al* 1974). They can also display a listener's competence and even affiliation. In other words, the GP may have been following the patient's talk so closely that he was able to anticipate what she was going to say next and said it for her. This display of independently achieved knowledge can show alignment.

However, an examination of the content of the interruption indicates that the utterance is more complicated. One possible hearing of the GP's rabbit-remark is as a *joke*.<10> By initiating a joke, the GP has not only taken over the floor, but he has initiated a 'story' of his own. The patient has been placed in the complementary position of recipient, where she is being invited to laugh along with the GP. Thus, she is not only prevented from continuing her troubles-telling, but she must now respond with laughter to his initiative.

In addition to interrupting the patient's troubles-telling and shifting their membership categories in terms of the form the talk takes, the GP's utterance introduces an *ambiguity* around the trouble itself. On the one hand, the joke can be heard as reassuring. Such a hearing depends on the patient being able to perform one or both of the following steps. She must be able to view the fact that the GP could even make a joke in the course of a troubles-telling as an indication that her trouble can't be all that serious. Or, she can find the content of the joke itself reassuring. Everyone knows how prolific rabbits are, so if they are able to reproduce in such awesome numbers, despite having a double uterus, there may be hope for the patient's daughter as well.

On the other hand, the patient may find herself confused that the GP begins, out of the blue as it were, about rabbits. Or, she might take offense at her daughter being compared with a rabbit at all. In that case, she will probably not be reassured. The remark then becomes a trivialization of her daughter's present distress, as well as an indication that the GP does not take the patient very seriously either.

It is this ambiguity, which makes the rabbit remark, interactionally speaking, difficult to respond to. If the patient laughs, she will acknowledge that a joke has been told, thereby establishing herself as someone who is competent at 'getting the joke,'<11> as well as the interaction as a friendly, pleasant affair. At the same time, however, she will relinquish her position as troubles-teller and implicitly lend credence to the notion that her troubles were a laughing matter. Her reason for coming to the GP in the first place won't have to be taken seriously.

If, however, she treats the GP's remark as a rude interruption or as an insult ('Are you calling my daughter a rabbit, or what?'), the GP may quite legitimately retreat behind the other (Janus)face of his joke; namely, that he was a) only trying to reassure her or b) just making a little joke. She is then placed in the position of having to account for being upset about something so patently well-intentioned. Let us look at how the patient deals with this particularly complex obstacle to her troubles-telling.

To begin with, *she does not laugh*. After pausing and thus minimally acknowledging the interruption, she continues where she left off (line 21) with 'and well ....' Her utterance recapitulates the point of the trouble. 'You work yourself up' is not only a description of how *she* feels at the moment, but how *anyone* would feel under similar circumstances. This is accomplished by means of the generalized 'you,' whereby shared background assumptions about how people normally feel in certain situations are implied. A recounting of an additional troublesome event follows; namely, the daughter's pending kidney examination. Whether the patient is upset because the examination is unpleasant or because she is worried that it can lead to more far-reaching difficulties,<12> she links it to her own medical experiences, thereby providing a rationale for her being upset at present. This point is reiterated, once again, in lines 35-36. And, finally, she presents the crux of the matter: it is she who would like her daughter to have children. Despite her mitigating laughter and assurances to the contrary, the patient is worried about whether she will become a grandmother (lines 38-39).

In this way, the patient carefully constructs a case for the seriousness of her present distress by explaining what the trouble is and why it is so upsetting to her. In this way, she re-embarks on the business of aligning the GP to the point of her story about trouble, which she may have reason to believe is unclear to him. By repeating her point, the GP can - despite his previous blunder - be brought to understand what the problem is all about.

This is, however, to no avail. If the GP did not appreciate the seriousness of the trouble before, he doesn't seem to now, either. He interrupts her with a medical question about heart palpitations (lines 41-43), thus shifting the topic away from the patient's daughter and her own distress. In the extended Q-A sequence which follows (Chapter 7, pp. 181-182), various other medical problems, including the pills the patient is presently taking, are discussed. Her initial trouble, however, has vanished completely from the conversational stage.

The patient's second attempt to engage in troubles-talk about her daughter's double uterus occurs at the end of the consultation. The GP has just provided the patient with an ointment for her muscle pains and reiterated his preference for such remedies rather than pills.<13>

EXCERPT 11

125 D: Otherwise your life will start to be one
continuous string of pills-
126 P: ((loudly)) No, no thank you. Because uhh-
127 ((laughs)) I'd rather not. ((laughs))
((3.0))
128 D: I'll just help you into your coat, just
129 like at the hair dresser's.
130 P: Yeah. ((laughs))
Would- yeah- would it uhhh- well yeah-
131 would that really have an effect on - with
my daughter?
132 D: With your daughter?
133 Uhh- it has- uhhh in fact it does probably
134 effect the possibility of getting
pregnant. But that doesn't mean that if
135 she has a double uterus like that that she
136 for sure won't be able to get pregnant.
137 P: Pecu:liar.
138 D: Yeah, huh? Well it kind of has to do uhhh
139 with the way that your uterus is uhh
140 formed in your development as a little
baby inside you <14> *(But everyone-*
Yeah:
141 everyone has at a certain point a double

142 *uterus like that but that fuses then*
at [a certain moment
143 *[ahh, of course, yeah:*
144 *and- uhh yeah- she just stayed behind*
145 *there a bit. You do often see ano- ano-*
146 *ANOmalies like that [difficult word is*
[Mm
called.

At this point, the consultation has come to a close. The patient is being helped into her coat. This is obviously not the time to embark upon further troubles-talk. The GP's remark 'just like at the hairdresser's' constructs the talk, retrospectively, as something, which wasn't very serious. Although troubles certainly get talked about at the hairdresser's, this talk is more likely to be regarded as gossip than talk about matters meriting the serious attention of a medical professional. Whether the patient hears this remark as a friendly or humorous signal that the consultation is over or whether she regards it as an assessment of their interaction up until now, it does not, in any case, assist her in re-embarking on more troubles-talk.

However, that is, in fact, what she does. Despite its inappropriateness at this particular juncture of the talk, she returns to her original trouble: her daughter's double uterus. After four starters, she requests some information and is, for the first time, successful in getting it. At long last, the GP provides an explanation for the 'anomaly,' as well as the consequences it can have for the patient's daughter getting pregnant (lines 133-136).

The GP's explanation is a rather curious amalgam of medical and everyday understandings of the events in question. Whereas the intricacies of uterine development are delineated (lines 138-146) and the problem graced with an appropriately technical label ('anomaly'), due account is taken of the layperson's world where such words are apparently regarded as 'difficult' (line 146) and where 'little babies' are thought of as being 'inside you' (line 140) or daughters as just 'staying behind there a bit' (line 144). All in all, the message is clearly that a double uterus is quite routine and nothing much to worry about.<15>

The patient does not seem entirely convinced, however, as evidenced by the stretch of troubles-talk which follows:

EXCERPT 12

147 P: Sad I think - yeah ((laughs)) well then
wouldn't at all uh well

148 yeah, (honey <16> ((complaining)) it's
(Yes- of course
149 sooooo: hard Yeah- (I uh-
150 D: (Yeah. (Yeah-
151 ((rapid)) It's (terribly bothersome-
152 P: (Should I know that?
153 D: I can imagine that that- that fear of
154 whether or not you'll be able to have
155 children (that that can make you really
(Yes
very tense (on the other hand, I tend
(Yeah
156 to say, yeah(hh): ((rapid)) you can be
better off with a double uterus than with
157 just one arm, but, yeah- (
158 P: (((snorts)) YEAH:
159 yes but, you can go without children, too,
160 of course, but, (that
(Yes
161 IS NOT (the way it's supposed to be, of
162 (That's - it's annoying-
course, right?
163 D: that the ones who do want to
have them- (they can't (have
164 (Yeah, they- (Yeah-
165 them the uh- the ones who don't want to
have them they (get them anyway.
(Yeah
166 Yeah but (let's
(And what can I do now-
look a little more at the sunny side for
the time being- (
167 P: (in your opinion about my
168 nervousness?
169 D: Uhh- uh I would like to do nothing about
170 it for the time being. Look, I can give
you a pill (because uh- but that ner-
(Yeah:
171 vousness, you do use that to work through
172 the problem with your daughter.
173 P: Yeah(hh).
174 D: Yeah: All that worrying, you'll keep doing
175 that of course (and well I could sweep
(Yes

176   that under the table with a pill⌈ but I
                                      ⌊No:

Having received an explanation of what causes a double uterus and its consequences for getting pregnant, the patient begins to evaluate the experience. It is 'sad,' 'so-o-o-o- hard,' all of which is rendered in direct speech as though she were commiserating with her daughter about it (lines 148-149). At first, the GP displays affiliation with her point; he appears to have finally grasped the seriousness of the problem (lines 151;153). Appearances are deceptive, however, as evidenced by the following utterance, which is not unambiguously affiliative: 'You can be better off with a double uterus than with just one arm' (lines 156-157).

There are several ways this utterance can be understood, each of which displays it as an obstacle to the patient's attempts to establish her present difficulty as something serious enough to warrant being told. The first hearing is that the GP's assertion is simply irrelevant to the talk at hand. It concerns parties, who have nothing to do with the patient's problem. The patient has, after all, been talking about a concrete person (her daughter) as well as a specific problem (a double uterus). The GP suddenly produces a statement about a generalized other (you) with a completely different sort of problem (one arm). Since conversations are generally supposed to proceed in an orderly and coherent fashion, the remark would then have to be viewed as constituting an interactional breach of conduct.

This is only one possibility, however. The utterance can also be heard as a suggestion that the patient's daughter should draw a parallel between herself and those individuals unfortunate enough to have only one arm. Since the GP has informed us that this is less desirable than having a double uterus, this could work as a reassurance. Whether the patient feels reassured will, of course, hinge on her being able to make the connection between her daughter's predicament and that of other anonymous, one-armed individuals. Furthermore, it will depend on her being able to see the appropriateness of having to make such comparison in the context of a medical consultation.

Understanding the GP's remark is additionally complicated by the form in which the remark is made. By comparing the number of body parts in the context of a commonplace, the GP can also be heard as making a pun (Scherzer 1985).<17> Joking is a useful conversational activity. It not only can provide a welcome time-out from troubles-talk (Jefferson 1984a), but it is a way of establishing rapport between participants without things getting overly emotional or personal (Gordon 1983). Joking allows members to let off steam; it reassures them that things are not as bad as they seem.

Unfortunately, puns or jokes are not necessarily affiliative, particularly when the object of the humor is also the person with whom rapport is sought. Humor requires participants to be relatively uninvolved in the object of their mirth (Korsemeyer 1978), as the slapstick slipping-on-a-banana-peel-routine attests. We can laugh, as long as the misfortune belongs to someone else. To work as a joke and, consequently, as affiliative response, the patient must be able to laugh with the GP at the expense of her daughter. In addition to distancing herself from her daughter, she must distance herself from herself as someone, who has, up until now, found the problem anything but a laughing matter.

In view of these contradictory demands, it is perhaps not surprising that the pun 'flops.' The patient interrupts the GP with a snort and re-formulates her trouble as a serious matter (lines 158-161). To this end, she makes use of shared background assumptions (Polanyi 1979) about having children being 'the way things should be.' Heavy intonation is employed and emphasis is provided by means of 'naturally' and the generalized you. Thus, the GP is appealed to as a member, who should be expected to know this. Unfortunately, whereas a woman might, he does not. Having provided the GP with another formulation of the point of her story, the issue now becomes: will he, at long last, confirm it or will he stick to his version, which seems to be that there is really nothing to worry about? In the segment which follows, this is precisely what is being negotiated.

The GP begins by responding with affiliation (it is upsetting), but follows it up with a commonplace about women always wanting things they can't have (lines 163-165); i.e. an irritating, but common dilemma for which the best course is to simply take an optimistic view (line 166). Thus, despite his initial lip-service to the upsetting character of the problem, his version is upheld; namely, that the daughter's predicament is not something meriting serious attention. It would appear that the patient has not been very successful in getting the point across.

Her next utterance indicates that she has given up trying to establish her point or even to talk about her trouble altogether. Whereas this marks a capitulation in terms of her original stance, it is worth noting *how* she goes about it. She interrupts the GP mid-stream with a request for advice on a previously-mentioned trouble: her nervousness. In and of itself, this is an unusual occurrence, interruptions being, empirically speaking (West 1984), the prerogative of physicians rather than patients. Her request does more than interrupt, however. The GP has just provided an advice of sorts; namely, the patient should try looking at the sunny side. An appropriate response for advice would be an acceptance or even a rejection, thereby at least acknowledging that advice has been proffered. The patient's utterance does not fit in either of these categories. By requesting advice herself, she negates that the GP has, in fact, just provided some. Moreover, her referral to a previously-mentioned topic demonstrates that, up until now, no suggestions for that problem have been forthcoming. Given that advice is what professional helpers are supposed to do, her request could be heard as an indirect challenge or criticism of the GP in his professional performance (Labov and Fanshel 1977:94).

The GP responds to the patient's request with a new suggestion; i.e. he would 'rather do nothing' (line 169). This somewhat disappointing piece of advice is softened by a gratuitous 'for the time being,' as well as an extended stretch of accounting (lines 170-177). In a commendable display of some of the more progressive trends in modern medicine, which stress non-technical solutions for health problems and individual responsibility (Berliner and Salmon 1980), as well as integral approaches to illness (Menges 1986), the GP indicates his reticence in prescribing pills (psychotropic drugs, presumably) for nervousness. These only 'sweep' problems 'under the table,' creating an 'unnatural situation.' The 'natural' and preferable one would be to 'talk about one's problems' (line 177). That is exactly what he would like to do, albeit at some later date and only when things really 'get out of hand.' This implies not only that her present problems are not 'out of hand,' but also that he is willing to talk about problems in general - something which has not been born out by the interaction, up until now.

In addition to justifying his advice, the GP's account performs an additional function, however. A retrospective look at the consultation will show that the patient has not actually requested tranquillizers. Whereas she may, in fact, want them, her request concerns what she might *do* about her nervousness.<18> At the same time, she has implied that the GP has, up until now, not provided any advice or help for that problem. By treating her request as though she *had* asked for tranquillizers, the GP is able to establish his own position as a reasoned stance against the prescription of tranquillizers for life-problems. In this way, the patient emerges as someone who wants something, which not only runs counter to her best interests, but is unreasonable as well. The GP comes to the fore as rational champion of her best interests. Attention is thereby directed away from both the regrettable fact that he has not provided help for her current distress and from the potentially ticklish issue that she has just criticized him to this effect. By indicating help will be forthcoming at some later date, should it prove necessary, he can cover up his present inability to help, presenting it as a matter of preference, based on medically-sound criteria, rather than his own powerlessness or carelessness. Having re-confirmed his position as medical expert, as well as her incompetence in matters concerning her own health, the consultation may be brought to a friendly ('don't forget your bag') and unproblematic close.

STRUGGLE FOR POWER

[the 'double-uterus' tape]

In the 'double uterus' tape, the patient encountered obstacles to her troubles-talk throughout the consultation. She had difficulties getting the floor, recounting her troubles, and establishing her trouble as terrible enough to be story-worthy. It is to this last obstacle that I shall now direct my attention in order to show how the struggle for control over the ongoing interaction was being waged.

Power in interaction

As in the previous consultations (the 'classic case of gall stones' tape and the 'a little bit nervous' tape), both the GP and the patient drew upon interpretative schemes available to them in order to influence the interaction. Since GPs and patients have unequal access to professional systems of relevance, the outcome of any interactional struggles ensuing in the course of the consultation tended to be skewed in favor of the GP. In contrast to the other consultations, the GP and the patient in the 'double-uterus' tape did not simply employ resources, which were (asymmetrically) available to them within a medical frame. They also made use of skills and resources, available to them as ordinary members in order to establish (or obstruct) the point of the story.

The patient employed various interactive procedures open to any storyteller, enabling her to display her trouble as unusually bothersome. She continually repeated the point, underlining its significance and re-establishing it as *raison d'être* for her story. She employed dramatization devices like heavy intonation and direct speech to emphasize the seriousness of the problem. Direct appeals were made to the GP as someone, who could be expected to share cultural understandings about what not being able to get pregnant means for women, both potential mothers, as well as grandmothers

(Excerpt 10, 12). The patient made use of resources taken from a friendship frame (Chapter 5), where storytelling will ordinarily yield an affiliative response from the recipient.

In fact, it was only at the very end of the consultation that the patient, having failed to elicit any interactive displays of reciprocity from the GP, reverted to the medical frame and the resources it had to offer her. This did not enable her to establish the seriousness of her present predicament, but it did allow her to challenge the GP in his authority as medical expert (Excerpt 12).

The GP in the 'double-uterus' tape, like the patient, did not remain securely oriented to the medical frame either, but also exercised control by means of mundane conversational skills and resources. Not only did he refrain from making use of specialist or technical knowledge available to him as medical expert, but he even rather conspicuously withheld medical information he had at his disposal. Since this information could have confirmed the seriousness of the problem of the double uterus, his silence might be viewed as implicitly undermining the patient's efforts to establish her trouble as especially bothersome. Ample use was made of a friendly interactive frame, up until the end of the consultation. It was only when the patient, half-way out the door, returned to the topic of the double uterus, that the GP treated it as something deserving of an explanation. Even then, the explanation was geared primarily toward establishing the double uterus as something not worth worrying about (Excerpt 11).

The GP's lack of affiliation with the point of the patient's story was expressed primarily by means of utterly routine, everyday conversational devices: jokes, puns, commonplaces of the 'that's-life-variety,' and myriad reassurances to the effect that the patient had nothing to fear.The only time that he made use of resources specifically available to him as professional was to sanction the patient's criticism of his competence (Excerpt 12). By providing a treatise on modern medical ideology, replete with the value of 'working through' problems rather than resorting to tranquillizers, he disqualified any suggestion that he might have been remiss in helping her. Thus, as was the case with the other two GPs, he employed his professional resources selectively and specifically in order to maintain his authority and ensure that his was the final word concerning the assessment of the patient's current troubles.

Power and agency

The exercise of power will be linked to agency. Sometimes this will entail influencing the interaction towards specific or intended outcomes; other times, however, the exercise of power will be a routine or 'for all practical purposes' affair, bounded by conditions of which the participants are not aware, as well as consequences, which may not have been intended.

The patient in the 'double-uterus' tape displayed herself as a knowledgeable and competent social actor, oriented to interactional procedures appropriate for a medical consultation. She embarked on the consultation with a suitably medical entree (blood pressure), following it up with a request for information, which she might reasonably expect the GP as medical expert to have (the consequences of having a double uterus for a normal pregnancy). Moreover, she ongoingly displayed the appropriate deference to the GP in his position as medical expert, never challenging his assessments openly or becoming less than polite, even when he, as was repeatedly the case, treated her current troubles as laughing matter. If the interaction proceeded in a smooth and friendly fashion, then it was to a large degree due to the patient's often remarkable adroitness at keeping it that way. She provided, once again, a good example of the reputedly female talent for performing the 'interactional shitwork' (Fishman 1978) necessary for any conversation, if it is to proceed in a well-oiled fashion.

This was, however, only part of the story. In addition to doing her share in keeping the interaction friendly and harmonious, the patient managed to resist and even combat the GP's ongoing trivialization of her current troubles. This was not done in-so-many words and it did not have an entirely successful outcome, but, when put 'under the microscope,' as has been done in the present inquiry, her activities provided a convincing demonstration of how social actors can (and do) perpetually 'nag at the bases' (Giddens 1984) of the powerful.

By tirelessly underlining the gravity of her daughter's double uterus, the patient not only stuck to her guns in terms of her own feelings about the situation, but she even made a valiant attempt to instruct the GP in what not having children might mean, generally speaking, for a woman. Direct reference to such cultural assumptions proved necessary as they did not seem to be shared by the GP whose gender may have prevented him from empathizing with an experiential perspective, which would have been more readily available to a woman.<19> Whereas the patient's use of shared background assumptions concerning pregnancy might have been perfectly appropriate for establishing the evaluative point of her story in a chat with another woman, it might have been inappropriate for a gendered context like a medical consultation. In other words, the GP's gender provided an unacknowledged condition for the patient's inability to make her point, as well as his to get it.

The GP in the 'double-uterus' tape, like his colleagues in the previous consultations, skillfully steered the consultation along appropriate medical channels, using his expertise to interrupt the patient's troubles-telling with medical-type questions (Excerpt 10) or to close off the conversation with an assessment of the patient's problems and what should be done about them (Excerpt 12). His activities were, however, by no means limited to the production of orderly medical talk. He put considerable interactional effort into displaying the encounter as a friendly one as well. This began with an attentiveness to ritual conventions like greetings or helping the patient into her coat, reminding her not to forget her bag, etc. Not stopping at the minimum politeness conventions, he produced jokes, platitudes, and reassurances, all of which provided the consultation with a friendly and even intimate touch.

In part, this friendly construction work allowed the GP to *not* get the patient's point without causing disruption to the ongoing interaction. After all, who could get angry at someone so obviously bent on doing his best to bring a bit of 'social cement' (West 1984) to the chilly bricks of institutionalized medicine?

Power and the dialectic of control

The exercise of power will involve a dialectic of control with both parties attempting to influence the course of events within the interaction. In the 'double-uterus' tape, the struggle for power centered around whether or not the patient's problem warranted a troubles-receptive hearing in the context of the consultation. Along with the trouble, the patient's credibility as sufferer was at stake. Was her concern about her daughter and the attendant problems of insomnia and heart palpitations legitimate worries? Or, was the patient simply making a mountain out of the proverbial mole-hill?

Judging by the GP's response, he was not entirely convinced that the patient's trouble was a matter of much consequence. In all fairness, as medical complaint the double uterus itself left him few options for providing service of a professional sort. To begin with, the patient's daughter was not his patient. Moreover, a double uterus was, by his own admission, 'just' an anomaly, a kind of birth defect. Aside from providing some general information about the implications of a double uterus for getting pregnant and having a normal birth or supplying some rather depressing statistical information about the probabilities that the patient's story might have a happy end, the GP was more or less relegated to lending a sympathetic ear to the patient's problem. Of course, this presupposed that he was, in fact, able to sympathize with it. In view of his response to her troubles-telling, there is some reason to suspect that this was not the case. Thus, faced with a complaint that he could do nothing about, medically-speaking, and for which he had little sympathy as fellow human being, he had two options. He could do nothing at all, explaining to the patient that he couldn't help her. A bitter pill to swallow, perhaps, but truthful. On the other hand, he could equivocate, pretending that he had, in fact, done something to help her. Like his colleague in the 'a little bit nervous' tape, the GP chose the latter route.

By providing a medical explanation for the double uterus, which served primarily to downplay its seriousness as problem (Excerpt 11), as well as myriad jokes and platitudes, the patient's trouble was gradually reduced to little more than a bagatelle. Since the patient's problem was nothing to worry about, the GP could quite legitimately come up with advice like 'look at the sunny side for the time being.' For the eventuality that this proved inadequate, he could set himself up as champion of the patient's other medical interests (Excerpt 12). By showing that preventive activities were warranted in order to keep the patient from behaving in ways detrimental to her own well-being (taking pills, sweeping problems 'under the table'), he could come to the fore as someone benevolently engaged in the all-important activity of protecting her against herself.<20> This could be accomplished, even though closer inspection reveals that he, in fact, did nothing which was particularly helpful, including, just parenthetically, the lending of a sympathetic ear.

The patient did not confirm the definition of her problems as 'nothing to worry about.' By engaging in various subtle and not-so-subtle attempts to align the GP's sympathy, she worked tirelessly to establish the seriousness of her present problems. When her troubles-telling was countered by a joke, she did not laugh. When the problem of the double uterus was treated as triviality, she emphatically underlined its seriousness, drawing upon cultural assumptions which, if shared, might have enabled him to share her distress. When the GP persisted in trivializing her problems, she snorted and even more emphatically made her case (Excerpt 12). She succeeded in initiating the double uterus as topic a second time, even though she had already been helped into her coat and was being slowly ushered out the door. When placed 'under the microscope,' such an act of resistance takes on almost heroic proportions. The patient in the 'double-uterus' tape was able, just barely and with one foot out the door, to get some attention for her original problem and, at long last, a medical explanation of sorts. Once the double uterus seemed to be exhausted as topic, she reverted to the medical frame, servicing the GP with a (metaphorical) slap in the face for not having helped her in his capacity as medical professional (Excerpt 12).

In short, although the patient never really managed to establish the point of her story, she clearly did not go down without a struggle.

Power as enabling and constraining

Interactional power struggles will exhibit both constraining, as well as enabling dimensions. The interactional control exercised in the 'double-uterus' tape had as its most distinguishing feature an intertwining of enablement with repression. The outcome of the negotiations around the point of the patient's story did not fall in her favor. Despite her often heroic attempts to establish the seriousness of her trouble, the GP generally managed to steer the talk in the opposite direction, constructing the problem as something too trivial to merit serious attention. Because this was accomplished by means of friendliness or even benevolence, it became increasingly difficult for the patient to respond adequately.

In ordinary conversations, jokes require that the recipient laugh, indicating that she has, indeed, 'gotten' the point (Sacks 1978). In a medical conversation, however, a joke as response to a story about trouble requires that the recipient uphold the harmonious quality of the interaction at her own expense. By laughing, she simultaneously affirms that her own problems are not worth being taken seriously or that they are nothing more than a laughing matter. Seen in this light, the GP's jokes in response to the patient's trouble were something of a mixed bag, combining social pleasantries with a trivialization of the patient's experiences.

The GP's jokes and platitudes (Excerpt 10, 12) were ambivalent in terms of their content as well. A comparison of a person's experience of suffering with the experience of others can work as a reassurance by producing a comforting sense of community or 'we're all in this together.' Such reassurance entails combining a respect for the sufferer's experience with an attempt to help her see her difficulties as part of the human condition. In this way, she may be helped to cope and, ultimately, get back to the business of living her life or, as Jefferson and Lee (1981) put it, 'continued functioning, regardless.'

In the 'double-uterus' tape, the potentially reassuring dimension of the GP's responses also rested on the patient being able to compare her daughter's plight with that of others: not only with other women who want (or don't want) children, but also with rabbits and one-armed individuals. It was not surprising that the patient might have had some difficulty feeling reassured, wondering, perhaps, about the grounds for making such comparisons in the first place. Nevertheless, a reassurance, even a rudely formulated one, displays the speaker's good intentions, making it no easy matter, interactionally speaking, to reject them. An appropriate response to the trivialization of her experiences along the lines of 'Are you calling my daughter a rabbit, or what?' would have had the unavoidable consequence of calling the GP's benevolence into question.

The GP's use of jokes and reassurances as response to the patient's story about her trouble could constrain the kinds of activities in which the patient could engage more effectively than would have been possible with more openly authoritarian forms of control. Confronted with two, in part, mutually exclusive definitions of the situation, a kind of interactional double bind was created, thereby impinging on the patient's ability to act at all. The ambiguous quality of the interaction curtailed her opportunities to identify, let alone, do something about the way her problems were being treated in the consultation.

Power as process

Power will be processual, whereby negotiations around the obstacles to the patient's troubles-telling slowly, but surely, produce and reproduce relations involving structures of domination. In medical encounters, physicians will invariably be faced with the interactional task of diagnosing and treating their patients' complaints. To this end, they will need to ensure at least some degree of compliance from the patients. Authoritarian methods offer several immediate disadvantages. Being more bald and overt, they are, at least in theory, more available to being called into question by the patient, providing she is daring enough to do so. Moreover, they may well be unsuited to problems for which there is no clear-cut diagnosis or course of treatment. In that case - and this is the case for the average GP (Chapter 1) - subtle and benevolent forms of control may turn out to be the most effective.<21> The 'double-uterus' tape provided a particularly good illustration of why this is so.

By introducing activities normally found in a friendship frame to the medical interaction, it was possible for the GP to exercise a subtle, but effective form of control over how the patient's troubles were defined and, ultimately, treated. At the same time, a social relationship could be constructed, which cemented the GP's position of dominance, while undermining the patient's authority and making her even more dependent upon his services than she was at the outset of the consultation.

As in the previous consultations, professional control over the process of defining or diagnosing <22> the patient's current distress had to be ongoingly produced and sustained. This control was not accomplished in a straightforward fashion, with the GP openly baring his teeth as medical expert as was the case in the 'classic case of gall stones' tape, but rather in a paradoxically friendly and even intimate fashion. By means of a covert and stealthy 'iron fist in the velvet glove,' the patient's troubles could be defined as basically trivial and inconsequential in terms of her life experiences. At the same time, she could be instructed to look to the GP for authorization about what was wrong with her and what she should do about her life in order to enhance her health and well-being.

Thus, the 'double-uterus' tape provided a bird's eye view of the moment-by-moment process by which the cooptative control of medicine as an institution could be interactionally reproduced. By demonstrating the difficulties involved in identifying, let alone resisting or combatting such control, the consultation also provided a convincing display of what makes 'soft' and friendly power more effective than its harder or more openly repressive manifestations.

OBSTACLES TO GETTING THE POINT II: POWERLESSNESS AND THE NEED OF HELP

[the 'one-continuing-story' tape]

In the consultations described up until now, the troubles-talk has often not gotten much farther than a fragmentary beginning. The following consultation, is, however, notably devoid of obstacles to getting the floor, recounting the trouble, or establishing the trouble as especially terrible. Not only does the GP readily relinquish the floor, but he allows the patient considerable leeway to engage in storytelling. He consistently commiserates with her current distress, often, in fact, indicating that he finds her difficulties worse than she herself does.

Although this consultation seems to be a veritable paradise for storytelling, the troubles-talk does not proceed without the occasional snag. The obstacles are primarily centered around the second dimension of the point of a story about trouble: the patient's powerlessness and need of help (the 'skandalon'). This is where she needs to establish herself as a victim of circumstances (Rehbein 1980), who is not only helpless in the face of her present suffering, but has, at the same time, done everything she could to help herself. The following excerpts will illustrate how difficult it is for the patient to establish this dimension of the point and the kind of negotiations which occur in the process.

After producing two medical entrees, the patient embarks on an extended storytelling about her anxiety, including recountings of anxiety attacks experienced while walking down the street, vivid descriptions of her physical symptoms (diarrhea, heavy perspiration), as well as dramatic evaluations of how terrible everything is (Chapter 7, pp.186-187). She is in therapy and her therapist has provided the following 'advice': she should 'choose' whether she wants to go on ruining her life for the next 20 years or finally make a break-through (Excerpt 6). In the following excerpt, the patient begins negotiating the 'skandalon' with the GP:

EXCERPT 13

77 D: Well, small wonder. It would seem really
78 crazy to me, too, if you didn't find that
scary. Uhh- yeah? When you start seeing
79 your world getting smaller and smaller
and- and you're being notified from the
80 other side that, well, listen here, it's
your choice and you have to do it uhh-
81 uhh- it means actu- well
82 P: (Well yeah-
83 D: that- that- but-
84 P: (it is my life that I have to do it
85 and I wouldn't know how- (do you
86 D: No. No
understand?
D: but that you so you seem to
be heading towards a kind- a kind of
87 crisis (like heading toward one,
(Yeah that's
88 you know, like you're world is small (
(Yeah (Yeah
and you keep getting told by others that
89 you have to make the choice yourself so -
uhhh well ((so there your are))
90 P: (Yeah- that's right because nobody
91 can tell me uhhh- whether I well - how I

can get rid of this (
Mmhmm
92 That's impossible, I can uhh- yeah- I
93 don't kno- I can take a pile of pills, I
can drown myself or just anything
94 else (you know. Then I do feel a
Yeah(hh)
95 bit calmer but it isn't the solution
96 either (
No:
97 I've already tried too ((but))-
98 D: Listen you- you
-you were talking about the psychologist
99 who- who who like uhhh- who had also been
saying that like-

After displaying affiliation with the point of the patient's story; i.e. that it is very serious, the GP indicates that he doesn't find the therapist's advice particularly helpful either (lines 77-81). Whereas the patient heartily agrees with his assessment, she produces a slight correction. She emphasizes that it is *her* life and she is the one, ultimately, who will have to decide what to do about it. The only problem is that she doesn't know how to help herself (line 85), all attempts on her part to do so having been to no avail.

The patient's utterances perform two interactive tasks. First, by reiterating her present powerlessness, the patient displays her problems as, by no means, 'over.' The therapist's advice is established as unhelpful, thereby making further troubles-talk a legitimate activity. Accepting advice automatically transforms troubles into problems-to-be-solved, requiring a different form of talk (Jefferson and Lee 1981). If the advice is demonstrated as unacceptable, however, continued complaining remains the order of the day. This interactional strategy is employed regularly by the patient in the 'one-continuing-story' tape.

Second, while establishing her right, conversationally speaking, to engage in more talk on her problems, she, at the same time, counters the notion that others can tell her what to do or make her decisions for her. In this way, her autonomy is established. She needs help, but is only willing to receive it under conditions of her own making.

The negotiations around this point are interrupted when the GP shifts the topic back to what the psychologist has to say, eliciting a new story about the patient's experiences in therapy (Chapter 7, pp.

186-187). These experiences are displayed as not only singularly unhelpful, but demeaning and somewhat ridiculous as well. The therapist, for example, tends to side with the patient's husband in therapy. This is characterized rather drily by the patient as maybe being 'good for your self-confidence, but ...' Advice like getting up at 9 o'clock sharp or having to call in if she over-sleeps are presented as unrealistic or laughable ('you never know ahead of time, of course, whether you might want to stay in bed until 10 one morning'). In the following excerpt, the point of the story is summed up:

EXCERPT 14

139 P: ((complaining) But I don't see any-
140 pretty soon I do have to walk back and I
141 don't s- see any help there, you know.
142 I- I (I can get so mad about it,
(Mmhmm.
you know (
(yeah
143 And then I can then talk about it there
144 and then maybe that man thinks like, yeah,
what am I really doing here, but that's
145 what I think too.
((1.0))
D: Yeah(hh).
146 P: And I think like- tha- would I ever like-
put me under hypnosis or or something. Or
147 beat it out (((what could make it
(Yeah. (Mmhmm (Mmhmm
worse)). Yeah.
148 D: You sound- you really sound a little bit
worked up, (I think
149 P: (But I'm
150 mad about it, ((complaining)) yeah, you
guys, I have to get through it again
151 pretty soon, not him. (
(Yeah. Yeah.
152 D: Do you feel a- like you've been left on
your own. Or do you feel like he's
153 failed you a little bit uhm- is that it?
154 P: Yeah. He can hardly walk around holding my
155 my hand for ((all these things))- (
156 D: (No: but
157 ((2.0)) did you expect more help too, or

that you-
158 P: *Yeah, I did expect more but*
159 *that's maybe dumb too because when you've*
160 *been suffering for ten years and then all*
of a sudden uhh- it's no wonder.
Yeah-
161 *That's impossible of course.*
((2.5))
162 *But I AM so afraid that that with the*
163 *talking it won't t- won't help. That's*
what I am really afraid of, yeah.
((2.0))

As in the previous example, the point here is a double one. To begin with, the patient's suffering is displayed as continuous, unremitting, and very probably even insoluble. It is not just that no solution has, as of yet, been forthcoming, but she can only imagine being (passively) placed under hypnosis or having 'it beaten out' of her (lines 146-147). In view of such suffering, complaining is clearly in order. The unsuccessful attempts to help her only underline this fact.

As in the previous excerpt, the patient is quick to point out that her present powerlessness does not extend to her general competency as an adult. *She* is the one who must walk home after an anxiety attack, not the therapist (lines 150-151). *She* is angry (lines 149-150). *She* can bring it up in therapy - for all the good it will do her (lines 143-145). However powerless she is at the moment, her autonomy and ability to cope with her life are ongoingly displayed as well.

In the negotiations which follow, it becomes clear that only the first dimension of the point has gotten across. In the GP's formulations, it is the patient's helplessness, which is underlined. She is 'all worked up,' feels 'left on her own' by her therapist, has 'expected more help.' The connection between being upset and needing help is emphasized rather than her basic ability to help herself. In each of her utterances, the patient corrects the GP's formulation of her problem. She is angry, for example, rather than worked up.<23> She does not expect the therapist to 'hold her hand,' thus contradicting the notion that she feels abandoned by him. Whereas the expectation of help is confirmed, it is immediately qualified as being out of proportion (no wonder, after ten years!) and, in fact, a bit stupid. Her final conclusion is that she is afraid she can't be helped at all (lines 162-163).

In this final excerpt, the negotiations around this point are summed up by the GP as follows:

EXCERPT 15

```
        ((3.0))
175 D:  I was just saying like, I have the feeling
176     that you are working up to a kind of
        crisis. Uhh- do you have that feeling
177     yourself too?
178 P:  Yes but I don't know what will be happen-
179     ing. It's uhh- yeah- ((1.5))
180     Then there are- very bad things too which
        have been happening lately uh- I don't
181     know - uhh all of that I could -
182     apparently could put up with it very well.
183     My husband thought I would get hysterical
184     bu- but I didn't. (
                          (No:
185     That's what I had expected too but it- it
186     just didn't happen at all so uhh- (
                                           (Mmhmm
```

Once again, the GP provides a reading of the patient's present distress as a 'kind of crisis.' If she confirms his reading, then it can work to establish it as the outcome of negotiations over the definition of her current problem. This would enable the talk to move on to a new topic: for example, what to *do* about the crisis. Or, the conversation as a whole might be terminated. The GP's request for confirmation (lines 175-177) is organized in such a way that, should the patient agree with his reading, it can be displayed as what she herself also feels. This picks up quite economically on the patient's much-reiterated point that she is the one who, ultimately, must decide what to do about her present distress.

In spite of the GP's careful organization of his formulation of the patient's point, she does not confirm it. After prefacing her utterance with an obligatory agreement (Pomerantz 1984), she produces a counter-example (lines 181-186), indicating how she has been able to manage quite well, despite all the terrible things that have been happening and her husband's belief to the contrary. Thus, she implicitly disagrees with an assessment of her present situation as a crisis or that her own activities might be leading in that direction. In short, the patient's present powerlessness and need of help goes hand in hand with a request for respect for her autonomy and competency.

STRUGGLE FOR POWER

[the 'one-continuing-story' tape]

In the 'one-continuing-story' tape, the patient was able to engage in troubles-telling over long stretches of talk, obtaining a highly affiliative response for the seriousness of her present distress. Obstacles emerged, however, as she attempted to establish her present need of help without having to forfeit her claim to respect as someone, who could, under normal circumstances, take care of herself. It was here that the interactional struggle for power between the GP and the patient became apparent.

Power in interaction

The exercise of power in interaction requires that participants draw upon skills and resources available to them in order to influence the course of events. These resources are organized as interpretative schemes or as normative sanctions. In the 'one-continuing-story' tape, the patient made use of skills and resources open to any troubles-teller for establishing the 'skandalon' of her suffering. With considerable adeptness, she managed to underline both her own powerlessness and the powerlessness of others in the face of such suffering. By means of dramatization and other evaluative devices, she could show how each attempt to alleviate her problems only met with failure (Excerpt 13, 14). In this way, her position as a victim of circumstances could be established; i.e. someone who, through no fault of her own, has been unable to help herself.

At the same time, the patient was careful to demonstrate that this (temporary) state of powerlessness did not mean that she had relinquished all autonomy. Ultimately, she would have to be the one to help herself. This was accomplished by means of repeated contrasts between her present powerlessness and assertions that she didn't expect anyone 'to hold her hand' (Excerpt 14) or counter-examples indicating that, however serious her 'crisis' might be, she was still able to cope, at least to some degree (Excerpt 15). In this way, she managed to establish both the legitimacy of continued rounds of troubles-telling and have some say in the conditions under which she would be prepared to receive help. Features inherent in storytelling could be employed as resource for exercising control over the way her experiences were being defined.

Unlike his colleague in the 'double-uterus' tape, the GP in the 'one-continuing-story' tape displayed considerable affiliation with the first dimension of the point of the patient's stories; namely, that her troubles were 'unusually terrible.' He was, however, less proficient at aligning himself with the second dimension of the patient's point and, in fact, actively obstructed it. Drawing upon resources like listening skills and the ability to express empathy, which are standard ingredients of the modern therapeutic arsenal (Rogers and Kinget 1960; Sigling 1982), the GP could display commiseration with the seriousness of the patient's current distress. Ongoing encouragement was provided in the form of listening noises (mmhmm's, yeah's) throughout the patient's storytelling. He reformulated her evaluative point in terms which upgraded her own rendition, thereby demonstrating not only his own sympathetic stance, but showing that he found her difficulties perhaps even more terrible than she herself did (Excerpt 4).

This same GP, however, who was otherwise a paragon of responsiveness to the patient's troubles, turned a deaf ear to her protestations that she had to eventually help herself or, for that matter, that she had been doing everything she could to help herself all along (Excerpt 13, 15). By consistently omitting affiliation to these elements of the patient's story, he made it more difficult for the patient to establish the conditions under which she could acknowledge her present powerlessness without losing face (Goffman 1955; 1967). This omission was especially marked by virtue of its contrast to the GP's abundantly reciprocative responses to the first part of her point; i.e. that her troubles were unusually terrible. The contrast itself not only helped to establish a case for the patient's need of outside, and presumably professional, assistance, but it served as an interactive sanction, restricting her in her attempts to establish her fundamental, if impaired, ability to take care of herself. In this way, the expression of empathy could be used as a resource: employed to *encourage* the patient's troubles-telling or withdrawn to *discourage* it, depending upon the tasks facing the GP in his capacity as professional.

Power and agency

Agency will be central to the exercise of power in any instance of interaction. Activities of the interactants are likely to be bounded by unacknowledged conditions and unintended consequences. The patient in the 'one-continuing-story' tape was, of all the patients up until now, the least concerned with embedding her troubles-talk within the dominant medical frame. In fact, after producing the obligatory medical entree concerning the potential side-effects of birth control pills (Chapter 7, pp. 184-185), she did not seem to pay much attention to it. Instead she embarked on an unceasing 'tale of woe,' with each new story about a trouble following hot on the heels of the previous one. As troubles-teller, she showed great expertise, managing to capture the GP's attention and affiliation for no less than 15 minutes and his time for nearly 45 minutes-something of a feat in general practice which average 7-8 minutes.

Not only was she particularly skillful at dramatically depicting the crippling quality of her current distress but, at the same time, she displayed herself as someone who was, in spite of it all, still a 'fighter' (Excerpt 6). Thus, she could avoid the pitfalls inherent in any admission of powerlessness regarding activities belonging to the behavioral repertoire of a competent adult (Goffman 1961).

The GP in the 'one-continuing-story' tape demonstrated that he, too, had all the makings of a knowledgeable and competent social actor. Clearly well-versed in the social and therapeutic skills belonging to the professional repertoire of the modern GP,<24> he displayed unflagging empathy, encouraging the patient to 'explore' and 'clarify' her problems *à la* 'Methodisch Werken' (Chapter 1). Considering the conspicuous absence of anything even vaguely resembling a medical complaint in her troubles-talk, this was not an unwise course to take. Furthermore, by making the provision of sympathy a stable feature of the interaction, something the patient could come to count on as response to her troubles-telling,<25> the GP paved the way for a cooperative consultation in which the patient might also be expected to display affiliation with *him* and *his* interactional tasks, when the time arose.

As it turned out, it was the medical frame of the consultation as a service encounter, which, ultimately, formed the unacknowledged condition of the interaction, bounding the activities of both participants. The patient's efforts to establish her autonomy regarding her present troubles would have been well-taken in a routine instance of troubles-talk between friends, where one of the aims is to return the troubles-teller to her position as 'member of the fold' (Jefferson and Lee 1981). In a medical consultation, however, these same efforts tended to conflict with the business at hand. Providing service requires an object to be serviced (Chapter 6). By emphatically stressing her ability to help herself, the patient put the GP as service provider in a difficult position - a kind of unintended consequence of this aspect of her story's point. Whether or not he actually had a diagnosis to make or help to provide, he still remained under the constraint, at least initially, of having to assess the patient's troubles in that light. This state of affairs could account for the GP in the present consultation not being able to hear the patient's protestations. Whereas it may have been the point of *her* story, it was not the point of *his*.

Power and the dialectic of control

The exercise of power will involve a dialectic of control with relations involving autonomy and dependence being produced, sustained, and undermined in both directions. In the 'one-continuing-story' tape, the struggle for control emerged most clearly in the negotiations around the patient's powerlessness in the face of her present troubles and the implications this had for the provision of help in the context of a medical encounter. Whereas the GP was aligned as recipient throughout her troubles-telling, performing all the required interactive tasks with equanimity, he contested her story when it began to interfere with the constraints imposed by the medical frame and his own professional systems of relevance. Although the boundaries of the medical frame within this particular consultation were more flexible than those in operation in the previous tapes, here, too, they had, upon occasion, to be defended. Thus, the 'one-continuing-story' tape provided just one more instance of how storytelling and its attendant friendship frame, will, sooner or later, conflict with the constraints imposed by the institutional setting.

The struggle for control over the ongoing interaction did not ensue immediately, but emerged gradually in the wake of the patient's efforts to establish the second dimension of the point of her troubles-telling. Whereas these efforts would probably have been rewarded in a chat between friends with a reassurance along the lines of 'sure, you're suffering, but you'll manage in the long run,' here they clashed with the requirements of the medical frame where professionals are supposed to provide help and laypersons to be helped. As long as this institutional frame could be expanded to include 'problem exploration' or 'clarification' along with more traditional diagnostic fare, the GP in the 'one-continuing-story' tape could remain securely oriented to his own professional schemes of relevance, while, at the same time, responding affiliatively to the patient's troubles. Allowing the patient to get the floor and recount her troubles coincided unproblematically with the therapeutic mandate that problems need to be clarified before a diagnosis can be made. Displays of empathy not only constitute an appropriate response from any troubles-recipient, but they have therapeutic value as well. Thus, the GP's alignment with the patient's troubles-telling could often be accomplished without running into conflict with the medical or service frame.

It was only when the patient's storytelling did collide with the constraints of the service frame that the difficulties began and the GP was forced to adopt a somewhat less affiliative stance. This was not done abruptly or even overtly, but rather by omission; he simply withheld commiseration. Taken together with the fact that sympathy had been so freely given at other points in the interaction, its absence served as an obstruction to the patient's story about her troubles. Instead of supporting her attempts to show that she could cope or, at least, expected to be able to do so at some later date, he underlined her powerlessness. In such a state of affairs, the obvious and only solution was professional help.

Unfortunately, part of the patient's current troubles centered around the fact that professional help (therapy) had, as of yet, not been particularly helpful (Excerpt 13, 14). Her protestations, however, were, ultimately, to no avail. Unlike everyday encounters, commiseration cannot be provided within a medical consultation unless the patient is willing to admit that she is no longer able to help herself and is, therefore, willing to hand herself over to the professional helper - lock, stock and barrel.<26> The patient's reiterations of her autonomy, however essential for establishing her orientation to normal appearances, were dysfunctional here in that they served to unleash a power struggle around the GP's professional hegemony. This occurred, despite the fact that the GP in this particular consultation had set himself up, more consistently than any of the other GPs, as emphatically 'on her side.'

Power as enabling and constraining

Interactional struggles for control will display constraining as well as enabling features. This Janus-faced quality of power was never more evident than in the 'one-continuing-story' tape, where empathy and good listening skills became intertwined with the curtailing of the patient's attempts to present herself as not only the victim of various and sundry difficulties over which she had no control, but, at the same time, as someone who had not yet given up the fight. On the one hand, the patient's masterful efforts to get her point across facilitated the ongoing troubles-telling, eliciting an affiliative response to her troubles for a period far beyond the time normally allotted in a general practice consultation. In spite of institutional constraints, which can make storytelling an unwelcome activity

(Chapter 5), she was able to do a great deal of it, thereby exercising control over the interactive frame for long stretches of the consultation. In terms of immediate control over the course of the conversation, the patient seemed to even have the upper hand.<27>

Whereas the patient's skillful presentation of the skandalon and her own fundamental ability to help herself facilitated her control over the interaction in one sense, it worked, paradoxically, to undermine it in another. In order to pull off a successful story about trouble, she needed to ensure that the GP was receptive to her troubles throughout her story. By stressing her autonomy, she lost some of this affiliation, provoking, at the same time, a contestation of her right to the conversational floor.

In a similar vein, the GP's expressions of emotional reciprocity with the point of the patient's story enabled the patient to talk about her troubles in a form of her own choosing. These very same displays of empathy, however, elicited protestations from the patient that her troubles weren't so serious that she couldn't help herself (Excerpt 15). Although sympathy might be expected to 'produce' a highly compliant and grateful patient, in the present consultation, it had quite a different effect; namely, a patient who, with increasing vehemence, appeared intent on not allowing herself to be helped. In this respect, the GP's therapeutic skills created additional troubles in the course of the interaction; most notably, *his own*. Thus, the GP's empathetic displays had both a facilitating and a limiting dimension in terms of his ability to exercise control over the consultation. Likewise his omitting to show empathy restricted the patient's attempts to establish her autonomy, while, at the same time, playing an essential role in the constitution of the consultation as professional service contact.

Power as process

The production and reproduction of relations involving domination and subordination will not be a simple matter of structural determination. Power is exercised as part of a moment-by-moment process, whereby asymmetrical relations are ongoingly constituted and sustained, but also and, in fact, simultaneously, undermined. In the previous consultation (the 'double-uterus' tape), we saw how friendly power practices could be especially effective in the production and reproduction of asymmetrical relations between a GP and patient. This is, however, not a hard and fast rule. The 'one-continuing-story' tape demonstrated that friendliness is not necessarily an effective power practice and, in fact, too much friendliness can even become something of a liability for the GP, if an (asymmetrical) relationship was what he was after.

Having been drawn into an interactive frame bearing considerable resemblance to a chat between friends, he not only appeared to have lost control over the ongoing interaction, its topicality, and the appropriate membership categories, but he was faced with the attendant matter of having to approach the patient as an interactional equal.

Storytellers are able to control the way a conversation proceeds, whether the stories are produced within a friendship frame or an institutional setting. By engaging in a continuing tale of woe, the patient in the present consultation was able to keep the interactional reins firmly in hand. She determined which troubles became topics, as well as how they were to be defined (as insoluble) and treated (with sympathy rather than professional help). Moreover, she could exercise considerable influence over the GP's activities, keeping his attention and sympathy for surprisingly long stretches of talk. In this respect, the 'one-continuing-story' tape provides some empirical backing for Quasthoff's (1979) argument that one of the reasons storytelling is so popular is that it is the *only* way that patients or clients can determine the kind of interaction generated in an institutional encounter.

Staying within the friendship frame set by the patient, the GP had various courses of action open to him. He might have supported the patient's 'business-as-usual' attitude, indicating that it was, indeed, remarkable that she had managed as long as she had with so many serious troubles. Or, he might, at some juncture, have engaged in the conversational practice of linking some stories of his own to her. For example, a story about how full his waiting room was with a dramatic rendition of his worries about getting home in time for lunch might not have been amiss. These actions, whatever their vicissitudes, would also have entailed him taking leave of the last vestiges of his professional authority, however.

The line between sympathy as part of a story about trouble being told in the context of a chat between friends and sympathy which goes along with an expert display of good listening skills in the context of a medical encounter appears to be a thin one. When the difference between the two becomes cloudy, as was the case in the 'one-continuing-story' tape, confusion concerning the nature of the interactional frame, as well as the respective positions of the interactants can emerge, leading to some rather powerful negotiations.

The interactional control open to patients through the temporary switching of interactive frames is, ultimately, doomed to failure (Quasthoff 1979). Even the patient in the 'one-continuing-story' tape, who, of all the patients up until now, exercised the most control over the unfolding consultation, including how her complaints were dealt with, did not achieve the outcome she presumably sought; namely, sympathy *and* respect for her basic ability to take care of herself.<28> The outcome was somewhat ambivalent, with the GP promising help rather than respect, and the patient unwilling to accept help unless coupled with an acknowledgement of her capacity and desire to take care of herself.

In this chapter, I have dealt with power struggles, which can emerge when the double point of a story about trouble is obstructed. In both the 'double-uterus tape' and the 'one-continuing-story' tape, power practices could be delineated in obstacles to the point of the patient's story. Now a few words are in order concerning the quality of the power practices themselves. Despite the differences between the two analyses, the power practices in each of the consultations demonstrate that not only evaluative points to stories about trouble are two-sided; so is power itself. It is to this Janus-face <29> of power that I shall now direct my attention by way of a synthesis.

A SYNTHESIS: THE JANUS-FACE OF POWER

The evaluative point of a storytelling is its most essential ingredient, what makes a story a story (Chapter 6). In a story about trouble, this point is a double one, each dimension being an integral part to the story being told. Whereas both dimensions are crucial for the point, they introduce, at the same time, an ambiguity into the interaction. It is an ambiguity, which informs the power practices occurring within the interaction and which ultimately accounts for a kind of control more intensive, subtle, and effective than would have been possible by more straightforward or authoritarian means.

Neither patient was successful in establishing the point of her story about trouble. The patient in the 'double-uterus' tape could not convince the GP that her trouble was serious enough to be worrying about in the first place. Instead she was constantly reminded that things weren't all that bad and, ultimately, invited to forget about her problems and return to the fold; i.e. a member once again, but without a leg to stand on, trouble-wise.

The 'one-continuing-story' tape provided a mirror-image of the first consultation in terms of obstacles to the point. Here the seriousness of the patient's troubles was not at issue, but rather her position as 'member of the fold;' i.e. someone who could, generally speaking, be expected to take care of herself. In other words, she was invited to forget about her membership to the category of competent adults and devote herself to her trouble.

Whereas this double-edged quality of the point of a story about trouble can lead to some interactional confusion at the best of times (Jefferson and Lee 1981), it becomes particularly difficult to manage in the context of a service encounter. In a chat between friends, for example, it is not problematic to commiserate with how terrible things are, while, at the same time, expressing sentiments, which indicate a respect for the storyteller's basic ability to come to terms with them. A story about trouble, no matter how dramatic, will eventually come to end with a drying of tears and returning to life as usual. This, however, is precisely where the problems arise in the context of a general practice consultation.

Within a medical frame, it will be the GP's task to assess the patient's problems and, if he is able to come up with a diagnosis, to provide some sort of help for them. In the above consultations, however, the patients introduced troubles for which, medically-speaking, there was little to be done. Thus, as professional service providers, the GPs found themselves in that decidedly unpleasant condition of being powerless, unable to *do* anything as professionals. What they could do, of course, was to lend a sympathetic ear to the troubles, much as a neighbor or family member might. By temporarily participating in the shift from the medical frame to a friendship frame, whether patient- or self-initiated, and trading in their membership category as professional for that of interactional equal, the GPs could at least have provided their patients with the opportunity for engaging in troubles-telling with all of its attendant advantages.

Whereas this might have been an option in the consultations above, it was, in fact, not the one chosen by the GPs. Instead of taking the inevitable loss of professional face in stride, they took another path. If they couldn't be 'just friends' or 'just professionals,' they could perhaps try being a little of both. Here Janus fortuitously reared his head, enabling the GPs to look toward the friendship frame, while keeping an eye on the medical frame at the same time. Thus, the GP in the 'double-uterus' tape readily engaged in conversational practices normally belonging to a chat between friends while, at the same time, producing definitive assessments of the patient's trouble as (medically) irrelevant and, therefore, nothing to worry about. In the 'one-continuing-story' tape, the GP behaved much as a friend might with his enthusiastic displays of partisan concern and his unflagging alignment through such unprofessionally long stretches of troubles-talk. The 'friendship' came to an end, however, at the moment that he became caught in his own dilemmas as professional service provider. A friendship frame was employed and/or acknowledged by the GPs just as long as it coincided with the requirements of the institutional encounter. When it did not, it was withdrawn.

In both consultations, power struggles emerged around storytelling as an activity for friendly talk between interactional equals and the constraints on the GP to maintain his position as professional in the context of a service encounter. The interactional ambiguity produced by engaging in both frames at the same time enabled control to be exercised over how the patient's problems were being defined (as nothing to worry about, as 'something for the therapist') while, at the same time, making it difficult for the patient to contest this definition, should she wish to do so. By responding to one of the Janus-faces ('Are you calling my daughter a rabbit or what?' or 'What do you think I am, a basket-case?'), the patient would not only run the risk of disrupting the friendly quality of the interaction, but she would be sure to lose the confrontation. In perfect conscience, the GP would only need to jump behind the other face ('Can't you take a joke?' or 'Sorry, just trying to help'), leaving the patient, conversationally speaking, without a leg to stand on. Taken together, the 'double-uterus' tape and the 'one-continuing-story' tape demonstrate how 'friendly' power practices work to produce expansionary or cooptative forms of social control, which, if we are to believe current critics (Illich 1976; Ehrenreich and Ehrenreich 1978; Conrad and Schneider 1980), are on the upsweep in medicine today.

NOTES

1. For a detailed typology of evaluative devices, the reader is referred to Polanyi (1983). She distinguishes between phonological, lexical, syntactical, and discourse levels. Whereas I found many of the same devices in my own material, the focus of the present inquiry made it unfeasible to deal with evaluation in such a comprehensive way.

2. I 'discovered' this phenomenon in the following way: in the initial phases of the present inquiry, I would spend hours listening and re-listening to the taped consultations. Often I listened to the entire tape in one sitting. Other times, I stopped to take notes about what I was hearing. Sometimes I retracked and checked out earlier observations by re-listening to parts of the tape. This was all part of the general process of getting acquainted with my data.

Much later on, I decided to listen to the tapes 'just' for their rhythm; i.e. I attempted to block out what was being said. I know from previous investigations that this kind of listening can yield fresh insights into what is going on in the conversation. Thus, I closed my eyes, propped my feet up on my desk, and let it roll. The first thing I noticed was that there were regular shifts in the rhythm of the conversation throughout. They reminded me in many ways of a refrain of a song. This pattern occurred in each of the tapes. When I returned to these segments later on in my transcript, I discovered that they all had one thing in common. It was that the patient was *complaining*! The sing-song rhythm I heard was the intonation patients use when they want to stress just how terrible/painful, etc. the event was.

3. See McConnell-Ginet for an interesting account of intonation and its functions within a conversation. She suggests that intonational 'tunes' are the primary device employed by speakers to express emotional involvement and, as such, may be put to various strategic uses. For example:

> To sound highly emotional might enable one to attain ends not reachable by calm behavior if one is a subordinate, such as a child interacting with an adult or a woman with a man. And of course certain situations promote the expression of emotion: the care of a child puts greater demands on emotional expressiveness than repairing telephone lines.(1983:78)

This, I would suggest, also applies to doctor-patient contacts, where the patient may employ the emotional intonation of 'complaining' to make her point.

4. As Goffman has noted:

> Stories can call for persons to eat, make love, and be tortured, but as part of an inclusive human drama, not as an isolated display or a matter of interest to examine closely in its own right. (1974:56)

I would agree that to be dramatic, something must also be human and, therefore related to the mundane world of daily life. Normalcy and drama go hand-in-hand.

5. There is no English equivalent to the Dutch 'hè' which occurs frequently at the end of an utterance. It is accompanied generally by a rising intonation and functions as an elicit of affirmation. This is similar to tag-questions in English ('isn't it?') or 'you know?,' 'right?' I have used these possibilities interchangeably in the translations.

6. The patient began the consultation with an ostensibly medical request for information about the advisability of stopping birth control pills. It becomes gradually clear, however, that she would, in fact, quite like to have a baby and it is her husband who doesn't want her to. This is, in and of itself, a 'trouble'; i.e. her husband finds her unstable and is 'dissatisfied with her (see Chapter 7, pp. 184-185).

7. The reader will possibly sense that there are other ways of hearing this utterance than as a reassurance. There is also good reason, i.e. on the basis of the ongoing

talk, to assume that the patient was not reassured. At this point, however, the construction of the point is being treated as a cooperative, and, in that sense, harmonious venture. A bit later on, I shall be examining what happens when meanings about what makes the trouble troublesome are *not* shared by participants.

8. At this point, I am only discussing alignment in the form of agreement with the troubles-teller's point. Disagreement is, of course, also a way to acknowledge the point. For example, the recipient might question the grounds upon which the claim of helplessness is based (Turner 1976). I shall be returning to this point later on in this chapter.

9. Once again, 'tablets' appears in the diminutive.

10. I am using the term 'joke' rather loosely here. Scherzer (1985) maintains that there is a distinct difference between 'joking' (nonserious communicative behavior) and a 'joke' (full-fledged discourse form, complete with set-up and punch-line).
Whereas other interpretations of the utterance are possible (and will be suggested further on in the text), this one is a particularly benevolent one as far as the GP's intentions go. Others suggested to me by less kindly readers, are: he didn't know what else to say, maybe that's what they learn in medical school, or it was the first thing that came to mind when he heard the complaint 'double uterus.' My own inclination here was to give the GP the benefit of the doubt.

11. As Sacks (1978) has demonstrated, the recipient of a joke does not display the same kind of response as to a story. Affiliation or a second story are not appropriate responses. The recipient shows understanding by laughing, indicating not only that she has 'heard' the story and gotten all the information, but she is aware that it is a joke; i.e. funny or at least meant as such. Failing to respond with laughter opens the recipient up to possible criticism - that she has no sense of humor, that she is somewhat dim, or, as in the case of a dirty joke, that she is not up to par in terms of her knowledge of sexual matters.

12. This segment is ambiguous. Kidney examinations entail an injection, an IVP (intravenous-pyelogram), followed by an x-ray. The connection with a double uterus is missing, since this would not require such an x-ray. It is possible that the patient's daughter is being examined for a host of problems, i.e. also her kidneys. If the patient herself has had any hospital experience, she may be wary of examinations as instances of 'being turned inside out.' The GP, however, refers to it 'all coming out.' Taken together with the subject of the double uterus, he may be implying that the patient is worried about her daughter having a hysterectomy.

   Despite these 'factual discrepancies,' the participants appear to be understanding one another in a way, which matches the problem-at-hand, if not the correctness of their knowledge of medical procedures. As such, it provides an excellent illustration of Garfinkel's (1967) et-cetera clause in action. This is the practice whereby members bring unforeseen instances or occurrences under a rule-in-use and regard them as complying to it. In this way, talk can appear remarkably orderly and understandable even when it, upon closer scrutiny by a disinterested observer, is quite deficient in both areas.

13. In this segment, the GP initiates a lengthy Q-A sequence in which he checks up on various medications that the patient has been taking for her ailments (a sore toe, water retention, muscular pain). Subsequently, he provides her with some helpful suggestions (taking longer walks), an ointment for her muscular pain, and a 'handful of pills' for water retention. Although the patient has not requested advice for any of these complaints, she gets it anyway, with the GP coming to the fore as someone who has her best interests at heart. For a more comprehensive analysis of this segment as an example of how medical paternalism may be constructed, the reader is referred to Davis (1988).

14. Interestingly, the GP reverts from a generalized 'you' when discussing the way 'your' uterus develops to a direct 'you' here, thereby explicitly instructing the patient to imagine her own uterus with her daughter the 'little baby.'

15. It is worth noting that, whereas the GP's information is correct, it is also somewhat incomplete. Uterine malformations can cause a host of clinical problems like, for example: dysmenorrhea, renal troubles, infertility, repeated miscarriage, premature labor, uterine rupture, transverse or breech births, retained placenta, prenatal and postpartum blood loss (Stolk 1975). Thus, whereas a double uterus does not necessarily mean that a woman can't get pregnant or have a normal birth, it would be going a little too far to say she has 'nothing to worry about.'

16. Literally: 'maid, girl.' It can be used as 'good girl' or 'old girl,' but in this case, where the reference is to the patient's daughter, a term of endearment is what seems to be meant.

17. A similar (both in form and content) English proverbial equivalent is: 'Better a bird in the hand, than two in the bush.'

18. In view of the patient's reticence to take pills for other ailments as in the following:

    113 D: Otherwise your life will start to be
    114 continuous string of pills-
    115 P: ((loudly)) No, no thank you. Because uhh
    116 I'd rather not.

    a convincing case might be made for the fact that the patient would actually prefer *not* to take pills. In any case, she has not requested them up until now. 'Doing' something for nervousness could entail taking pills, but it is, by no means, the only course of action.

19. This is, of course, strictly speculative. It would be impossible, on the basis of a conversation analysis, to make any conclusive statements concerning participants' motives. Nevertheless, we might assume that a woman would be more likely to understand some of the problems associated with having children for the simple reason that she, having that option herself, will probably have given the matter some

thought. This is true whether or not she actually has children. See, for example, Dowrick and Grundberg (1980).

20. In their discussion of the features of the social control of women, Hutter and Williams (1981) suggest that even 'normal' women are regarded as having a somewhat childish incapacity to govern themselves and a corresponding need for male protection. Regarding this as part of the female nature 'renders' women a 'fit subject for the care, protection and control of medical experts' (Hutter and Williams 1981:19). The 'double-uterus' tape provides a good look at this process of 'rendering.'

21. This is, by no means, an original point. In their analysis of medicine as an institution of social control, Ehrenreich and Ehrenreich (1978) distinguish between two forms of control. The first is *disciplinary*, aimed at keeping people out of the medical system and on the job. The poor and ethnic minorities are particularly subject to this kind of control. *Cooptative* or expansionary control, on the other hand, is characterized by low barriers to entering the medical system. People are even encouraged to make use of medical services in situations when they are not sick (contraception, marital problems, preventive health care).

22. In the 'double-uterus' tape, the patient's troubles did not receive a diagnosis, as was the case with complaints in the previous tapes. Nevertheless, the GP had to weigh what he heard in terms of his professional schemes of relevance. In the case of this patient, his conclusion was clearly that no diagnosis was in order since nothing was really wrong with the patient.

23. The difference between 'upset' and 'angry' is not simply a matter of semantics. As any therapist will attest, being angry is, by far, less passive than being upset, possibly a reason that this particular patient insists that she is the former rather than the latter.

24. He was far more competent in applying these skills than his colleagues in the previous consultations. This is not surprising considering that he was the only GP in my corpus of tapes involved in training programs for general practitioners. Since his job was to teach social skills to medical students, it is reassuring to note that he possessed and employed the skills in his own practice.

25. As is evidenced by her expression of appreciation at the end of a protracted round of troubles-telling about her therapist:

199	P:	But I- I think I'm better able to talk with you
200		than with that man. Listen I-
201	D:	Mmhmm
202	P:	-don't know why
203	D:	Mmhmm
		((2.0))
204	D:	Yeah(hh).
		((4.0))

205 P: H-he does always let me talk of course. He does
206 do that. ((1.0)) But-

This expression of appreciation on the part of the patient was not an entirely welcome one, leading to the unfortunate 'Doctor's Dilemma' described in Chapter 7. However, it indicated that the patient, up until that point in the consultation, was quite satisfied with the response she had received from the GP to her troubles.

26. This is strictly old hat in medical sociology. Parsons (1951), in his functional description of the 'rights' and 'obligations' of the patient, included the duty to seek the competent help of a physician and follow his or her advice forthwith. Interestingly, this had as corollary within the Parsonian medical social system that the physician have as primary obligation a returning of the patient to her normal position as full-fledged member of the workforce or her everyday concerns. My material seems to indicate that the patient advocates this return to normal functioning more strongly than the physician.

27. Appearances can be deceptive, however, as we shall see in the next chapter. In the long run, the patient lost control over the interaction, as well as the frame in which it occurred.

28. Once again, this is strictly speculative, as we have no way of knowing what her motives were. Based on what she actually said, as well as empirical studies on storytelling and its functions, I have assumed this is what she was aiming at. A case might be made for some other 'motives' like, for example, a spiteful wish to make things difficult for the GP or a grudge against professional helpers in general. Whereas this kind of speculation could open up possibilities for uncovering potentially pathological behavior on the part of the patient, it would also bring us even more rapidly into the realm of the speculative than the above reading.

29. Janus, the Roman god of gates and beginnings, was famous for his two faces; one facing backwards, one forwards. This anatomical aberration accounted for his being celebrated on New Year's Eve when a certain amount of ambiguity in terms of backwards- and forwards-looking is inevitable. Since then, he has come to stand not only for ambiguity, but also hypocrisy and insincerity. I am using the term in both senses as well.

12

POWER PRACTICES: RELINQUISHING THE FLOOR

RELINQUISHING THE FLOOR

Ordinarily, it will be up to the troubles-teller to alert the recipient that her troubles-telling is coming to a close, so that routine turn-taking patterns can be resumed. There are various devices available for this: summarizing statements, making arrangements, or conversational closing sequences.<1> In troubles-talk, this task is not an easy one. As Jefferson (1984b) has pointed out, it is not only difficult to get a troubles-telling started, but it is hard to get off it as well. Certain topics, like jokes or pleasantries, cannot be embarked upon without creating an appropriate conversational environment first. This may be accomplished by means of 'other-attentiveness;' i.e. remarks which exhibit attention for the recipient ('So, how are things going with you?').

Getting off troubles-telling did not follow this pattern in the consultations comprising my data, however. To begin with, it was rarely the patient who changed the subject.<2> GPs did this by introducing another kind of talk or closing down the conversation altogether. This is not surprising. Control over topicality is one of the primary ways that power is exercised by professionals in institutional encountering (Scheff 1968; Hughes 1982; West 1984; Fisher 1986). Moreover, it is a control which, sooner or later, will have to be reinstated in professional hands. This can be accomplished in various ways.

The simplest is to ask a medical-type question:

EXCERPT 1
[the 'double-uterus' tape]

```
81 P: Yeah(hh) - well yeah- you're actually uhh
82    ((complaining)) thinking constantly(
                                         (Mmhmm
      about it(
              (yeah.
      ((2.0))
83 D: Let's see. What were all those delicious
84    things you're taking at the
      moment?
85 P: (Mmm- ehm- Oh dear!
```

In this excerpt, the troubles-telling is transformed without further ado into a medical checking operation. Having been marked as topically disjunctive ('let's see'), no special sequence is needed before being introduced. Unlike ordinary transitions, medical matters do not require accounting when initiated by the GP. He can bring new topics up, even abruptly, without doing damage to the ongoing interaction (West 1984).

Physicians are protected by virtue of their institutional authority from sanctions, which would be conventionally operative in talk between equals. It is this authority, which enables them to take over, legitimately and without disruption at any point within a consultation.

Another way of closing off troubles-talk is to provide a *summary statement*, assessing the trouble and then moving into a 'making arrangements' segment. This, like the medical question posed above, was a frequent occurrence in the consultations of my data. For example:

EXCERPT 2
[the 'a little bit nervous' tape]

```
129 P: But my husband says, just go to the doctor
130    ((laughs)) with that headache.(
                                     (Yeah(hh)
131 D: Well. You can tell your husband that it
       really isn't uhh-(
132 P:                  (not uhh
133 D: isn't physical-
134 P: Ohhh:
       ((2.0))
```

```
135 D: ((clears throat)) No. I really thought
136    that uhh- the
       problems     ((clears throat))     - were
137 P:         [Yeah, that's what I thought too
138 D: just getting caught in your [throat
                                   [would that-
139 P: Could that be it? Did you think.
140 D: Yes.
141 P: Because I got those seresta pills from you
142    the last time and ((animated)) when I did-
143    it's really crazy - when I take one every
       once and awhile, then I do feel uhhh- do
       feel fitter.
144 D: Yeah.
       ((2.0))
145 P: Well, I guess that I have gotten myself a
       little bit ((laughs)) too upset-
146 D: Just ((use)) this [     it's a little-
                         [Yeah-
147    little lighter than seresta [       and
                                   [Oh yes.
148    you'll be less likely to get used to it
       too [       Yeah.
           [Oh yeah.
149 P: ((soft)) well. [We'll just try it out.
150 D:                [((a little while ... on ...
       and then))
```

Following a protracted round of troubles-telling, the GP supplies an assessment of the patient's current distress; i.e. it is *not* a somatic problem. A summary statement is also provided, which has a remarkably diagnosis-like format (lines 135-136). With the patient's compliance, they can move unproblematically into the treatment phase of the consultation. He writes her a prescription for a mild tranquillizer, which she accepts. The trouble having received a medical solution, the patient's troubles-telling can be brought to an end and the conversation as a whole terminated.

In addition to introducing other (in this case, medical) matters to the ongoing talk, thereby allowing the troubles-telling to be concluded, the recipient may also simply *initiate a closing of the conversation* as such. The next excerpt follows a series of attempts on the part of the GP to bring the consultation to a close. He has repeatedly summed up the patient's difficulties (Chapter 7, p.191), with an accompanying 'Listen, I wish you luck' or 'Well, good luck with it.' The third time around, this is successful:

EXCERPT 3
[the 'one-continuing-story' tape]

515 P: Well yeah, I'm just scared to death to go
516 back that way [but yeah, you don't get
[Yeah.
517 any place. [((sniffs))
[Yeah.
((2.0))
518 I'll make it, you know.
519 D: I think so.
((2.0))
520 I'll give you a steady hand.
521 P: Well, I'll make it.
522 D: Listen. O.K.? Good luck.
523 P: I hope that this helped towards uhh-
((laughs)) [
[Yeah:
524 a good impression of
((me with you)) - [
525 D: [Don't know, but [
[Uhh-
526 it doesn't matter anyway, O.K.?
527 P: Merci beaucoup. <4>

Whereas this particular return to turn-by-turn topic talk proves problematic, as will be demonstrated shortly, it has, in any case, been successful in bringing the ongoing troubles-telling to a halt. This is accomplished in primarily two ways. First, the GP as recipient does not align himself to the troubles-telling being initiated in line 515. He is either silent or simply fails to produce a suitably reciprocative response (lines 519-520). Second, both he and the patient make references to conventions belonging to closing segments of conversations; i.e. hand-shaking, wishes for well-being, thank-you's. However unhappily, both participants do use closing conventions, thereby enabling the troubles-telling as well as the consultation as a whole to come to an end.

In conclusion, since stories about trouble entail a temporary disruption of the normal turn-taking machinery, interactional work will be required to re-engage in turn-by-turn talk and reinstate the GP's authority as medical expert. In a consultation, this entails returning to medical Q-A sequences or terminating the interview altogether. In the previous tasks involved in troubles-telling, the

GP's cooperation as recipient was required and obstacles occurred when this cooperation was not forthcoming. Now it becomes the patient's turn as storyteller to cooperate by relinquishing the floor. When this does not occur, obstacles emerge and need to be negotiated. By refusing to give up the floor, the patient undermines the institutional authority of the physician, thereby giving rise to a struggle for power.

OBSTACLES IN RELINQUISHING THE FLOOR

[the'one-continuing-story' tape]

In the 'one-continuing-story' tape, returning to turn-by-turn talk proved the greatest interactional obstacle within the troubles-talk, making it difficult to close down the consultation as a whole.
Whereas the patient had ample opportunity throughout to engage in stories about troubles, it could not go on indefinitely. Mid-way through the consultation, the GP begins to abandon his affiliative stance, indicating that he may be having some troubles of his own:

EXCERPT 4

```
288 D: Listen, uhhh- that uh- uh-uh- you are in
289    contact with that with that psychologist-
       I think that uhhh uh you really have to
290    watch out a bit- of course you
       can (come here to talk things through
291 P:     (Well I think that I've been acting too
292    negative, you know, (about that man maybe.
293 D:                     (Well-          that-
294    that's something you have to decide your-
       self, I guess.
       That can be pretty logical. (
295 P:                              (But that
296    doesn't help naturally very well.
297 D: WELL: in any case - that's something
298    too - those feelings you have them now
       after all- (        There's nothing wrong-
299 P:            (YES! I feel that way myself too
300 D: at all either that you wi- come out with
301    that. I uh- I just find that that I really
302    have to watch out not- uhh not to get in
       the way (                 do you understand?
303 P:         (that doesn't have to happen at
       all n-
```

304 D: So, I'll listen to it
305 P: No, I'm not f-forcing it on you at all either-
306 D: No, but I also like to- to just
307 talk about it but- I will be staying
308 neutral myself beyond that
Yeahh.
309 Uhh- just to prevent that uhhh- that you get two therapists you know?
Mmhmm!
310 ((telephone rings)) I just have to get the
311 telephone, sorry-

((interruption, telephone call))

312 D: Yeah? ((2.0)) Uhh-
313 P: ((soft)) No, I don't need two.
314 D: Uh, does the psychologist know that you take reapam?
315 P: ((coughs)) No, he doesn't know that
316 because last week uhh- I had to come here.

By engaging in protracted storytelling about her troubles with her therapist, the patient has brought the GP into conflict with his professional code. Moreover, he will, sooner or later, have to terminate the consultation and she shows no signs of running out of story-material. The problem has now become how to get off troubles-talk so that the consultation can be brought to a close.

To begin with, the GP attempts to enlist the patient's sympathy or, at least, cooperation in dealing with his dilemma. He describes his problem, indicating that he needs to 'watch out' (lines 289-290; 302) that he doesn't 'get in the way' (line 303) and that he wants to remain 'neutral' (line 308).

The GP does not simply state his case, however. Taken together, his utterances can also be heard as a criticism of the patient (Labov and Fanshel 1977:97). It is her activity; i.e. talking about her problems with him, which has led to the GP's present predicament. By referring to this state of affairs, he indirectly challenges that behavior. All of this is done in an extremely mitigated fashion. In case the patient may have gotten the idea that he doesn't want to talk to her, he hastens to reassure her that *of course* she can come back and talk any time. In fact, he even 'likes' chatting with her. Despite these laudable attempts to forestall the potentially disruptive effects that his remarks may have produced, the desired effect is not forthcoming.

The patient clearly 'hears' his remarks as critical. Whereas her first response is an admission that she might have been too negative about her therapist (lines 291-292), she responds later with a flat denial that it was her intention to 'force' him to talk to her. *She* doesn't want two therapists either.

This segment provides a good illustration of how attempts at mitigation can backfire. By being indirect and reassuring, the GP has not openly taken a stand about his present discomfort in discussing the patient's therapist. Moreover, he has not indicated that he would like to terminate the consultation, which is very probably also at issue here. Instead he obfuscates his own predicament with reassurances and indirectness, while, at the same time, making it clear that the patient's behavior *is* problematic.<4> When she responds to this criticism, he does not validate that either. He shifts the topic by referring to the patient's tranquillizers (lines 314-315). In this way, the treatment phase of the consultation can be initiated and the consultation brought to an end.

The patient isn't having any of these reassurances, nor does she seem prepared to relinquish the floor. What follows is a veritable tug-of-war, with the patient attempting to re-align the GP as troubles-receptive listener to her talk and the GP, just as valiantly, trying to dis-engage himself from that position and wind up the consultation. I shall examine several instances in more detail, taking them chronologically. It will become clear that the struggle intensifies as the talk progresses, with both participants resorting to ever more extreme interactional procedures. The GP is attempting to reinstate his institutionally-based dominance, while the patient resists it, every step of the way. For example, here is the first 'hurdle,' taken by the patient:

EXCERPT 5

334 D: Would you like to continue with it or are
335 you saying, like, well, no, let me just
try it again on my own?
((3.0))
336 P: ((soft)) Well, yeah, I have to, yeah, on
my own uhh- Ehm.
((5.0))
337 Yeah. You know, I been getting terribly
338 upset now thinking about the way
back. (Plus that afterwards I also
(Yeah.

339 still have to go to the pharmacy, you
know (
(Yeah(hh).
I'd most like ((laughs)) to say, no-:
340 but only because I wouldn't have to do it
341 then (You know, ((to withdraw into
(Mmhmm
myself (like-)) I'll go right away
(Yeah
342 If necessary I'll call a cab and go
343 straight home and I'll just have to see
344 ((1.0)) and uhh- even though I know it
or - so, le- ju- say, like, yeah(hh)- uhh-
345 ((do I really need them, question mark)).
346 Those conflict situations come back every
day and-
((2.0))
347 Or you just don't go uhh- your shopping
at all.

The GP requests that the patient decide what she wants to do about her tranquillizers: continue or stop. The latter option is formulated in a way, which refers back to prior talk and the patient's repeated assertion of her autonomy. Instead of taking the GP up here, however, she puts off his request (Labov and Fanshel 1977:86) by remarking, once again, that she has to do things on her own. As an assertion about a desirable state of affairs, she can remain 'off the record,' until the GP provides some additional information (for example, how she can go about it). In this way, she can, politely, but legitimately, refuse the GP's request that she decide right now.

In contrast to the GP's usual affiliative behavior, this utterance merely evokes a long silence. He refrains from providing even a minimal response.

Undaunted, the patient takes the floor, embarking on another round of troubles-telling. After providing a preface ('I'm getting terribly upset now thinking about the way back') and a cursory attempt at alignment ('you know'), she proceeds to recount the problems that getting a prescription for tranquillizers filled would entail. Thus, she utilizes the GP's request itself as topic for further troubles-talk about her agoraphobia. It is transformed into *just one more instance* of the problem-at-hand, thereby eliciting a further round of storytelling (lines 346-347).

In this excerpt, the patient recounts her trouble, using a series of contrasts: she would like to say 'no' to the pills, but laughs this off as an illegitimate action (line 339). Her first inclination is to dash right home (lines 341-343), but she knows that she really needs the pills (line 344). The point of her story, summarized in line 346, is a familiar one: her suffering is unending ('conflict situations come back *every day*') and, at the same time, she will be the one, in the final analysis, who has to do something about it.

Her use of contrasts not only serves to focus the GP's attention, but has a phatic function as well; i.e. the troubles-telling can be prolonged (Quasthoff 1980). No matter which way the GP responds, the patient can fall back on the other part of the contrast. In this way, she has him locked into his position as troubles-recipient. Her laughter and 'you know' soften, but do not otherwise alter this rather constraining method of alignment.

Throughout the troubles-telling, the GP continues to refrain from displaying affiliation. His responses are limited to minimal Mmhmm-ing. Unlike previous storytellings, he does not re-formulate her talk, marking the significance or indicating that he has been listening, nor does he fill in the occasional silence with an appropriate 'mm' or a 'yeah'. Despite being noticeably less aligned as troubles recipient than he was earlier on, the GP's passive stance is not enough to disrupt the patient's story. It is only many turns later that he actually manages to regain the floor, once again, recycling the psychologist as topic:

EXCERPT 6

359 P: *All these- these horrible things, you know*
360 *((gagging sound)) Bahh. Bohh.*
361 D: *And what kind of arrangement do you have*
362 *at the moment with the psychologist?*
363 P: *I go uhh- Monday evening - to*
him (again at 8 o'clock.
364 D: *(Monday evening, 8 o'clock. Yeah. Yeah.*
365 *Do you know what? If I would just give you*
uhhh- a prescription- and you ((1.0)) can
366 *simply see for yourself afterwards whether*
367 *you want to get it filled? Whether you*
have the nerve (Uhh- and you have it in
(Mmm
368 *any case in the house.*
369 *Do you see that as a kind of compromise*
uhhh-?

Interrupting the troubles-telling with a question, the GP returns to the matter of the prescription. Since the patient has put off making any decisions about it, he now frames his request in a way which will make it more difficult for her to refuse. He will write it out and she can decide later. He goes to considerable effort to mitigate his action: by putting prescription in the diminutive form ('a little prescription') and by suggesting that she can 'simply' see whether she wants to get it filled or not. His suggestion is presented as a 'compromise' rather than something she has to do. All of this can work to align her to the position of advice-recipient - something she has, up until now, refused.

Whereas success appears to be just around the corner for the GP, the patient does, in fact, manage to prolong her story-telling, and, with it, the consultation, once again. This time, however, she has to resort to more drastic measures, interactionally speaking:

EXCERPT 7

378 D: I'll give you a prescription.
379 You have reapam ten?
380 P: Yeah.
((3.0))
381 Last time I said, just give me librium.
382 Well, he said, then I don't know what
you're going to be doing with that.
((laughs))
383 D: ((surprised)) OH?
((2.0))
384 Were you planning to uhh- do crazy things
with it then? Or something.
385 P: Sometimes I do a bit yeah.
386 D: Yeah?
387 P: Yeah. Yeah. ((is th-)) crazy? ((laughs))
388 It's actually not crazy, you know.
Absolutely not crazy- No: (in
(((starter)) (No-
389 situations (it comes- (sometimes
(Mmm (Yeah-
it's real normal. Won't do it of course,
no. ((2.0))
390 I'm crazy but not that crazy.
((3.0))
391 D: Yeah.
392 P: Yeah(hh). Just like that- it's not just

393 *because- yeah of course it is, I have to solve it but-*
(Mmhmm
394 *there's my husband too who does all of*
395 *these things which uhh I step over too, you know?*
(Yeah.

Up until now, the patient has used a variety of techniques for holding the floor for continued troubles-telling: series of stories, repetition of the evaluative point, dramatization, and contrast constructions. In this segment, she produces what might best be called a threat. It is a veiled one, but, nevertheless, unmistakable.

The GP is now writing out the prescription for reapam, having checked on the dosage (line 379). The patient picks up on this, beginning a new story in which she indicates that the last time she asked for librium,<5> her therapist protested. She might 'do' something with them, i.e. presumably commit suicide. This is presented very indirectly, with the patient not actually threatening to 'do something,' but suggesting that this is her therapist's worry. In this way, she can enlist the GP's sympathy, without having to be held accountable for her actions.

It has the desired effect. The GP breaks off writing the prescription, and reacts very affiliatively with a surprised 'Oh?' This is followed by several elicits, inviting her to provide further accounting for her utterance (lines 384,386). The patient's triumph, interactionally speaking, resides in her having managed to break the medical frame and force the GP to react to her as a peer; i.e. with shock and concern. After providing an account for her utterance (387-390), she is then able to launch still another round of troubles-telling, this time about her problems with her husband (394-395).

Thus, despite the fact that the consultation is in its final stages, the patient has gotten the GP's attention and re-aligned him as troubles-recipient. Just how receptive a recipient he is going to be, however, is open to debate, as the following excerpt demonstrates:

EXCERPT 8

462 P: *Well yeah. ((soft)) And uh just keep fighting, I know that myself too of*
463 *course, but HOW LONG can you- how*
(Yeah-
464 *long can a person fight? I wonder? How long can it uhh-*
(Yeah(hh) <6>

465 D: *You have been able to for ten years, I understand.*

At this juncture, the GP's patience has clearly come to an end. He no longer displays passive resistance to the patient's attempts to enlist a troubles-receptive hearing from him. Although he does not assert his unwillingness to listen in-so-many-words, his stance is definitely a less than friendly one and it remains so until the end of the consultation.

In the above excerpt, the patient has produced a rhetorical question. With considerable dramatic flourish, she asks how long a person (any person) should be expected to keep up the fight. This serves, once again, to underline the seriousness and, in fact, the tragedy of her present situation. As is generally the case in indirect communication (ironies, metaphors, understatements), the actor leaves herself 'out,' allowing the addressee to decide how to interpret it (Brown and Levinson 1978:216). Since the utterance is made in context, it can be assumed that there are enough clues for the other person to be able to interpret it correctly. Certainly this is true for the rhetorical question above. The GP has had ample opportunity to discover how serious the patient's present situation is, as well as become familiar with her extensive efforts to 'fight' it.

Instead of honoring this interpretation, however, the GP treats it as though it were not a rhetorical question at all, but rather an actual request for information. 'How long' is answered by 'ten years, so I understand.' This serves not only to undermine the seriousness of the patient's problem (if you've been at it this long, why not a little longer?), but it casts doubt on the propriety of the question as well.

This segment is followed by two unsuccessful attempts on the part of the GP ('Listen, good luck') to terminate the consultation (Chapter 7, pp.188-191). When he does finally succeed, the ending proves anything but a happy one:

EXCERPT 9

503 P: *Now - I'm just thinking now what shall I*
504 *do? Uhh- ((3.0)) Shall I sit here for a*
little bit, in the waiting room, or shall
505 *I hurry on immediately? Or - you know -*
506 *I've already been sitting- t-t- it's*

already spinning [around again. [
[Yeah [Yeah.

D: Yeah.

507 P: Will I ge:t it while I'm on my way back

508 or- will I-I- start hyperventilating,

((rapid)) that doesn't do any harm of

509 course, when I start [thinking like

[Yeah

this. [((2.0))

[Yeah.

510 Or shall I just simply say to myself like,

girl, head up. You are just [going to

[Yeah.

511 go to the [pharmacy.

[Yeah.

512 D: What are you telling this to me? Also a

513 little bit because you expect an answer

from me? Or a- a - little support

or so- [

514 P: [More support really, because

515 P: Well yeah, I'm just scared to death to go

516 back that way [but yeah, you don't get

[Yeah.

517 any place. [((sniffs))

[Yeah. ((2.0))

518 I'll make it, you know.

519 D: I think so.

((2.0))

520 I'll give you a steady hand.

521 P: Well, I'll make it.

522 D: Listen. O.K.? Good luck.

523 P: I hope that this helped towards uhh-

((laughs)) [

[Yeah:

524 a good impression of

((me with you)) - [

525 D: [Don't know, but [

[Uhh-

526 it doesn't matter anyway, O.K.?

527 P: Merci beaucoup.

In this final segment, the patient makes a last ditch attempt to engage in troubles-talk. She pursues the topic of her agoraphobia, this time linking it to the present situation. As usual, she not only

recounts the trouble itself, but places emphasis on her own efforts to overcome it (lines 508-511). The GP has, on various occasions, become aligned to the seriousness of her problems. He has, however, up until now refrained from confirming the second dimension of her evaluative point; i.e. that's she's doing her best to alleviate them.
It is, of course, unlikely that the patient will be able to achieve this alignment so late in the game.

Here, the GP, in accordance with *his* interpretative professional schemes hears her account as a request that he *do* something. He formulates this (lines 512-513), allowing the patient to decide. Unfortunately, this is precisely what she does not do. She provides an account which, on the one hand, serves to answer his request. She would like a 'little support' because she's 'scared to death' to go back. On the other hand, she indicates that she already realizes that none of this will really help (lines 516-517) and that she'll make it on her own (line 518). This is the second dimension of the evaluative point, once again, still awaiting confirmation. Unfortunately, as answer to the GP's request, it places him in an interactional double bind. If he provides the support, she can retreat to the position that it doesn't help. If he agrees that nothing helps, she has every reason to continue complaining. Whereas there are ways out of such double binds, (Haley 1963; Watzlawick *et al* 1967), it would require some heavy duty therapeutic work on the part of the GP. A more practicable solution might be simply backing out of the interaction immediately and on no uncertain terms. This could be done by stating his position without mitigation ('Lady, I'm not going to put up with this stuff any longer'), something he has, up until now, refrained from doing. This is not what he does now either. The interaction remains indirect, with both participants managing to stay politely off the record. What follows demonstrates how politeness strategies are not only necessary for face-saving or preserving the harmonious quality of an encounter, but how they can be employed as weapons in interactional struggles as well. Being polite entails not saying what one means. Considerable effort, in fact, goes into *not* being entirely honest and packaging one's talk in terms, which are willfully vague or obscure. It is left up to the speaker to make inferences as to what is *really* meant (Brown and Levinson 1978).

In the above segment, the speakers use (or abuse) normal conventions for politeness by taking each other at their word and refusing to make inferences. In the context of her fears about going down the street, the patient's claim that she'll 'make it' (line 518) is designed to emphasize a stiff-upper lip attitude. By taking her remark at face-value, however, the GP disregards this contextual interpretation. His statement that he thinks she's get home all right indicates that he is not particularly impressed by her anxiety attacks. This challenge is off the record, since, on the surface, he is only agreeing with what the patient herself has said. Similarly, offering her a 'steady hand' is not an unambivalently supportive gesture. It is also a suggestion that she should leave.

Just as the GP remains politely off the record, so does the patient. Although she has some reason to suspect that he is less than happy with her, she formulates this indirectly, as a 'hope' that she's made a 'good impression' on him. Politeness conventions would require a confirmation that this has, indeed, been the case. The GP is not forthcoming, however.

The patient's final 'thank you,' set off in a foreign language, indicates something quite different than gratitude. The impression is unavoidable that thanking him is the last thing she has in mind.

Thus, a consultation which began as a paradise for troubles-talk grinds to an unsatisfactory halt - probably for both parties. Despite the fact that the GP has, throughout the consultation, downplayed his power, he does, ultimately, exercise it by bringing the consultation to a close.

STRUGGLE FOR POWER

[the 'one-continuing-story' tape]

It has been demonstrated that getting off troubles-talk, once underway, is by no means an unproblematic affair. Like the other tasks involved in the accomplishment of a troubles-telling, relinquishing the floor requires negotiation. The 'one-continuing-story' tape provided a look at what can happen when the storyteller does not make a move to terminate her story nor the GP to reassert his control over the interaction strongly enough that the consultation could be brought to a close. This led to an interactional vacuum within which a kind of tug-of-war ensued. As the consultation went on (and on), each party began to employ stronger and less-than-kosher means to maintain or regain his or her control over the conversation.

Power in interaction

The exercise of power in interaction will entail that actors draw upon skills and resources structurally available to them in order to influence the course of events. In the 'one-continuing-story' tape, the GP employed skills available to him as professional service provider, as well as competent conversationalist in order to regain the conversational floor and terminate the consultation. As was demonstrated in the last chapter, he continued to disalign himself from the patient's troubles-telling by refraining from displaying affiliation. He was notably silent at junctures in the ongoing storytelling when a listening noise would have been in order (Excerpt 5). Whereas his initial strategy was one of passive withdrawal of alignment, he later adopted a more straightforward stance, explicitly asserting his reticence to engage in further talk about the patient's therapist. When this was not sufficient to impede the patient's troubles-telling, he initiated various 'making arrangements' sequences (Heritage and Watson 1979), making welcome use of schemes of relevance available to him as professional. He summed up prior talk, discussed the possibility of a prescription, referred the patient back to her therapist and indicated that the patient could always come back for a follow-up consultation (Excerpts 4,5,6). In this way, the troubles-talk and, subsequently, the consultation could be brought to a close. When this did not succeed, he challenged the patient in a way which, while couched in polite terms, could be heard as an unmistakable criticism of her behavior (Excerpt 8). Ultimately, after all else had failed, he simply initiated a closing sequence anyway, replete with farewells, thereby bringing the consultation to an abrupt, but definite end.

The patient worked equally hard to ensure that her troubles-telling could continue, overcoming each hurdle as she came to it. To this end, she drew primarily upon resources available to her as storyteller, including dramatization, repetition of the evaluative point, and contrasts. All of these devices served to prolong her storytelling. As the consultation progressed, the content of the stories became even more dramatic, with veiled threats of incipient hyperventilation attacks (Excerpt 9) and even suicide (Excerpt 7). In this way, she made it particularly difficult for the GP to avoid expressing some commiseration or at least acknowledging her suffering.

In addition to these storytelling devices, the patient prolonged her troubles-telling by mobilizing topics introduced by the GP himself in order to initiate new stories about troubles. His suggestion of a prescription for reapam, rather than serving to move the consultation toward its final stages, could be mobilized as a trigger for continued rounds of troubles-telling about the perils to be encountered on the way to the pharmacy (Excerpt 9). And, last but not least, when the troubles-talk was finally brought to an end by a somewhat abruptly iniated farewell sequence, the patient managed to supply the parting shot, indicating that, whereas she was leaving, it was definitely not as a satisfied customer.

Power and agency

Even in contexts characterized by institutional hierarchies, the exercise of power will remain tied to agency. This does not mean that the actors will obtain outcomes entirely in accordance with what they intended, however. On the contrary, the exercise of power will display a mix of intended and unintended activities.

The patient in the 'one-continuing-story' tape was, of all the patients up until now, the most skillful in constructing her stories about trouble. She not only re-enacted the circumstances of her suffering in such a way that the recipient could relive her experience along with her, but she set the tone of the interaction for the greater part of the consultation. In addition to being able to display her experiences in her own terms, she successfully countered the GP's attempts to redirect or cut off her troubles-talk. To this end, she not only employed increasingly dramatic storytelling devices with great expertise, but also made use of routine conversational procedures for putting off the GP's efforts to 'make arrangements' (Excerpt 5), thereby obtaining more time to make her point. In short, the patient, managed to overcome each obstacle, obtaining the GP's compliance with remarkable adeptness.

Despite the patient's success in acquiring almost unlimited space for troubles-talk during the consultation, this did not mean that she had unlimited affiliation as well. In fact, one of the unintended consequences of such extended rounds of troubles-telling in the context of an institutional encounter may well be the progressive loss of emotional reciprocation on the part of the troubles recipient. Whereas the GP in the 'one-continuing-story' tape embarked on the consultation with a notably affiliative stance regarding the patient's troubles as well as her troubles-telling, it appeared to have diminished rather markedly towards the end of the consultation. His final responses defied even the most routine politeness conventions (Excerpt 8, 9).

Power and the dialectic of control

Any interactional struggle for power will dialectically produce and reproduce relations involving autonomy and dependence in both directions. In the 'one-continuing-story' tape, the struggle began over the issue of the patient's helplessness in the face of her present distress (Chapter 11). The conflation of the point of a story about trouble and the constraints imposed by an institutional context led to an interactional dispute as both participants attempted to maintain their foothold within the interaction. The patient put all her effort into continuing her troubles-telling and the GP to reassert his professional authority, which, though dormant, had not been actively employed to curtail the patient's activities until mid-way through the consultation. Whereas the patient's complaints about her therapist served to underline her point (this is terrible and I am powerless in the face of it), it also worked to put the GP as recipient and service provider in a difficult position. Not only would commiseration with her implicit critique of one of his colleagues conflict with his professional code (Chapter 7,p.188, 190), but it would undermine his own position as professional helper, faced with the unfortunate task to having to provide help for someone, who so manifestly cannot be helped. Thus, acquiescing to the patient's claim that she is 'the only one who can help herself' would have required him to enter another interactive frame; i.e. a friendship frame, at least temporarily, where he was not required to provide help. More importantly, however, it would have forced him to commit an act of disloyalty to the medical frame, which entails respecting other colleagues' territory and professional competence (Excerpt 8).

It was here that the power struggle began, with the GP gently withdrawing his affiliation from the patient's troubles-telling and the patient, somewhat less gently, trying to regain it. Drawing upon his training in social skills, he could maintain the appearance of being empathetic and interested in the patient's plight, while, at the same time, disaligning himself and, in the process, stepping on politeness conventions left and right (Excerpt 8, 9). Whereas politeness is essential for greasing the wheels of any conversation, the inevitable indirectness accompanying it did not contribute to the task at hand. The GP's ambiguity simply provided the patient with enough affiliation that she could launch onto another troubles-telling.

Although the patient did not take up the GP's indirect attempts to close off her troubles-telling, she did not ignore his dilemma altogether. In fact, at various points in the conversation, she even offered a formulation of his motives, engaging in a rare - at least, for patients - display of 'meta-linguistic' listening.<7> In Excerpt 4, her formulations of the talk-thus-far incorporated the GP's implicit criticism that she might be 'too negative' about her therapist or, worse yet, guilty of trying to transform the GP into just another therapist. At the end of the consultation, she suggested that he might not have a very 'good impression' of her (Excerpt 9). Each of these instances not only gave the GP the possibility to disconfirm negative readings of the talk, thereby smoothing over potential interactional disturbances, but they could have provided openings for him to discuss his own troubles as well. Of course, such talk would have been produced at the patient's instigation, leaving her with the conversational reins in hand.

Both the GP and the patient in the 'one-continuing-story' tape used indirection to terminate or prolong the ongoing troubles-talk. The patient didn't take the GP up on his attempts to terminate the conversation; the GP refrained from getting the patient's point. She did not, ultimately, get her point across or maintain his affiliation and the GP did not succeed in re-establishing his professional authority and regaining control over the consultation. This was not only, interactionally speaking, unfortuitous for him, but it had some rather unpleasant consequences for the patient as well. The consultation ended on a decidedly unhappy note; i.e. a kind of interactional draw with neither part really coming out on top.

Power as enabling and constraining

Power practices will exhibit enabling as well as constraining dimensions in terms of the construction of the encounter itself as well as the balance of power between the participants. The kind of control exercised in any instance of interaction will involve an interplay of both features. In the 'one-continuing-story' tape, the patient exercised interactional control primarily by telling stories. Depicting her troubles in a highly vivid and dramatic fashion, she could facilitate a sympathetic response on the part of the GP. As recipient to these tales of suffering, he could hardly avoid experiencing the events, once again, along with the patient.

However, when a troubles recipient, for whatever reason, wants to disengage himself, the same devices, which produced emotional reciprocity earlier on, can become constraining, putting him under pressure to display alignment. The patient's refusal to abandon her storytelling only aggravated this lack of alignment. The GP's affiliation wore thinner and thinner until it amounted to little more than an empty phrase, albeit a politely formulated one. Thus, it would seem that the patient's efforts to achieve a sympathetic response to her troubles through repeated renditions of them in story-form, in the long run, backfired. Ultimately, the scope of the patient's control was bounded by her ability to achieve an affiliative response from the GP and the boundaries became increasingly clear as the consultation progressed.

Just as the GP in the 'one-continuing-story' tape could use

displays of empathy to facilitate or constrain the patient's storytellings as needed (Chapter 11), he could employ politeness to end the consultation. Politeness has both enabling and repressive dimensions, which can be employed to gain control over the ongoing interaction. When it became clear that the patient was not going to cooperate with a conventional means of terminating the consultation; i.e. a prescription or referral, the GP began to resort to more repressive means. Rhetorical statements like 'how long can a person fight?' (Excerpt 8) or 'I'll make it' as brave announcement after a dramatic rendition of the difficulties involved (Excerpt 9) required the GP as recipient to fill in the context in which they were made (Brown and Levinson 1978). Refusing to do this by taking the remarks at face value was tantamount to an insult or denial that the patient had, indeed, been grappling with her problems or even that the problems themselves were particularly grievous. On the surface, his responses were nothing more than a reiteration of what the patient herself said, expressing disalignment, while enabling him to remain conveniently off the record. Thus, surface friendliness or attention to conventions ('I'll give you a steady hand') blended intimately with an insulting stance toward the patient, allowing the GP, at long last, to reassert his authority and end the consultation. At the same time, he could maintain an appearance of being sympathetically on her side. Once again, it becomes clear what can happen when Janus rears his (or her) head in the course of a medical encounter.

Power as process

Power will be exercised in a moment-by-moment process through which asymmetrical power relations can be constructed, sustained, and undermined. In the 'one-continuing-story' tape, these relations expressed autonomy and dependence in both directions. In this particular consultation, however, it was demonstrated perhaps more clearly than in the previous ones just how mutual this dependency is, as well as some of the difficulties, which can emerge when neither participant manages to establish her or his autonomy in a decisive way.

In a process which began with disputes around the point of the patient's stories (Chapter 11), the interactional frames, which had been in operation simultaneously throughout the consultation, had gradually reached the end of their compatibility. One of the frames would have to be establishing decisively, if the consultation was to proceed or become terminated with some semblance of harmony. As the patient persisted in talking about her troubles, it became apparent how storytelling is not only a powerful, but also, to some extent, a doomed activity in medical interaction. In accordance with Quasthoff's findings (1979), the patient's storytelling endeavors in the present consultation inevitably crashed on the rocks of the institutional encounter where, no matter how friendly the GP had been during the consultation, he was, in the final analysis, a professional and *not* the patient's friend. To be so would have entailed abandoning his position of institutionalized authority and meeting her on equal footing. Whereas some GPs seem inclined to do this,<8> my material provided ample evidence that this willingness was strictly temporary. Thus, the control exercised by the patient through her storytelling was, at best, a time-out from the medical frame and not its replacement. Ultimately, the medical frame with its institutionalized relations of authority took precedence over a friendship frame set by the activity of storytelling.

Lest the conclusion be drawn that the patient in the 'one-continuing-story' tape had her frames confused, let us not forget the GP. He deployed various social and therapeutic skills, which were not only useful for displaying commiseration as any troubles-recipient might do, but precisely these interactional skills came in handy for eliciting information about the patient's current distress or clarifying her problems in order to come up with a solution. The same skills are functional in both a medical and a friendship frame. Any sympathetic listener, when confronted with a story about trouble, will produce listening noises, formulate the problem from time to time to display understanding, express commiseration with the terrible-ness of the trouble, etc. In short, the conflation of membership categories, where the GP was both a sympathetic troubles-recipient and a professional helper, well-versed in the therapeutic skills of modern medicine, may have accounted for the GP's difficulties in re-directing or otherwise stemming the flow of the patient's talk. He did not have the consultation under control.

This did not have to be problematic, of course. The analysis, however, indicates that it was problematic for the GP in the 'one-continuing-story' tape. He appeared to be caught between two interactional frames, unable - in contrast to his colleagues in the previous consultations - to mobilize his professional authority to combat the problem of 'fuzzy boundaries.'

The result of this frame confusion was, in all probability, unpleasant for both parties. Whereas the patient had plenty of opportunity to talk about her troubles, it was clear by the end of the consultation that she had forfeited the GP's sympathy. By the same token, the GP could withdraw his alignment as recipient, but he still remained locked in the conversation, unable to terminate it in a way satisfactory to either of them - a decidedly unhappy end for all concerned.

NOTES

1. I am excluding the possibility here that the recipient engage in some troubles-telling of his own. This is done frequently in everyday conversations where speakers link stories to previous ones, producing series-of-stories (Ryave 1978). With one exception, this was not done in the consultations investigated here. This corroborates Quasthoff's (1979) finding that storytelling is limited to clients. Professional helpers do not engage in this activity. Interestingly, the exception in my data (preliminary analysis) involved a GP, who gave a patient some advice on what to do about her child's bedwetting. He began this advice-giving by recounting that he'd recently been to a conference on bedwetting. Although it lacked the other features of story- or troubles-telling, it was unusual in that it involved something he had experienced.

2. I am referring here only to topic shifts from a troubles-telling to something else and not the other way around. The only example in my data of a patient-initiated shift from troubles-talk to turn-by-turn talk was in the 'double-uterus' tape (Chapter 11, Excerpt 12). The GP responds to the patient's troubles-telling with a platitude, suggesting that 'we' should look at the sunny side of things. The patient initiates a new topic at the first possible completion point by asking him a medical question. This marks a shift in the troubles-telling and moves the entire consultation towards closure.

3. This last utterance was in English in the original. To get the same effect, I have translated it into French. I selected French as I assumed that an English speaker would be about as likely to know this expression as a Dutch native speaker to know English.

4. As Turner (1976:248) has pointed out, complaints may be touchy affairs in that they stand to be reformulated by recipients as to whether the complainer has sufficient grounds for making the complaint ('I should have such problems') or with respect to the complainer's character ('what did you expect?'). Similarly, reassurances do not always work unproblematically, but stand to be re-formulated by the recipient. This can happen when the credibility of the giver of the reassurance is in doubt, giving rise to the suspicion that he or she is engaging in a 'selling approach,' whereby the goods in question are not what they should be (Byrne and Long 1976:107).

5. *Reapam* is a benzodiazepine; i.e. a mild tranquillizer which decreases anxiety and works as a relaxant. *Librium* is also used to decrease anxiety, but can be given for psychoses as well (Kuiper 1978:312). Unless the dosage is known, there is, however, no

way of knowing whether or not librium was actually stronger than reapam.

6. 'Tja' in Dutch has a slightly different connotation than 'yeah.' It suggests some speaker hesitation, almost as if to say: 'Yeah, what am I supposed to say to *that* one?'

7. By engaging in 'meta-linguistic listening,' an utterance can be labelled by the person hearing it in terms of what it does rather than what it actually says. This having been accomplished, a possible motive for the action can be sought and defined as neurotic, making this kind of listening a useful tool for professional helpers (Schwartz 1979; Davis, 1986). Apparently, the skill alone is not enough. When employed by a patient, as was the case here, it does not seem to have the same effect at all, indicating that many conversational manoeuvres are only powerful when performed by those who are already in power.

8. There are differences, as demonstrated even by my limited corpus of tapes. Of the consultations included in the analysis, the GP in the 'classic case of gall stones' tape appeared to be the least willing to enter a friendship frame. Of course, the patient was also the most heavily oriented toward the medical frame, making the question of responsibility something of a chicken-and-egg question.

IV

EXPLAINING POWER:

TOWARD A GROUNDED THEORY

13

GROUNDING POWER IN MEDICAL ENCOUNTERS: SOME CONCLUSIONS

INTRODUCTION

The present inquiry has provided an in-depth analysis of how power works at the level of face-to-face interaction between male physicians and female patients in general practice consultations. I have approached the problem of power from various angles, creating a kind of microscope through which I (and the reader) could look in order to discover and delineate how power works. Just as a scientist prepares a cell specimen for observation under the microscope, I had to get my data ready for further analysis by transcribing audio-tapes of general practice consultations. When treated with the appropriate dyes, even the most minuscule slice of cell tissue will display different structures. So it was with the consultations which, having been prepared for further investigation, revealed various organizational structures. For example, the first dye exposed the structure of the overall conversation itself, indicating what the participants were doing as well as the topics being discussed. Later on, these same specimens could be put under the microscope again and another structure investigated - the structure of troubles-talk. This yielded a wealth of additional information about what was happening in the consultations as well as a few surprises, particularly for those of us who thought that storytelling was limited to casual talk between friends. Apparently, patients do it, too, and are sometimes even aided and abetted by their physicians. Had I been a linguist concerned with storytelling for its own sake, I might have left the investigation at that. However, since I am a social scientist and- more to the point - a social scientist with a 'hidden agenda,' I put my eye to the microscope, once again, this time, however, using a stronger magnifying glass. With the help of this lens, I could see how storytelling in consultations came to be interrupted or even obstructed. Finding these obstacles and discovering how they worked required painstaking observation, but the result was gratifying. A good look at the various kinds of obstacles emerging in the course of patients' stories about their troubles could be obtained.

This was only the beginning, however. Having uncovered what appeared to be a moment-by-moment struggle around patients' storytelling endeavors, another look had to be taken, this time through the lens of power. This final step enabled a view of how asymmetrical relations of power were being produced, sustained, but also undermined in the course of the consultations. Being able to observe these relations in action required using various dyes to ensure that the patient remained visible, that the interactional power struggle could be viewed dialectically, and that the processual and multi-facetted character of both the power practices as well as their outcome would remain discernible.

The result of this investigation was a blow-by-blow account of what power looks like 'under the microscope' and, more generally, how it works in the production and reproduction of asymmetrical relations between male physicians and female patients. Having reached the end of the inquiry, the remaining task is a two-fold one:

First, the *empirical* results of the analysis need to be tied together in the light of the research problem formulated at the outset of the book. These results will be summarized and conclusions drawn concerning how women's problems are defined, diagnosed, and treated in the course of a general practice consultation, as well as how power is being exercised in that process.

Second, some *theoretical* conclusions are in order regarding the relevance of the research findings for the construction of a more general theory of power and gender in medical encounters. This will be the subject of the next chapter.

FROM TROUBLE TO COMPLAINT: DEFINING, DIAGNOSING AND TREATING WOMEN'S PROBLEMS

The first part of the research problem concerned how women's problems are defined, diagnosed, and treated in the course of consultations with male GPs. Traditionally, the path toward uncovering this process has entailed starting with a complaint or some reasonable facsimile and uncovering the interactional work which goes into shaping it up for an appropriate medical diagnosis and, subsequently, subjecting it to further examination or treatment. This means that the investigator begins her analysis, much as the physician does, by sifting through the initial stages of the consultation to see if she can find a likely candidate for a complaint and then looking over the physician's shoulder as he attempts to do something about it. Whereas taking this professional vantage point (Murcott 1981) can be very useful for analyzing how consultations come to be organized and recognized as medical events, it had some drawbacks for a research problem like my own. Since the patient's 'voice,' as Mishler (1984) calls it, already tends to be somewhat muted by the institutional arrangements (the 'voice of medicine'), it seemed more sensible to take the patient's position as vantage point for observing the interaction.<1> By situating myself behind her shoulder, I hoped to uncover potential contributions on her part to power struggles around the business of how her problems were being defined, diagnosed, and treated in general practice consultations.

Taking the patient's vantage point meant, among other things, focussing on what *she* did in the consultation, how she talked about her troubles and what happened to them in the course of the consultation, rather than on how her complaints were diagnosed and treated, which is immediately what emerges when the focus is on the physician's contribution to the interaction. Thus, the original formulation of the research problem remained implicit, diagnoses and treatments often being subsumed under the heading of an obstacle to the troubles-talk.

I shall now return to the matter of how the patient's problems were being defined, diagnosed, and treated and attempt to draw some general conclusions based on the empirical findings of the present analysis.

Defining the trouble

There are various forms of talk which can be selected if one wants to define a problem. Although the patients packaged their problems as descriptions of symptoms or illness progress reports, they invariably engaged in some troubles-telling as well. In fact, even when the consultation was short and routine, running smoothly and in accordance with professional systems of relevance, they still managed to squeeze in a little bit of the experiential quality of their troubles. It was this emotional experience of suffering, which pervaded even the most business-like rendition of a complaint, whether the trouble concerned a physical complaint, mental discomfort or simply a problem of living. It appeared as a groan or gasp, embedded in an otherwise objective description of the location of the pain or as the complaining tone, which provided the refrain for a whole list of problems. If given the chance, it could easily unfold into a lengthy re-enactment of the events leading up to the present distress: a full-fledged story about troubles. Several conclusions can be drawn from this.

First, the patients apparently did not simply have complaints, but they tended to complain about them as well. In fact, they eagerly took every opportunity to engage in troubles-telling, even when it did not meet with a welcoming response on the part of the physician. To this end, they had to do considerable interactional work, displaying an ongoing awareness of cultural constraints, which require that troubles-talk be given special handling. GPs routinely received ample warning that a trouble was on its way and were reassured that the troubles-teller would not be making an undo fuss or taking up the recipient's valuable time with unnecessary complaining. The patients did not insist on talking about their troubles, but manoeuvered them adeptly around the institutional constraints of the medical encounter. Thus, whereas they engaged in troubles-talk, they were careful to organize it within the dominant medical frame.

Second, by recounting their troubles in story-form, the patients defined their own problems as part of their everyday life contexts rather than as a complaint, hygienically presented for further diagnosis. The activity of troubles-telling generated whole series of problems, running the gamut from sore-throats and stomach-aches to worried requests for information to an ongoing lamentation about life in general. A patient often had more than one complaint and, in the course of a troubles-telling, these troubles displayed the bewildering tendency to insinuate themselves into her body and mind as well as the various contexts of her everyday life. Even in the consultations where she began with a suitably medical complaint, it soon became only too apparent that her troubles were neither limited to nor encompassed by these entrees. Inevitably, other troubles appeared to have been lurking just around the corner. Sometimes the troubles were additive as in the 'just one more thing...' variety. At other times, they provided the contextual elaboration for the initial trouble, indicating that the first complaint was just the proverbial tip of the iceberg. In short, troubles seem to come in two's and three's in everyday life and that's the way they are talked about in medical consultations as well.

Third, by engaging in troubles-telling, the patients underlined their feelings as part of how the problem was to be defined. These feelings were, without exception, characterized as terrible, bothersome, painful or awful. Not a single instance was available in my data of a patient describing her problem as 'nothing to worry about.' Stomach-pains were retrospectively groaned about and remembered anxiety elicited an exclamation of despair or worried anticipation of future anxieties. No matter what the nature of the complaint or the GP's assessment of its seriousness, the patient's definitions of their troubles treated them as serious enough to warrant, at the very least, a sympathetic reception.

Fourth, the patients defined their problems in such a way that their powerlessness to help themselves was displayed as a temporary lapse from normalcy, thereby entitling them to the respectful treatment conventionally due any competent adult. Even the most serious trouble did not impair the patient's ability to function, 'just a little bit nervous' in the face of life's daily grind. Or, no general 'crisis,' however large it loomed on the patient's horizon, could prevent her from expressing the desire for recognition that she, and she alone, would be the one to pull herself together in the long run. Helplessness was temporary and the patient, in the midst of her suffering, was already anticipating her return to the normal flux of everyday life where she could go about her business like everyone else, competently able to take care of herself without having to enlist outside help.

In conclusion, if the patients defined their problems as part of the activity of complaining rather than as a complaint, as an experience rather than a diagnostic category, as something serious enough to feel bad about, and themselves as persons deserving both sympathy and respect, the next question becomes whether a similar definition of their troubles could be found in the diagnoses produced by the GPs in the course of the consultations.

Diagnosing the trouble

In medical encounters, troubles are not simply talked about. Any trouble presented in this context will be subject to assessment as a potential candidate for a medical intervention. The process of diagnosis begins as soon as the patient starts talking about her troubles (Turner 1972). However receptive the GP may be to the patient engaging in troubles-telling for its own sake, he will primarily be concerned with assembling the information he needs to fit her troubles into his professional schemes of relevance in order to produce a diagnosis. It is this diagnostic process, which marks the beginning of a transformation, but also of a distortion of the patient's definition of her troubles.

In the present inquiry, each of the GPs attempted to assess his patients' troubles in accordance with his specialist knowledge and technical skills. As it turned out, this was not an easy or straightforward task. The majority of the complaints bore little resemblance to a medical complaint and even those that did had to be ferreted out after previous complaints had been rejected as unsuitable for further attention. Symptoms were presented for which nothing could be found upon examination; information requested for problems which did not seem worth worrying about at all (at least, to the GP); or, perhaps worst of all, a complete palette of complaints was displayed, making it difficult to know where to begin or, for that matter, where to stop.

Leaving the interesting question of representativity aside for the moment,<2> it would seem that the GPs in this inquiry were primarily confronted with troubles for which a conventional diagnosis, theoretically derived by means of clinical reasoning, was simply unavailable. In such a situation, three courses of action are possible:

1. The GP could inform the patient that there is no medical diagnosis for her present distress and proceed to terminate the consultation.
2. The GP could lend a sympathetic ear to the patient's troubles without attempting to diagnose them.
3. The GP could look around for another complaint to be diagnosed.

In this inquiry, the GPs exhibited a marked preference for the third option. This was the case, even when it meant having to resort to decidedly common-sense assessments of the patients' problems as a general 'crisis' or a sore-throat being the result of problems 'having gotten stuck in your throat,' or uterine malformations described reassuringly as something which 'rabbits have, too.'

Despite the fact that his *professional* repertoire often left him in the lurch when it came to supplying an appropriately medical diagnosis, the GP could use it indirectly as a resource for keeping the consultation going until something (anything) could be done. Thus, additional troubles could be elicited, should the first one prove unsatisfactory. 'Nervousness' proved a valuable helpmeet for diagnostic dilemma's by virtue of its felicitous tendency to accompany all sorts of other troubles. Moreover, it could easily be transformed into a problem in its own right, a psychological problem of sorts. Of course, the GP could refrain from diagnosis altogether without loss of face, simply by devoting himself to 'clarifying' the patients' troubles - an activity which, in view of the contextual messiness of most troubles, could keep him busy indefinitely. And, last but not least, he always had the option of assessing the troubles as 'nothing to worry about,' a judgement which, by virtue of his expert knowledge, required no further legitimation.

Obviously, these were only a few interactional possibilities. On the basis of my findings, however, two general conclusions may be drawn concerning how women's troubles may be diagnosed in the course of a general practice consultation.

First, there was a discrepancy between the patient's version of her troubles and the diagnosis provided by the physician. The initial versions produced by the patients displayed the troubles as part and parcel of their daily lives. Even while they were emphasizing how unusual their present suffering was and how extreme their need of help, they remained throughout with their feet planted squarely in the 'normal appearances' of everyday life. In the diagnoses that emerged in the course of the consultations, however, their troubles were invariably removed from their context and transformed into a problem, which the GP could label and, subsequently, treat. Whether the complaint was organic (gall stones) or psychological ('nerves') or psycho-social ('a crisis'), the physician separated one aspect from the patient's overall experience of suffering and organized it into a diagnosis. Diagnoses, whatever their ilk, moved in the opposite direction of a story about trouble.

Second, diagnoses did not correspond unproblematically to what the patient 'had' at the outset of the consultation, but had to be carefully constructed, often starting from scratch, in the course of the consultation. Since the patients did not always have complaints that lent themselves to diagnosis, the GPs often had to begin by finding another candidate and building it into a complaint suitable for a diagnosis. In the present inquiry, they seemed to find it less acceptable *not* to diagnose, than to diagnose illness when the patient wasn't actually ill (Scheff 1968).<3> The motto was clearly: when in doubt, diagnose something! Despite the interactional struggles waged by the patients in an attempt to exercise some degree of control over how their problems were being defined, their own rendition of the problem rarely survived the rigors of the diagnostic process. Whether a patient's assessment was tossed out abruptly ('we don't do that here') or, as was more frequently the case, simply left dangling, it remained an unheralded and unauthorized version of the problem at hand.

In conclusion, if the patient's versions of their problems were subject to distortion in the course of their being diagnosed, often having to be constituted on the spot as the consultation proceeded, the next issue becomes how the problems, which, ultimately, came to be diagnosed, were treated. Did this treatment coincide with the patient's request for help or were there distortions to be found here as well?

Treating the trouble

Once a problem has been assessed and diagnosed in the context of a medical consultation, the next step becomes some form of intervention - further examination, referral to a specialist, or therapy. In the present inquiry, the provision of help took a variety of forms. It could be anything from a referral for x-rays and a diet recommendation to a prescription for tranquillizers or a suggestion that the patient 'look at the sunny side' for awhile. In some cases, the patient was advised to talk about her problems, either with her husband or with her therapist. If that failed to help, she could always return to the GP for another chat.

Just as each of the GPs managed to find a diagnosis for their patients' troubles, they also were able to provide some kind of help. Whereas a treatment could consist of a straightforward warning to stay away from fatty foods, in most cases the GPs could not rely on textbook information to help their patients. Problems like 'nervousness' had to be first established as a bonafide complaints before they could be treated. Moreover, the treatment itself (tranquillizers) was rarely welcomed with open arms. On the contrary, the GPs demonstrated considerable reluctance at having to resort to this solution at all. In a similar vein, the solution of 'talking troubles out,' whether at home, with a therapist, or in the GP's office, was not entirely unproblematic. The GPs had little control over what happened on the home front, could do nothing about a patient's dissatisfaction with other professional helpers, and often had little to offer besides the rather lame suggestion that the patient might come back 'any time' for another chat.

On the basis of these findings, two conclusions might be drawn:

First, even when the GPs had little to offer their patients in the way of medical or therapeutic help, they invariably provided it anyway. Just as new problems could be generated if the original ones proved unsatisfactory for diagnosis, the need for some therapeutic activity could be established as well. After a few cursory attempts to indicate that they couldn't 'find anything' or that the problem went beyond their expertise, the GPs in the present inquiry did eventually and without exception provide some form of help. This often required some extra interactional work like eliciting new troubles or directing attention away from the unfortunate fact that no help had been provided for the patient's original problem.

Second, the GP's powerlessness to help the patient led, paradoxically, to an increased effort to establish her need of help. In the present inquiry, the patients not only expressed their own powerlessness in the face of their current distress, but the GPs seemed to be grappling with their own powerlessness as well. This issue of mutual powerlessness proved a tricky one in the context of the medical encounter, leading to a dialectic power struggle around the provision of help. Despite the gravity of their current problems, the patients remained with their feet firmly anchored in the ordinary business of everyday life where suffering does eventually come to an end (if only because the sufferer has other things to do) and where sufferers, bolstered by the sympathy of their peers, are expected to dry their tears and, once again, shoulder responsibility for themselves. Whereas this orientation is well adapted to dealing with troubles in everyday situations, it tended to run counter to the professional requirements of the consultation as a service encounter.

In attempting to cope with their own powerlessness, patients, unwittingly perhaps, seemed to stimulate the GP's powerlessness; i.e. his sense that, in fact, there was very little that he could do. One way of coping with unpleasant feelings of personal powerlessness is shift the blame, which is what the GPs in the present inquiry did. Thus, the physician's discomfort at his inability to help could be transferred from his shoulders to the patient's, thereby obfuscating the limitations of medical solutions for many routine problems of health and well-being.

In conclusion, the results of the present inquiry showed that women's problems as presented in their initial version underwent a transformation in the course of being diagnosed and treated in general practice consultations. Whereas patients were often allowed or took conversational space to talk about their troubles, this did not automatically guarantee them a sympathetic or respectful hearing for their troubles. In the process of selecting problems for diagnosis and treatment, the GPs attended to or rejected complaints depending on whether something could be done about them within a medical frame and not because they were problematic for the patient.

At this point, a few words are in order considering some alternative courses of action - courses of action, which might have been taken in the consultations of the present inquiry, but weren't. This will provide an elaboration of the significance of the findings, paving the way for some final conclusions concerning the process of defining, diagnosing, and treating women's problems in general practice consultations.

Faced with problems for which he could do nothing about, the *first* possibility for the GP would have been to simply *lend a sympathetic ear*. Retrospectively, this course of action seems a highly plausible solution to the situation at hand. In view of the kinds of troubles encountered by the GPs in the consultations, it would have made sense to listen for as long as time allowed without attempting professional intervention. Adopting this stance would not only correspond to the kinds of problems being talked about, but it would put the GP in the position of a conversational peer, having nothing more (but also nothing less) to contribute than any other sympathetic troubles-recipient. In addition to 'just listening' to the patient's troubles, the possibility for some troubles-telling of his own would have been open, for example, about being busy or worried about his overly-full waiting room or even resenting having unreasonable demands made on him.

Of course, for the GP, who does not want to abandon his position as expert, even temporarily, there is another option. The *second* possibility would entail remaining securely within the medical frame and *taking stock of his expertise by considering what he can and cannot provide by way of service for the patient*. As any service provider will attest, not all demands from the client can be met. For example, even the most skillful plumber is bound to stumble across the occasional recalcitrant pipe, which evades her skills. The most well-trained mechanic will, from time to time, encounter a rattle in an automobile muffler, defying repair. For such eventualities, the service provider will have to learn to say no, girding herself for the client's inevitable disappointment.

Both options would have been possible and, in fact, plausible in the consultations investigated in this inquiry. Moreover, each course of action could have been facilitated by means of social skills. In the former case, the GP could have relied on the skills available to him as member, performing the tasks necessary for the reciprocative reception of a troubles-telling. In the latter case, social skills might have helped the GP to express the limits of his expertise, cope with his own feelings of powerlessness, or meet the patient's disappointment head-on.

In the present inquiry, the GPs did not take either of these courses of action. Whereas they did use social and professional skills as a resource, it was for neither of the above-mentioned tasks. On the contrary, they employed skills and resources available to them as professionals as well as ordinary members toward quite a different end. They went to great pains to assert their professional authority, protecting it zealously against even the smallest nudge on the part of the patients, while, at the same time, attempting to extend the boundaries of their service to encompass any trouble, which crossed their path. Thus, the conclusion can be drawn that the interactional activities of the GPs in the present inquiry were directed, first and foremost, at establishing (or re-establishing) their position of authority over their patients' problems. The patients, whether by virtue of their activities in the encounter itself or simply due to the intractability of their problems, had apparently managed to undermine the GP's authority to such an extent that its rehabilitation and the re-establishment of asymmetrical power relations became the order of the day.

This brings me to the second part of the research problem; namely, how the process of defining, diagnosing, and treating women's problems in consultations with male GPs reflects or constructs asymmetrical relations of power.

ASYMMETRICAL POWER RELATIONS IN MEDICAL ENCOUNTERS

The assumption at the outset of the inquiry was that the process of defining, diagnosing, and treating women's problems in the context of general practice would involve asymmetrical power relations. This starting point differs from the one generally taken in studies of medical interaction, which tend to focus on medical talk as a, more or less, cooperative endeavor. Based on the feminist critique of the health care system, it was assumed in the present inquiry, however, that participants would not necessarily entertain similar definitions of the patient's problems and how they should be diagnosed and treated; nor would they inevitably agree about their respective roles within the medical encounter. On the contrary, I expected that their understandings on each of these issues would diverge from time to time, thereby giving rise to the occasional interactional struggle. And, finally, I assumed that the GPs and their patients would not be interacting as peers, but would have differential and, more to the point, unequal resources for influencing the course of the conversation. In short, a medical encounter would involve power asymmetries between the participants, as well as an ongoing struggle for control over the interaction.

To know that power will be implicated in the interaction between a male GP and a female patient does not mean that it is immediately available for further analysis. The analysis of power required following the patients' troubles-talk until it became impeded or obstructed and subjecting those instances of talk to closer examination.

Obstacles occurred at various points within a story about trouble. The patient might find that she couldn't get the floor long enough to produce much more than a story-fragment or that her initiatives were continually being interrupted by medical-type questions. Or, having gotten the floor, she might discover that essential ingredients of 'what happened' were continually getting lost along the way. Obstacles could arise as the patient tried to convey the significance of the troubles. Sometimes the GP was not able to appreciate that the events in question were, in fact, troublesome. Thus, while the patient was allowed to talk about her troubles, she did not necessarily receive sympathy for them. And, finally, obstacles emerged when the GP failed to treat the patient with respect; i.e. as an autonomous and competent adult. Failure to honor the potentially face-threatening aspects involved in talking about troubles could become an interactive obstacle, even obstructing the successful completion of the consultation as a whole.

Obstacles were never simply thrown in the patient's path by the omnipotent GP, but emerged as part of an ongoing struggle where both participants attempted to gain (or regain) control over the process of defining, diagnosing or treating the patient's problems. The negotiations around these obstacles revealed interactional power practices which, in turn, served to construct, maintain, but also undermine structured asymmetries involving domination and subordination.

Based on the analysis of these power practices, some general conclusions can be drawn:

1. *In interaction, power was exercised when actors drew upon skills and resources structurally available to them in order to influence the course of events.*

   Patients employed skills and resources available to them as troubles-tellers. Use was made of the medical frame as well, most notably, in order to call the physician to task.

   Using troubles-telling as a resource, patients could employ various features of storytelling to influence how their problems were defined. Moreover, the activity of troubles-telling itself worked to establish the interactional frame of the consultation as a reciprocal one, where they, as well as the physicians, had something to say about the troubles at hand. With troubles-talk, they could construct a friendship frame within the dominant medical frame, thereby - at least, temporarily - defining the situation, as well as their respective positions within it in more egalitarian terms. They remained oriented to the medical frame as well, drawing upon shared understandings of what constitutes appropriate material for a medical consultation, in order to introduce and develop their troubles-talk. Interestingly, it was often when their troubles-talk was being the most strongly obstructed that they reverted to the medical frame, employing shared schemes of relevance as a sanction to remind the physician of his duties as medical expert. Thus, they could move between the interactional frames as needed in order to exercise power within the encounter.

GPs drew upon skills and resources available to them as medical experts. Use was made of a friendship frame as well, however - most notably, in order to exercise 'friendly' forms of control or sanction the patient's activity without disrupting the encounter. Professional schemes of relevance and technical expertise generally sufficed as resources for enabling the GPs to maintain or re-establish their control over the conversational floor, should the patient's troubles-talk get out of hand. Sanctions could be employed, based on institutionalized authority, which were not available to patients. The same activities performed by a patient did not serve to control the direction of the talk or sanction undesirable behavior in the same way that they did when employed by the GP.

Whereas the GPs employed resources available to them as medical authorities, they did not do so exclusively, as we might have expected. In fact, with one exception, the GPs all made use of a more mundane or friendly interactional frame as well. They also employed resources, which were, theoretically, available to any member to influence the conversation. In fact, it was often exactly at the moment that their control as medical expert began to wane, that they began to resort to these more routine measures. Jokes, platitudes or friendly reassurances could be employed in order to regain some semblance of control over a conversation which threatened to get out of hand.

2. *Power was always linked to agency. Although resources and skills were asymmetrically available to the actors, the weakest party still managed to exercise some control over the interaction by means of activities which were a mix of intentional action and action bounded by unacknowledged conditions and unintended consequences.*

Patients were not the helpless or hapless victims of omnipotent or oppressive GPs, but engaged in numerous subtle and not-so-

subtle attempts to undermine their undisputed authority in the course of the consultation, thereby exercising some control over how their problems were being defined, diagnosed, and treated.
Whereas the physicians had institutionally based resources at their disposal and used them when necessary, their control was not seamless. Patients could and did resist and exercise power on their own. Even though the patients, when all was said and done, did not have the final word concerning their problems, nor did they come out on top in terms of their conversational control over the consultations as a whole, they did not, by any means, go down without a struggle. Throughout the consultations, they worked at displaying themselves as not only competent helpmeets for the GP in the medical business at hand, but also as agents with a mind of their own. They did not acquiesce to being relegated to a position of passive recipient to the GP's ministrations, nor did they accept having their problems treated as trivial, strictly psychological, or so serious that they could no longer be expected to take care of themselves. Whereas their resistance to the physician's attempts to define their problems in such undesirable directions did not necessarily take the form of overt protest, it was, nevertheless, visible in more covert or subtle displays like not laughing at jokes or producing counter-examples to the GP's assessment of their problems ('yes, but...') or simply repeating their own version with the tirelessness of a broken record. When all else failed, patients did not necessarily shy away from more unorthodox methods like manipulation through somewhat histrionic displays of suffering or even veiled threats, designed to coerce him into seeing things her way.

These activities did not necessarily lead to an increased control on the part of the patients over how their problems were being defined, diagnosed or treated; nor were they invariably successful in eliciting a sympathetic or respectful response from the GP. In fact, they often had a boomerang effect, whereby displays of knowledgeability in medical matters produced a reminder that the patient did not know what she was talking about or especially dramatic renditions of troubles got her nothing more than a cold shoulder. Ultimately, and perhaps as we would expect, it was the GP who had the last word. Nevertheless, the patient's activities were an integral feature of the power practices in each of the consultations in this inquiry, part and parcel of the way relations of power were ongoingly constituted, maintained, but also undermined in these encounters.

3. *In each of the consultations, the exercise of power involved a dialectic of control with relations involving dependence and autonomy being produced and reproduced in both directions.*

Power was not a straightforward, top-down affair, but part of a moment-by-moment struggle for control with participants gaining and losing their foothold as the interaction proceeded.

Despite the patients' myriad attempts to influence the course of events or undermine the GPs' undisputed authority over their problems, it was, ultimately, the GP who had the final word - if not in the form of a tidy diagnosis, at least in terms of how the consultation as a whole was to proceed. He did not simply determine the course of events in accordance with his own schemes of relevance, however, but found himself embroiled in a dialectical tug of war with both parties gaining or losing ground as the consultation progressed. Control over one aspect of the interaction often went hand in hand with the loss of control over some other aspect. The authoritarian GP, who made all the decisions concerning the patient's problems and what was to be done about them, might have more control over the organization of the consultation itself, but less grip on what was actually wrong with the patient, endangering his ability to provide adequate health care. On the other hand, the modern GP with his arsenal of social skills and therapeutic techniques could move with greater skill among the patient's troubles. The activities of 'problem clarification' or empathetic listening could keep him respectably busy for the remainder of the consultation, enabling him to exercise control over the patient's problems without having to resort to an over-hasty diagnosis or an irresponsible course of treatment. Whereas this kind of control was more adequate, particularly for the kinds of problems presented by the patients in the present inquiry, it produced a simultaneous loss of control. By giving the patient unlimited access to the conversational floor as well as a sympathetic hearing for her problems, the GP ran the risk of seeming like a friend rather than a professional service provider - something which got him into trouble in terms of the constraints imposed by the institutional encounter like having to wind up the consultation, not taking the patient's side against other colleagues, etc.

Thus, even institutionally-based authority did not ensure immediate and absolute control over what happened in the consultation, but had to be ongoingly re-established. This was not always an easy task, but required constant surveillance and reinforcement against inroads made, unwittingly or not, by patients. It was this struggle to constitute or maintain professional authority, which remained at the forefront of the struggle for power. Despite having a venerable and powerful instance like the institution of medicine to back him up, the practitioner could not sit back and rest upon his laurels as expert, demonstrating that, indeed, even the armour of the most powerful will display the occasional chink.

4. *The power practices were, by and large, not of the repressive variety, but rather of a decidedly 'friendly' sort. It was precisely this friendly quality, which enabled a more effective form of control than would have been possible by more openly authoritarian means.*

In the present inquiry, control was sometimes exercised in the consultations straightforwardly with the GP drawing upon his professional knowledge and skills or applying sanctions legitimated by his authority as medical expert. This was, however, the exception rather than the rule. The power practices tended to be routine, subtle and, above all, ambiguously 'nice.' In general practice consultations, power turned out to have a Janus-face.

The consultations often bore more resemblance to a chat between friends than an institutional encounter with the professional bent on diagnosing and treating the patient's problems with a minimum of muss and fuss. This was, perhaps, understandable in the light of the unfortunate fact that many of the problems did not lend themselves to conventional medical diagnostic procedure. Both routine friendliness, as well as the more professional social skills of empathetic listening, could be employed to cover up the regrettable fact that the GP did not have much to offer by way of medical help and, when all was said and done, had, in fact, done little more than lend a (more or less) sympathetic ear.

Ultimately, this elusive, double-edged quality of 'friendly' power enabled a control which was more intensive and effective than would have been possible by means of more authoritarian means. Medical jurisdiction could be extended over areas of the patient's life, bringing them gently, but decisively under the physician's control. Sanctioning processes of the 'iron fist in the velvet glove' variety were notoriously effective due to their quality of continually seeming like something other than constraint or repression. The gruff doctor, who interrupts the patient midstream with some advice or who listens with half an ear while writing out a prescription may not be a pleasant conversational partner, but the patient will, at least, have a clear idea of who is in charge and what their respective positions are. This was not always possible with the empathetic GP, who appeared to be on her side, while, at the same time, behaving in accordance with institutional constraints no longer visible for the patient.

Thus, whereas social or therapeutic skills might seem, at first glance, to be a step in the direction of a more egalitarian encounter between GPs and patients, a look at these skills in action indicated that this was not necessarily their effect.

5. *Interactive power practices produced, sustained, but also ongoingly transformed social structures involving relations of domination and subordination.*

Social structure, including gender inequality or asymmetrical relations based on institutional or professional authority, did not simply determine the regularized or stable quality of interaction, but was dynamically reproduced in the structuring activities of the participants who, sometimes intentionally, but more often in routine, habitual ways, produced and sustained these social structures. Power practices ongoingly undermined and transformed these arrangements as well.

In each of the consultations, asymmetrical relations of power were established where the GP could exercise control over how the patient's problems were defined, diagnosed, and treated. In this way, both professional dominance over lay forms of healing, as well as male dominance over how women see themselves and their problems, were produced and reproduced at the micro-level of interaction. This control worked to undermine the patients' understandings of their troubles as well as their own authority to do something about it. Not only were women's problems shorn of their contextuality and forced into professional schemes of relevance, but the GP seemed unable, in many cases, to understand what made the problems problematic in the first place. Troubles like being overburdened with responsibilities as family caretakers, or being old and worried about not having grandchildren, or feeling generally powerless and angry about the lack of control over one's life seemed to escape the boundaries of what the GPs could sympathize with.

Not being able to sympathize with a particular difficulty did not prevent them from attempting to treat it, however. Sterilization decisions, marital problems, family difficulties were all grist for the medical mill. Psychologizing women's 'problems of living' was not just a convenient response to the interactional problem of having to find 'something' to talk about, but reflected structural tendencies within medicine as an institution to extend its control over ever more areas of everyday life. Expansionist or cooptative forms of social control have not been limited to modern medicine, but have historically been directed at women as well - a favorite means for channelling women's dissatisfaction with their lives into safer paths and teaching them to look to men for a solution to their problems (Ehrenreich and Ehrenreich 1978; Ehrenreich and English 1979).

Institutionalized relations of power were not simply produced and reproduced within the consultations, however. They were also ongoingly undermined and transformed as well. The 'chinks in the armour of the powerful' not only demonstrated that asymmetrical power relations are already being undermined in routine medical encounters, but, by pointing out the contradictions in the power relations themselves, they provide an indication of the direction social change might take. By separating themselves from patients as laypersons, professionals not only establish their own authority as experts, but they force their patients, robbed of their skills and self-confidence in matters concerning health care, to 'learn the ropes' of the medical encounter, taking on the medical jargon and routines employed by the professional. Social skills and empathy can extend the boundaries of medical control by bringing problems under professional jurisdiction, which were, formerly, not necessarily 'something for the doctor.' While increasing medical control, they serve, at the same time, to make the boundaries fuzzy between the encounter as an institutional one and the encounter as a friendly chat between equals. This conflict of interactional frames - medical and friendship - not only accounts for interactional power struggles, but it signals contradictions at a structural level as well- problems inherent in the expansion of medical control and the increasing intractability of patient's problems for medical solutions. This points in a new direction where practitioners might, indeed, become less professional and more likely to just lend a sympathetic ear or, on the other hand, more professional, but at the same time, more selective about what they can and cannot treat.

In conclusion, an attempt has been made to uncover some of the 'how's' of power at the level of interaction between male physicians and female patients in medical encounters. Having provided a résumé of my research findings, the question now becomes whether these results of have anything to offer of a more general nature. In the next chapter, I shall be directing my attention to the issue of power in gender relations at a structural level and how the findings of the present inquiry can contribute to the construction of a grounded theory of power and gender in medical encounters.

NOTES

1. Taking an analytic vantage point does not imply correspondence with how the patient or the physician 'see' the encounter. As has been previously discussed, I have no access to this. By vantage point, I am referring to my own position as analyst from which I will view the ongoing interaction; i.e. the focus of the analysis.

2. As discussed in Chapter 1, this may well be the usual state of affairs in general practice. Estimates ranging from 50% to as high as 80% have been made of complaints presented in general practice for which the physician can do little more than lend a sympathetic ear. See, also, van Es (1980). The issue of the generalizability of the present findings will be dealt with in more detail in the next chapter.

3. Scheff (1968) provides an analysis of medical decision-making, which is analogous to decision rules in the legal system. Type 1 errors involve failure to diagnose illness, similar to judging a person innocent who is, in fact, guilty. Type 2 errors concern diagnosing someone as ill who is healthy - an error to be compared with locking up innocent parties. In law, type 2 errors are regarded as something to be avoided at all costs, whereas medicine is more concerned in avoiding type 1 errors.

14

THEORIZING POWER AND GENDER: A CONTRIBUTION TO THE DEBATE

INTRODUCTION

My analysis has demonstrated *some* of the ways that power can work in medical interaction; it does not, by any means, encompass *all* of the ways that power works in medical encounters in general, or general practice consultations in particular. Although 'friendly' power practices received considerable attention in the present inquiry, there is no way of knowing whether they are representative for what can be expected in Dutch general practice as a whole. Another sample of GPs might well have yielded a different picture of power.<1>

This raises several questions - questions which regularly emerge in conjunction with an inquiry like the present one. Since my analysis of power was limited to one specific interactive constellation: a male GP and a female patient, the reader may find herself wondering whether power would be exercised in similar (or, indeed, different) ways in other constellations.

For example, if the GPs had been female, would we have found more egalitarian forms of interaction in the consultations? Or would the female GP have been just as oppressive as her male colleague, demonstrating that the professional relationship is the primary source of asymmetrical power relations rather than the gender of the practitioner?

In a similar vein, the question might be posed concerning the effect of the patient's gender on the interaction. Did the GPs exercise power as they did *because* the patient was a woman? Various authors have suggested that friendly power plays may, indeed, be typical of the interaction between (middle-class) women and male physicians (Ehrenreich and Ehrenreich 1978; Ehrenreich and English 1979), but does this mean that they are *specific* for encounters between women and physicians? Might not similar practices be found in the interaction between physicians and other patient groups - for example, the elderly, ethnic minorities, children, or, for that matter, men? And, what would happen if the roles were reversed entirely; i.e. a male patient interacting with a female GP? Would gender as a structure of domination prevail, giving the patient an edge over the female physician, even though she is the medical expert? West's (1984) findings, for example, indicate that physicians are more powerful than patients, 'except when the doctor is a lady.'

More generally, we could ask whether similar power practices might be found in all instances of interaction between the sexes or in institutional interaction in other settings.<2>

This is but a small sample of some of the questions, which might be raised on the basis of the present inquiry. My standard answer has been that, whereas they are all, in and of themselves, interesting questions and worth further investigation, it was not the aim of this particular investigation to do so. Qualitative research does not provide answers to questions concerning generalizability in the usual sense of the word. As an instance of this research orientation, the present inquiry is similarly unable to answer questions about how often the power practices delineated here will occur in which settings and between which (groups of) actors. What it can do, however, is to provide a way of looking at power - a kind of microscope. By looking through this microscope (or one like it), interactive power practices can be investigated in other similar, or even not so similar, situations. It is my hope that the present inquiry can help other scholars of power <3> to develop an eye for the phenomenon, enabling them to locate, identify, and delineate power practices in other contexts. They will know, to begin with, that the analysis of power in interaction requires, in fact, a microscope. Having received some instruction in how to use a microscope, other scholars can begin to identify power practices available in their own data. Thus, it is in this methodological sense; i.e. in developing a way to observe and investigate power, that the general relevance of the present inquiry needs to be sought, rather than in the empirical generalizability of its results.

In addition to this general methodological consideration, the research topic itself warranted setting the questions posed above aside in favor of an exploratory, in-depth investigation into *how* power works in the specific context of a medical consultation between a male physician and a female patient. The feminist critique of the health care system had already provided ample evidence *that* power is implicated in any medical encounter between a male physician and a female patient. With some exception, however, little empirical research had been done concerning how it works in the actual interaction between the sexes in medical encounters. Within medical sociology, studies of interaction have tended to ignore power and gender as a topic altogether, whereas sociological theories on power have, for the most part, remained at the macro-level of institutional or structural analysis and done little to ground their concepts in specific instances of social interaction. By showing how power works at the level of interaction in a limited number of cases, I was able to discover and delineate recurrent patterns, which can, ultimately, serve as a starting point for other research endeavors.

These arguments having been provided, I might, at this point, regard the issue of generalizability as closed <4> and conclude that I had accomplished the task I set out to perform with all that is left to do being to wind up this book with a few recommendations for medical practice or suggestions for further research. I am reluctant to do this, however.

In the course of thinking about the questions raised above, <5> it has become increasingly clear to me that they were, perhaps, not *empirical* questions after all - or, at least, not strictly empirical questions. In fact, I began to suspect that the 'real' issue behind the questions and the problem of the general relevance of my research findings might be a different one altogether. It is my contention that the question-behind-the-question of empirical generalizability is a *theoretical* one. It concerns what power is and why it is there to begin with. Apparently, showing how power practices work to construct asymmetrical relations involving domination and subordination between male physicians and female patients in medical encounters is not enough. It still leaves the question open: yes, but what exactly is this that you have demonstrated? What kind of power is it and, more to the point, where is it coming from?

If these are theoretical questions, then they cannot be answered satisfactorily on a purely empirical level. Providing an answer to them necessitates entering the domain of theoretical notions on power and gender and reconsidering the findings in terms of these assumptions. In a sense, this entails a somewhat reluctant, but nevertheless necessary return to the drawing board. I will need to take another look at my initial theoretical assumptions on power and gender - assumptions, which informed the research problem as it was formulated at the outset of the inquiry. As the reader will recall, my theoretical starting point was that power would be involved in any encounter between a male physician and a female patient. Moreover, I assumed that the exercise of power in medical encounters could not be separated from gender inequality as structural feature of all social life. These assumptions, based on the feminist critique of the health care system, provided both the rationale for the research problem, as well as a kind of sensitizing theory about power and gender for the investigation at hand. Whereas these assumptions stood me in good stead throughout the analysis, I have since come to entertain some doubts about them. Considered in the light of my results, I began to suspect that my theoretical assumptions on power and gender were, indeed, in need of some *theoretical* revision. This revision will not be a painless one for me, as the reader will see. Moreover, in view of current feminist theorizing on power and gender, it may not be an entirely popular one either. Nevertheless, the present inquiry points me unmistakably in this direction. Since scientific inquiry should ideally result in reviewing and, if necessary, revising notions which were available at the outset (otherwise, why should we bother?), this is, however, only as it should be.

Observations are never value-free. However painstaking and precise, they are always linked to theories or 'theory-laden' (Lukes 1979a). We do not simply 'see' what the empirical world has to offer, but observe it through our theoretical presuppositions. These theoretical presuppositions, whether of a 'common-sense' or scientific variety, help us make sense of and organize what we see. On the other hand, what we see, our observations, enable us to elaborate our original theories, altering them so that they can account for our empirical findings. Thus, whereas I - like any other investigator - secretly hoped that my empirical analysis of the 'how's' of power would provide the definitive bed-rock for a (grounded) theory of power and gender in medical encounters, it is not particularly surprising that it has not done so.

As retrospective consolation, it might be added that this would have been the case for any investigation. The scientist engaged in investigating how cancer cells multiply would also find herself faced with the problem of linking her observations to theoretical explanations of them. She can demonstrate *how* cancer spreads, how it can infiltrate healthy tissue and so on - in short, all of the 'how's' of cancer. However invaluable such information is for understanding cancer, it does not tell us where cancer comes from. By looking through the microscope, we are enabled to identify what we see as cancer and describe how it works. We do not know what is causing it. In other words, the 'why's' of cancer remain unanswered.

This can explain why an analysis of how power works, no matter how comprehensive or systematic, will still give rise to questions about the nature of power and, more specifically, the relationship between power and gender in medical encounters. The questions posed above refer back to my original assumptions: why it made sense to look for power in medical encounters between male GPs and female patients and, more generally, what accounted for power being there in the first place. Since the questions are primarily theoretical and, in fact, hark back to assumptions about power and gender, which had to be made before the investigation could begin, it makes sense to consider them in that light. This will involve re-considering the power practices observed in the general practice consultations between female patients and male GPs at a more general level. To this end, some structural explanations for asymmetrical relations in medical encounters will be discussed and evaluated in terms of their being able to make sense of the findings of the present inquiry. In particular, attention will be paid to the *kind* of structural explanation which is required in order to come to terms with power in gender relations in medical encounters.

SOME STRUCTURAL ACCOUNTS FOR INTERACTIONAL ASYMMETRIES IN MEDICAL ENCOUNTERS

Social actors do not simply produce social structures involving domination and subordination by means of power practices, but these structures exist as a virtual order by means of which these practices come to be constituted and reconstituted as asymmetrical power relations. In the present inquiry, considerable attention has been paid to how participants drew upon resources and skills available to them in order to influence the course of the consultation. In particular, an effort was made to resurrect the patient from her traditional position of neglect on the sidelines of the interaction and uncover the myriad, subtle, but often heroic ways she exercised power in the course of the encounter, attempting to establish her authority and interactional rights regarding her own health and well-being. Whereas both patients and physicians exercised power in the consultations, they did not employ resources and skills arbitrarily; nor did they embark on the interaction as conversational equals. Resources are structured by orders of domination, making them differentially and unequally available to participants. Since social practices, including power practices, and social structure, which routinely entails domination, presuppose one another, this 'fact' will need to be part of the analysis.

Whereas the problem of integrating structural social analysis with the micro-analysis of social interaction is a necessary part of any analysis of social life, it is, at the same time, a rather ticklish endeavor. It has been subject to considerable discussion within sociology in general and various strategies have been proposed for combining levels of analysis within social inquiry (Knorr-Cetina and Cicourel 1982). In recent years, the 'macro/micro problem' has also been specifically addressed within the field of medical interaction. For example, Fisher and Todd (1986:xiv) have stressed the need to integrate organizational, structural, and cultural factors in micro-studies of institutional interaction, proposing 'a contextual hermeneutic mode of analysis in which causality resides in a dynamic, reflexive process.'

In the present inquiry, which was also set up as a micro-analysis of power in medical talk, I began at the level of social practices rather than social structure. Whereas a certain degree of 'methodological bracketing' (Giddens 1984) is unavoidable in any inquiry, it remains an analytical rather than substantive distinction. In everyday social settings as well as institutional encounters, social practices and social structure are irrevocably intertwined, defying separation. Despite the fact that the social analyst will rarely have the scope within a single inquiry to deal with all levels of analysis in the same detail or depth, she will, nevertheless, do well to address the relationship between the various levels, regardless of her own point of entrance into the phenomenon being studied.

It is to this issue that I shall be directing my attention in the remainder of this inquiry. To this end, a look will be taken at several structural accounts, which have been put forth within medical sociology, to explain asymmetries in the interaction between patients and doctors. These accounts attempt to interpret interactional power imbalances in terms of the structural availability or non-availability of skills and resources in medical encounters.<6>

The first, and by far most common explanation, locates the asymmetry of power in *professional* structures of domination. In other words, any inequity within the interaction is attributed to the fact that the doctor is an expert and the patient a layperson.

More recently, this explanation has been expanded to include other possible sources for power asymmetries; most notably, *gendered* structures of domination. Power imbalances between physicians and patients are due to the fact that practitioners are, nearly without exception, male and patients, on the other hand, disproportionately female.

And, last but not least, some attempt has been made to bring these two structures of domination together, indicating that inequities inherent in medical encounters between male practitioners and female patients are *conflated*.

Whereas these are, by no means, the only structural accounts of power in medical encounters, nor do they provide an exhaustive account of all the possible structures of domination in operation within medical encounters,<7> they do give an illustration of how the power asymmetries, which have been delineated in the present inquiry in terms of their ongoing production, maintenance, and transformation, might be accounted for at a structural level. In view of the limited scope of the present inquiry, these accounts will not be dealt with in any comprehensive way; nor shall I be providing a full-fledged structural analysis of domination at the level of medicine as an institution.<8> On the contrary, my aim will be two-fold:

First, a brief description will be provided of three structural accounts of power in medical encounters between male professionals and female patients, demonstrating the kinds of explanations for my own research findings, which are made possible in each case.

Second, some conclusions will be drawn on the basis of these accounts concerning how power and gender need to be conceptualized at a theoretical level. A look will be taken at some of the missing pieces in the accounts. In particular, some of the problems inherent in the (implicit) assumption *that* structures of domination can be isolated *at all* will be discussed. I shall demonstrate how this attempt to separate structures of domination tends to produce a faulty account of power imbalances, both in medical and other encounters.

This critique will enable me to move on to the final section of this chapter, where a somewhat different approach to the business of theorizing power in gender relations will be proposed.

Professional Power

Within medical sociology, there have been basically two approaches to professional power. In the first, it has been treated as *unproblematic*; in the second, it has, albeit for widely different reasons, been regarded as *problematic*.

In the first tradition, power is regarded as an integral feature of any medical encounter, part and parcel of the division of roles between professionals and laypersons (Parsons 1951). Physicians automatically occupy a position of dominance *vis-à-vis* the patient. By virtue of their specialist knowledge and training, it is their task to discover what is wrong with the patient. This necessitates exercising control over the interaction. Since patients do not have access to these skills, they are, by definition, dependent on the physician for help and will, presumably of their own volition, take their rightful position as second fiddle within any medical encounter.

This social roles perspective on power in medical settings has had, and continues to have, an enormous impact on studies of medical interaction. It is, by far, the most usual way of accounting for any asymmetries within the encounter. Asymmetries in the interaction are treated, implicitly or not, as the inevitable concomitant to the provision of health care; i.e. the *sine qua non* of the therapeutic encounter. Thus, whereas power is clearly involved in the interaction, it is not regarded as problematic and asymmetries in the doctor-patient relationship tend to be treated as something of little concern either to the patient, the practitioner, or, for that matter, to the analyst. At best, it can require some extra 'negotiation,' most notably in helping the patient 'learn the ropes' or otherwise cope with her inadequacies as interactional partner (Hughes 1982).

It is hardly surprising that an explanation of power based on role theory is not particularly helpful for coming to terms with the kinds of power practices found in my own data. Whereas the consultations were collaborative endeavors, they bore witness to considerable struggle as well. It would be difficult to explain the interaction exclusively in terms of the complementary roles of doctors and patients, sharing similar normative commitments to the situation at hand. Such an explanation would leave those instances in my consultations unaccounted for where the participants seemed to entertain different interpretations of the nature of the encounter or their respective role obligations or gave the appearance of having opposing interests concerning the desired outcome of the interaction. Indeed, my initial assumptions concerning power and gender precluded anything as consensual as a social roles approach because it presupposed that power asymmetries would have decidedly negative consequences for the person with the smaller end of the stick. No structural account of power, which treats the power imbalance between doctors and patients as 'the way things should be,' would be adequate for explaining the potentially problematic dimensions of the asymmetrical power relations.

Fortunately, medical sociology has not followed unanimously in the footsteps of Parsons; nor has everyone agreed that the power asymmetry between doctors and patients is a desirable or unproblematic feature of medical encounters. For scholars like Freidson (1970; 1986), Zola (1972a,b), Illich (1976), Navarro (1976), and Waitzkin (1983) - just to name a few - asymmetries between medical professionals and patients are a matter of concern. This is true, whether professional power is dealt with it at the level of medicine as a profession, in terms of the ideological hegemony of medical over lay definitions of sickness and health, or in the broader context of the relationship between medicine and the social order in which it is situated.

Beginning with Freidson's (1970) opus on professions, the professionalization of medicine has been cited as a primary source of power imbalances within medical encounters. Medical expertise has become formal knowledge, situating it within the 'higher knowledge' of modern culture, justified and legitimized as distinct from both routine or everyday knowledge as well as non-formal, technical knowledge. By developing a body of formal knowledge, specialist training programs, and licensing for practitioners, medicine could corner the market for the provision of health care and develop a monopoly over all matters connected with illness and health.<9>

Medical expertise as marketable good can be regarded as the source of professional power, enabling structures of domination to be produced and reproduced in the interaction between doctors and patients. Professional power is employed to control communication in medical encounters, absorbing the patient's understandings of her problems into the incomprehensible framework of scientific medicine. Various authors have criticized the distorted and undemocratic quality of the resulting interaction, as well as pointing out that it is detrimental to the provision of adequate health care (Cicourel 1983; Fisher and Todd 1983; West 1984).

In addition to controlling communication on medical matters, professional power is increasingly used to address problems or justify decisions or actions, which are not technical or scientific at all and, in fact, have nothing to do with the professional competence of specialists. This is the 'medical imperialist' thesis; i.e. the tendency of medicine to 'colonize the lifeworld,' extending its jurisdiction to areas of everyday life, which formerly had nothing to do with medicine at all (Illich 1976; Conrad and Schneider 1980; Mishler 1984). Having cornered the medical market, the profession of medicine has a monopoly over all matters concerning *both* sickness and health, placing it in a position of hegemony with regard to other value spheres (Waitzkin 1983). Endowed with all the symbolic trappings of scientific rationality, medical assessments not only seem to be most 'correct,' but medical contexts have increasingly become the only legitimate place to deal with any problem of a person's well-being. Since professional power is often of a 'soft' or cooptative variety (Ehrenreich and Ehrenreich 1978), this paves the way for ideological, manipulatory forms of social control, which are particularly difficult for patients to identify, let alone combat (Fisher and Todd 1986; Davis 1988).

Like the social role perspective on professional power, this second perspective also views the professional relationship as a source of power asymmetries between patients and physicians. Whereas both perspectives explain power imbalances in medical encounters in terms of the difference between patients as laypersons and physicians as professionals, the second perspective, unlike the first, regards the imbalance as a problematic one. Moreover, by pinpointing potential areas of conflict, it opens up the possibility of investigating the ways that these asymmetrical power relations are contested.

As structural account, this perspective on professional power *can* help make sense of the kinds of power practices, which could be observed in my data. For example, the GPs in the present inquiry undeniably employed their medical expertise in order to gain control over the interaction. Myriad instances were available of the patient's 'lifeworld' being 'absorbed' into medical frameworks. Even when the patients' problems did not lend themselves to medical or bio-technical solutions, they were still invariably treated. In the course of trying to exercise power over how their problems were being defined, diagnosed, and treated, it was often this professional power, whatever its form, which allowed no dispute or sharing of authority. It was also this professional power, which was ongoingly being contested by the patients. In short, the power practices observed in my data could be explained in terms of how they produced, reproduced, but also undermined professional structures of domination.

Whereas a structural explanation of power imbalances in medical encounters based on asymmetries between the GP as professional and the patient as layperson can account for many of the power practices analyzed in the present inquiry, it also leaves some questions unanswered. Explaining structured forms of domination in terms of professional power does not address the issue of gender. It does not have anything to say about the sex of the participants-neither of the professional, nor of the patient. Medical professionals are usually men, just as laypersons in medical encounters tend to be women. In view of the negative consequences sexual divisions have had historically (Hartman and Banner 1974; Oakley and Mitchell 1976), and continue to have for women in nearly all walks of life (Mitchell 1971; Barker and Allen 1976) and within most cultures (Rosaldo and Lamphere 1974), it seems unwarrantedly optimistic to assume that sexual divisions within medical encounters will have no effect on the power relation between the participants. Moreover, the power relationship between the physician as professional and the patient as layperson cannot explain the kinds of problems women have encountered as patients in the health care system. Judging by the feminist critique of the health care system, there is, indeed, reason to believe that male and female patients do not have the same experiences with health care and that women as patients are less than satisfied with the way they fare at the hands of male medical professionals. In order to make sense of this, structural explanations are required, which take gender, both in terms of sexual divisions within the actual medical encounter, as well as gender inequality as 'social fact,' into account.

Gendered Power

Within feminist scholarship, gender is seen as fundamental to the way *all* social life is organized. It accounts for the way individual men and women behave, determines how meaning and value are assigned, and organizes human social relations in general (Harding, 1986). Just as gender has come to occupy a privileged position as central theoretical concept in feminist scholarship as a whole, it has also informed most of the research in the field of women and health care. Implicitly or explicitly, gender differences are regarded as playing a major role in the asymmetrical power relationship between male physicians and female patients and need to be part of any explanation of power in medical encounters.

A good example of a structural account, which locates the root of power imbalances in medical encounters in gender relations, is the classic study by Ehrenreich and English (1979). Beginning with a description of the historical usurpation of women's traditional skills and authority in the field of health and reproduction, the authors show how men set themselves up as experts in these areas. Not only did women's bodies and reproductive functions become subject to constant medical surveillance and intervention, but men began to busy themselves with the female mind as well. They delved into her 'psychology,' discussed how she should raise her children, clean her house, or keep her husband happy. Ultimately, this authority was extended to nearly every aspect of a woman's life.

The crux of '150 years of the experts' advice' was that women's experiences came to be defined by men rather than by women themselves and that these definitions were often derogatory, moralistic, and inaccurate, having, in some cases, dangerous consequences for women's health and well-being.<10> Even when medical interventions were not actually harmful, women were consistently treated as having nothing of consequence to say about their own lives and experiences. Robbed of their authority and lacking confidence in their own versions, women increasingly turned to men to tell them what was wrong and what better male than a medical expert? The physician became the obvious candidate to tell her how to view her difficulties - as a 'female problem,' a matter of 'nerves,' or simply as an inconsequentiality. This was not just a matter of defining women's problems in a particular direction, but concerned the construction of a specific kind of relationship between women and male professionals in particular, and men in general. It was a relationship of male domination and female subordination.

Although feminist explanations for the various inequities found in medical encounters vary, they do, by and large, share a common focus on gender as central to any account of power asymmetries between the sexes.<11> It is this structure of domination, which primarily contributes to the production and reproduction of asymmetrical power relations between male doctors and female patients. This is the case when sex differences between male and female patients are explained in terms of role behavior and expectancies (Nathanson 1977). It informs accounts of the stereotypic attitudes held by individual physicians (Fidell 1980) or their tendency to find and diagnose psychogenic disorder (Lenane and Lenane 1973), leading to a somewhat over-enthusiastic prescribing of tranquillizers for women (Cooperstock 1978). Gender is used to explain sexual divisions within the medical professions themselves, accounting for men disproportionately occupying the higher echelons of medicine with women in the lowlier ranks of the care professions as nurses and home help (Leeson and Gray 1978; Finch and Grove 1983). And, last but not least, gender is drawn upon as explanatory concept in the critique of medical ideology with its plentiful and notoriously misogynist lore on the 'nature' of woman (Ehrenreich and English 1979; Lowe and Hubbard 1983; Bleier 1986).

The assumption in each of these cases is that power in medical encounters is irrevocably gendered, historically and at present, from the face-to-face interaction between an individual male physician and his female patient to the organization of the health care system to the structural arrangements in which the institution of medicine is embedded. Gendered power accounts for sex-specific forms of control in medical encounters, as well as particular problems for women as patients in the health care system.

The question now becomes whether a structural account of structures of domination in medical encounters as *gendered* as proposed in much of the feminist critique of the health care system can provide a better explanation for the results of the present inquiry than would be possible with a structural account based solely on professional power. The answer to this question is yes, but, at the same time, it is also no. Beginning with the affirmative side of the matter, it is clear that the research problem of the present inquiry could only be formulated by virtue of the assumption that power would be implicated in any encounter between a male GP and a female patient. Taking the feminist critique of the medical system as a starting point automatically entailed assuming that gender would account, at least in part, for any power imbalances observed in the general practice consultations I investigated. This assumption was implicit in that I did not set out to prove that gender accounted for the power practices delineated in my data. Nevertheless, by developing a research problem, which could only be formulated by virtue of the assumption that gendered structures of domination will be produced and reproduced within any consultation between a male GP and a female patient, gender became an integral feature of the inquiry as a whole.

Not only did the research problem itself implicitly presuppose that power in medical interaction was gendered, but the results of the inquiry confirm the assumption as well. The power practices observed in my data resembled the same kinds of phenomena cited in the feminist critique of the health care system. For example, I found myriad instances of women's complaints being transformed into psychological problems or not being taken seriously. On the basis of my data, a strong case could be made for the claim that physicians usurp or, at least, severely undermine women's authority and control over their bodies and lives, providing instruction on anything from a sore-throat to a family crisis. Many of the patient's interactive efforts went into combatting these processes. By means of various subtle, covert, and, at times, not entirely kosher skills and resources available to them, they exercised their power to define their experiences and problems as matters warranting concern, as well as themselves as persons meriting a respectful and sympathetic treatment. By showing how processes like psychologizing women's problems, undermining their authority, or channelling their difficulties in unproblematic paths, actually work in medical encounters, as well as how these processes are continually countered by female patients, I was able to provide the feminist critique with some empirical grounding. At the same time, an implicit confirmation for the claim that health care is, indeed, gendered was given.

The question of whether a structural account of power in medical interaction based on gender is sufficient to explain asymmetries in medical encounters between male GPs and female patients cannot, however, be answered unproblematically in the affirmative. Whereas the assumption that power is gendered made, to some extent, the present inquiry possible, it is, at the same time, precisely this assumption, which makes it impossible to distinguish power resulting from gender from power resulting from other structures of domination. On the basis of the present inquiry, there is no way of sorting out what, for example, the GP was doing *because* he was a man (and the patient a woman) and what he was doing *because* he was a professional (and the patient a layperson). By setting aside empirical differences in power practices in other gendered combinations (female GPs/male patients; female GPs/female patients; male GPs/male patients), it became possible to tackle a research problem concerned with the 'how's' of power in one specific gender constellation. Whereas the results provide an adequate answer to the question of *how* power can work in interaction between male GPs and female patients, it also unavoidably begs the theoretical question of *what* is causing power to work the way that it works.

In conclusion, it would seem that structural explanations of power based on professionalization or on gender can be offered to account for the power asymmetries being produced, reproduced, but also undermined in encounters between male physicians and female patients. Power can be based on the professional relationship between GPs as medical experts and patients as laypersons. It can also be based on the gender relationship between GPs as men and patients as women. The question is how these structures of domination can be sorted out or, failing that, set into relationship in order to provide an adequate account of power in the specific context of a medical consultation. This is not simply an empirical problem, but an issue which, as I have suggested earlier (Chapter 4), needs to be addressed by *any* theory claiming to explain power and gender in the various contexts of social life.

Whereas a structural account, which treats power exclusively as *either* a matter of professional structures of domination *or* as a matter of gendered structures of domination, can make sense of many of the research findings in the present inquiry, it cannot explain how these structures of domination intersect in the concrete context of a medical consultation between a male physician and a female patient. Nor can it provide guidelines for how structures of domination, in general, whether they are based on gender, class, ethnicity, age, professionalization or whatever, might be set into relation with one another theoretically. Before we throw up our hands in despair, however, I will be taking a look at one recent structural account of domination in the context of medical encounters between male professionals and female patients, which at least attempts to weave gendered and professional structures of domination together. It is to this work that I shall now turn in order to see how it can help me make sense of my research findings and, more important, what it can offer of a more general nature concerning the way gender and power should be theorized.

Conflated Power: Professionalization and Gender

Having dealt with some of the difficulties involved in explaining asymmetries of power as result of isolated structures of domination based on professional power or gender, I shall now take a look at what happens when these structures are not separated, but treated, from the outset, as *conflated*. An example of this kind of approach to power has been proposed by Hearn (1982; 1987), who attempts to show how gender and professional structures of domination are interrelated in the context of medicine. Instead of trying to determine which structure is primarily responsible for power imbalances in medical encounters or, for that matter, figuring out what their relative contributions to the asymmetries might be, Hearn argues that the notion that professionalization and gender <12> *can* be separated is itself a faulty one and that any endeavor in this direction is doomed to failure. Since both structures of domination tend to be fused at all levels of medicine, it becomes impossible to consider one without taking the other into account as well.<13> Since Hearn explicitly tackles the problem of relating the various forms of domination- professional power and gendered power - at a structural level, it behooves us to take a closer look at how he goes about it, setting out some of his main arguments.<14>

Along with Ehrenreich and English (1979), Hearn maintains that the emergence of medicine as a profession is predicated on a male-dominated take-over of women's control over the sphere of reproduction, including areas associated with healing, sexuality, birth, spirituality, and death. The medical profession has historically and continues to be the cite of a 'battle between the sexes' - a fact, which has rarely received more than passing comment within theories on professionalization. If there has been any differentiation between groups able to establish and maintain power over other professional groups or groups of clients, then it has been in terms of class. Professionals are generally themselves members of the ruling class, giving them a vested interest in maintaining these interests (Ehrenreich 1978; Waitzkin 1983). Whereas professionalization concerns, as Becker (1961) puts it, how 'bright boys' become 'medical men,' the explicit link with gender is not made.

Even today, when the medical profession has officially opened its doors to women, (Western) medicine remains an indisputedly male preserve. The majority of physicians are still male, with women in the lower echelons of the health care sector. The terms 'profession' and 'semi-profession' only serve to obfuscate the fact that gender asymmetries are still integral to professional power (Hearn 1987).

Sex-segregation within medical practice is paralleled by the emergence of powerful medical ideologies, which reflect and help to construct women's subordinate social position. She is defined as dependent and, by nature, sick - the 'archetypic patient' (Hearn 1982). It would seem that if women were no longer qualified as healers, they had become all the more qualified to be patients, particularly for the male professional. Whereas medical ideologies have changed -most notably by locating female illness in her psyche rather than in her reproductive organs or general physical weakness, <15> women are still, numerically speaking, more likely to be patients. Thus, the most 'typical' combination in a professional encounter is a male professional with a female patient or client.

The conflation of professional and gendered power can also be found in the content of health care itself and, more specifically, in the central role medicine plays in the control of women's emotionality (Hearn 1986:137-146). The ideology of medical science rests on a duality of rationality and emotionality, which not only reflects and constructs professional hierarchies between physicians and patients, but the gender symbolism in how we think masculinity and femininity. As central values in medical science, technocratic rationality and instrumentality inform the individual practitioner's need to control his own emotions, while maintaining a facade of 'gentlemanly reasonableness' or 'affective neutrality.' In fact, this is more or less synonymous with 'being professional.' Patients, but, also, women, do not have to control their feelings and may even be expected to express them. With one foot in the 'life world,' patients are tied to the emotions, which inevitably accompany illness or distress, just as women, whether by nature, socialization, or a sexual division of labor which makes them responsible for mothering (Chodorow 1978), are more likely than men to give rein to their feelings.

This produces a situation where the professional may

> even talk *about* emotions, become 'expert' on emotions, yet not be seen as emotional. A degree of controlled friendliness may conjoin with this denial of emotion. (Hearn 1987:140)

Controlling one's own emotions, as well as setting the limits under which other people (patients) can express their emotionality, reinforces an implicit construction of a professional hierarchy whereby the experiences of others are objectified in the name of the provision of service. This public presentation of the ideal of non-emotionality, while, at the same time, dealing with the emotions of others, is not simply a feature of professionalism, however. It also helps to construct masculinity. Like professionals, men are also expected to keep their emotions at bay, setting themselves off from women, who are allowed to be emotional.

In conclusion, Hearn's central claim is that professionalization and gender are intertwined in the same process through which women's reproduction and emotionality came under the control of male professionals. If the emergence of professions is itself tied to structured relations involving male domination and female subordination, then no account of professional power will be complete unless the 'crucial connection with gender' is made (Hearn 1982:186). By the same token, any analysis of gendered structures of domination in medical encounters needs to take up and explore the relationship between professionalization and gender. In particular, the dichotomy between rationality and emotionality, so central to the construction of the relationship between the professional and the patient, requires analysis in relation to its links with gender symbolism around masculinity and femininity.<16> Thus, structural accounts of professionalization require gender, just as accounts of gendered structures of domination cannot be separated from the emergence of the helping professions.

Having set out some of the basic tenets of Hearn's theory, the question now becomes how a conflated approach to gendered and professional power can help me make sense of the results of the present inquiry and, more to the point, whether it can provide a more adequate structural explanation of structures of domination in medical encounters than is possible with a structural account separating these structures. In particular, I shall be focussing on two areas where a conflated approach to power at a structural level can prove beneficial for the present inquiry. The first concerns the thorny problem of the *sex of the GP and the patient* and its (respective) contribution to power asymmetries in medical encounters. The second pertains to the *quality of the power practices* delineated in the present inquiry and how to make sense of them. I shall be dealing with each of the issues briefly, drawing some conclusions at the end concerning what this can mean for theorizing power and gender in medical encounters.

Female GPs and male patients: do they make a difference?

By taking the approach that gendered and professional power are conflated, the selection of medical encounters between female patients and male GPs to the exclusion of other gender constellations (female GP/female patient; female GP/male patient; male GP/male patient) as material for the analysis of gender and power becomes unproblematic. If the profession of medicine is itself 'gendered' and, by the same token, gender relations in medical encounters thoroughly 'professionalized,' then there is no reason to assume that this structural conflation will suddenly vanish at the level of the *individual* practitioner or patient.

This has several immediate consequences. The notion that domination might be eliminated from medical encounters, if a female GP were substituted for a male GP, comes to rest on irretrievably shaky grounds. If the entire medical system, both historically and in terms of its present organization, is based on male domination over women, then it is unlikely that the individual female practitioner, however well-intentioned or even feminist, would be able to practice medicine in an entirely different way.<17>

Furthermore, a medical encounter where both GPs and patients belong to the same sex does not ensure that any power imbalance in their interaction can be neatly attributed to professional structures of domination. If medicine is, by definition, gendered, then any professional, regardless of her or his sex, will be drawing upon gendered structures of domination in order to exercise control over the interaction. Similarly, the patient, whatever his or her sex, will be caught up in structures of domination involving both professional and gendered power. This means that being a patient automatically entails being invested with attributes commonly associated with femininity like dependency and emotionality and, consequently, being treated within any medical encounter, at least to some extent, 'like a woman.'<18> Whereas this may shed some light on the several empirical events like, for example, the fact that male patients are reluctant to seek the services of a professional unless they are in the throes of acute illness or something requiring a 'technical' (masculine) solution (Nathanson 1977) or that male clients have a tendency to report their experiences rather than to tell them in a more emotional fashion, making them a somewhat less-than-ideal client for therapy (Wodak 1981), it does *not* help us to separate gender from professional power.

Thus, by treating power as conflated, any medical encounter becomes a potentially interesting starting point for the study of power and gender. Whereas different gender constellations would undoubtedly increase our knowledge on power and gender in medical contexts in a cumulative sense, it would not eliminate the conflation of power, thereby enabling us, once and for all, to separate gendered from professional structures of domination. Or, put a bit differently, we will simply never know whether the GP acts the way he does because he is a man or because he is a doctor (or because the patient is a women or because she is a layperson).

Gender, professionalization and the Janus-face of power

A conflated approach to professional and gendered power has some advantages in terms of explaining the specific character of the way power was exercised in my data as well. According to Hearn, the profession of medicine is predicated on the control of emotionality in general, and women's emotionality in particular. In the present inquiry, it has been demonstrated how the activity of troubles talk can become the cite for power struggles between GPs and patients. Treating power as conflated enables troubles-telling to be treated not only as a location for interactional power struggles, but also as an activity, which itself reflexively serves to construct professional power relations as well as asymmetrical gender relations. By definition, storytelling entails expressing emotions. When done in an institutional encounter, it will open the troubles-teller up to having her (or his) emotions subject to professional control. If men are, indeed, more likely to report their experiences rather than express them in story-form as women are inclined to do (Wodak 1981), it is not surprising that women's emotionality can become a frequent target for medical control. Male patients may be more likely to receive technocratic solutions because they present their problems instrumentally - as 'object to be serviced.' By the same token, if emotionality carries with it an implicit threat to rationality and masculinity, then a male professional may find himself doubly constrained to undermine the emotionality expressed in a story about troubles. Thus, gender and professional power would seem to be intertwined in the activity of troubles talk within medical encounters and treating it as such can help us make sense of how troubles talk and its obstruction are implicated in the production and reproduction of asymmetrical power relations between male GPs and female patients.

Not only is the conflation of gender and professional power involved in troubles-telling, but it can help account for the Janus-faced quality of the power practices as well. Emotions are elusive and do not lend themselves to control by open means of repression. Whereas straightforward expressions of aggression can be, at least for the time being, brought under control through incarceration, sorrow or 'nervousness' cannot. The soft, subtle, friendly forms of power are, perhaps, the most effective way of getting a handle on women's emotionality, just as punishment may be the most usual response to more 'masculine' expressions. In this respect, the cooptative or Janus-faced forms of control are not only more relevant to women because they are, numerically speaking, most likely to be the recipients of them, but the forms themselves may be 'gendered.'

In conclusion, professionalization and gender are not only conflated at the level of structures of domination in the context of medicine, but they are conflated at the level of the interaction between male GPs and female patients as well. A more comprehensive explanation of the power practices delineated in the present inquiry becomes possible with fewer threads left dangling, when power practices are not separated into those based on either professional or gendered structures of domination. In short, I would contend that the results makes more sense, when the conflation of gender and professional power is taken as a 'fact.'

GENDER AND POWER REVISITED

Gender, power and (feminist) theory

No social theory on power can ignore the problem of gender. Relations between the sexes are, and have been historically, power relations.[19] This means that any conceptualization of power will not only have to encompass the workings of power at various levels and in different areas of social life, but it will need to tackle the relationship between power and the specific (groups of) social actor(s) exercising it. Thus, in order to be worth its salt, any theory on power should be able to tell us how to conceptualize the most routine trappings of power as well; i.e. gender, social class, race, nationality, etc.[20]

Within feminist scholarship, gender has been the subject of considerable theorizing. As theoretical concept, it occupies a privileged position, indisputedly part and parcel of any study in the field. Gender as theoretical concept is not simply used to explain sexual difference, but, more importantly, to come to terms with how this difference constructs or maintains asymmetrical relations between men and women; i.e. relations involving male domination and female subordination. Thus, whereas gender differences may, in and of themselves, be of some academic interest, it is at the moment that gender embraces the issue of power that it becomes of particular interest and relevance within feminist scholarship.

Although power and gender are considered to go hand in hand by most feminists, the exact relationship between the two has not been subject to much explicit theorizing.<21> For this reason, I shall devote the rest of this chapter to setting out some thoughts on the conceptual relationship between power and gender, drawing upon the results of the present inquiry. I shall begin with the assumption that gender is essential for *explaining* how power works. This will be followed by a discussion of why the notion of 'gendered' power; i.e. specific forms of power operating in relations between the sexes, is a faulty one. And, finally, some suggestions will be made for how gender and power might (better) be conceptualized.

Some problems with gender as theoretical construct for explaining domination in relations between the sexes

Gender as theoretical concept made its entrance in feminist theory with Oakley's (1972) pioneering introduction of the distinction between (biological) *sex* and (social) *gender*. In an attempt to bury the notion, once and for all, that differences between men and women were natural or biological, she drew upon evidence on transsexuals to show that a person's sex is not determined by his or her anatomy, but is socially and culturally constructed. Further evidence for the relativity of biological sex was provided by anthropological findings on gender distinctions, which demonstrated that, whereas there have been differences between the sexes in every culture, the forms that gender identity have taken not only vary, but are contradictory from one culture to the next (Chodorow 1971).<22> In short, there seemed to be no biological bed-rock of sexual difference, but rather myriad social and psychological differences. Gender, rather than sex, was the appropriate construct for approaching differences between men and women.

Since these early formulations on the sex-gender distinction, feminist scholars have continued to elaborate our understandings of gender, developing ever more sophisticated frameworks for its conceptualization (Harding 1986; Hagemann-White 1987). Gender is no longer simply about the distinction between biology and society, but has since become a 'layered concept,' involving not only individual differences in behavior or sexual identity, but gendered structures in the division of labor or social activities as well as symbolic dichotomies in sexual meaning. Whereas these recent theoretical reformulations of the concept of gender undeniably contribute to our understanding of the variety of ways that sexual difference is reflected at all levels of social life, theories on gender remain, by definition, concerned, first and foremost, with *differences* between the sexes (or *similarities* within each sex). Within feminist scholarship, gender has not only been employed to come to terms with gender differences, but it is also used to explain the *asymmetrical* nature of these differences. The underlying notion is that if gender relations are always and everywhere asymmetrical, then we might expect to be able to uncover the core or cause of male domination by discovering what gender is all about. Thus, if we want to understand relations involving domination and subordination between the sexes, gender is where to begin.

Taking gender as a focus for understanding asymmetrical relations between the sexes unavoidably implies that power relations between the sexes can best be understood in terms of their uniqueness; i.e. what specifically differentiates them from relations of domination and subordination between other social groups. The emphasis on the gender as explanation for asymmetrical power relations between men and women also goes hand in hand with assigning structures of male dominance and female subordination a universality, which makes them a regular and stable feature of the social order. In other words, if relations between the sexes are essentially *different* than other power relations, then gender inequality becomes commensurable, structuring women's lives and situations in fundamentally similar ways regardless of the particularities of the moment, the setting, or the social actors (Smith 1983).

In recent years, the notion that asymmetries based on gender take universal or specific forms has born the brunt of considerable criticism within feminist scholarship (Rosaldo 1980,1983; Stanley and Wise 1983; Barrett 1987). Despite this critique, however, gender is still often regarded a useful theoretical starting point for *explaining* asymmetries between men and women.

It is my contention that this assumption is in need of some revision. To this end, it may be helpful to remember *why* gender was introduced into feminist theorizing in the first place and which theoretical problems it was expected to solve.

Originally, both feminist theory and practice were faced with the necessity of separating relations between the sexes from other relations involving domination and subordination. The emergence of a feminist movement required that women constitute themselves as a political group with, at least to some extent, common interests stemming from perceived regularities in situations over time and space. The constitution of women as a collective 'we' formed the precondition for female solidarity and an organized feminist movement (de Vries 1987). By the same token, sexual inequalities had to be established as a problem, something which is not simply part of the god-given, natural order. Within the academic world, gender relations were originally not even considered a topic. Relations between the sexes had to constituted as a 'social problem' before scholars could begin to study them.<23> Thus, the constitution of women as political subject and gender relations as theoretical problem was the essential first step *before* the more important task could begin; namely, uncovering how asymmetrical relations between the sexes came to be constructed, maintained, and, as we all hoped, transformed in the various contexts of everyday life.

Since the timely introduction of gender as theoretical concept in the halls of academe, things have, however, changed. Gender relations have achieved the undisputed status as problem, requiring attention both within, as well as outside the world of scholarship. Consequently, gender inequality is no longer something, which must first be constituted as topic, before it can be investigated. Whereas this has come as a welcome relief to many feminist scholars, who could finally get down to the business of investigating the 'how's' (and 'how-not-to's') of relations involving domination and subordination between the sexes, it also raises some question concerning the place of gender in feminist theorizing.

Gender as theoretical concept locates the problem; it does not explain it. Whereas it was essential for establishing relations between the sexes as a topic for research (as well as for social action), it is not the best theoretical concept for understanding asymmetries between the sexes. There are three reasons for this.

1. *Gender as a theoretical concept is about differences between the sexes; it does not explain why they are asymmetrical.*
   Even the most systematic or comprehensive gender construct does not, in and of itself, indicate how these differences can result in asymmetrical relations involving domination and subordination. Beginning with the early sex differences research, conclusions have always been drawn about social asymmetries on the basis of gender difference. For example, by showing that women and men were more alike than that they were different in terms of intelligence, capacities, traits, etc., it became possible to conclude that sex-specific educational or vocational patterns were not due to differences in ability, but were rooted in discriminatory social practices against women. In other words, uncovering differences (or lack of them) between the sexes could provide an indication of where to begin in analyzing the subordination of women in the various areas of social life. The differences themselves, however, could only point to a potential problem. Further analysis was always required in order to explain how these differences were implicated in asymmetrical power relations.

2. *Gender as a category can never be separated from other categories; nor can gender inequality be isolated from other structures of domination.*
Assuming that gender explains inequality, rather than locating a 'problem area,' goes hand in hand with the notion that power relations can best be explained in terms of their specificity. Power is regarded as working in fundamentally different ways in relations between the sexes than in other relations between dominant and subordinate groups.
This assumption is not without problems. It ignores the fact that unequal relationships have the disturbing tendency to be unequal in more than one way. Power inequalities

> ... cannot simply be discussed in terms of male domination; maleness is articulated within a set of power relations of race, class, and nationality. Similarly, femaleness is not a single thing but is also articulated within a set of power relations-again, class, race, and nation. These forms of domination are not separate, or exterior to each other, but are integrated. Thus, you cannot be a woman without being a woman of a certain class, race and country; similarly, you cannot be a woman without being a woman at a particular moment in history - and that moment in history will carry its own definition of what it means to be a 'woman,' what it means to belong to a certain class, race, and so on. All of these terms are continually being redefined in the process of ongoing political and ideological struggles; they are never static. (Fee 1986:53)

It is, therefore, hard to imagine *any* context where gendered structures of domination will operate in isolation of other structures of domination. If gender and other structures of domination are hopelessly and irrevocably intertwined, then understanding structures of domination between the sexes will inevitably entail understanding how these structures are interrelated.
3. *Gender is static; it is not about process.*
Gender inequality is not only inseparable from other inequalities, making it impossible to specify structures of domination based exclusively or even specifically on gender, but gender is not a concept which lends itself to uncovering the how's of asymmetri-

> cal relations in *any* context. Whereas gender can help us conceptualize sexual differences at various levels of social life, it is an essentially static concept. At best, it can delineate the 'layered' quality (Hagemann-White 1987), showing that gender differences pervade the entire social order. However, no amount of differentiation within the gender concept, nor the attempt to put it in context, can overcome that fact that it does not provide a way of uncovering the dynamics of power; i.e. how asymmetrical relations come to be produced, reproduced or undermined.<24>

After more than a decade of feminist scholarship, a convincing case can be made for employing gender as theoretical construct in order to explain differences between men and women as socially or culturally produced rather than biologically determined. As theoretical construct, gender can serve as an umbrella concept for the various individual, cultural, and social differences existing between men and women. If these differences are what one is interested in discovering, then gender is the preferred concept for explaining them.

Gender has, however, historically had another function. By establishing differences between men and women as social or cultural products, rather than the natural order of things, it became possible to locate sexual relations in the realm of power relations. The subordination of women could no longer be explained by her biology, but entered the realm of the political: as part and parcel of asymmetrical relations, which were socially produced and reproduced and, more important, could also be transformed.

This suggests that, whereas gender is essential for establishing relations between the sexes as asymmetrical, it is *not* an adequate theoretical starting point for coming to terms with *how* these relations are constituted, sustained and undermined in the everyday social practices of men and women. It cannot explain structures of domination in an inclusive way; nor can it deal with the constitution of these structures as part of a dynamic process.

If gender is not sufficient as theoretical starting point for coming to terms with the 'how's' of asymmetrical relations between the sexes, the question becomes: how about power? As theoretical concept, power does take the production, reproduction, and transformation of asymmetrical relations as its focus. Whereas conceptions of power vary greatly, with some conceptions being indisputedly more adequate than others for coming to terms with asymmetrical relations between different groups and in various social contexts, the focus of any theory of power is likely to be not only on *who* exercises the power, but *how* asymmetrical relations of power between individual actors or social collectivities are constituted in the first place, how they are maintained, and, of course, how they are transformed or might be transformed in the future. In order to come to terms with asymmetrical power relations, theoretical concepts are required which can explain the *process* and *dynamics* of asymmetries. This suggests that what we may need is simply a *good* theory of power; i.e. a theory, which is able to address the 'how's' of asymmetrical relations, no matter which social groups are involved in their constitution.

Things are not quite so simple, however. Despite its deficits as theoretical construct for explaining asymmetrical relations between the sexes, gender is not an easy concept to give up, even in part. With its emphasis on what makes relations between the sexes specific or different from other relations, it has provided feminist scholarship with her *raison d'être*. In order to demonstrate that relations between men and women were not simply a normal and unproblematic feature of everyday life, it was necessary to pinpoint the differences in women's and men's life chances, their everyday experiences, and their respective ability to exercise control over their living and working situations. By showing that women had specific problems, irrevocably tied to their social position, it became legitimate to study a whole host of topics relevant to women, which had previously not been regarded as suitable topics for scientific inquiry.

Questioning the privileged status of gender as central theoretical category seems tantamount to high treason to the feminist cause. Its adequacy as explanatory concept for coming to terms with asymmetrical relations between the sexes cannot be criticized without evoking, simultaneously, the fear that the feminist critique of these relations is coming under attack as well. A feminist perspective on relations between the sexes is predicated on the assumption that men and women have different experiences and, more importantly, that these experiences are characterized by male domination and female subordination. If the essence of these power asymmetries does not reside in gender, then various, decidedly uncomfortable questions arise. For example, does this mean that we have abandoned relations between the sexes as a topic in its own right? Will women's experiences, once again, become subsumed under the generic 'man,' part of the 'human condition?' If gender does not explain inequalities between the sexes, does this mean that the subordination of women has lost its structural base? And, finally, if gender was pivotal to the emergence of feminist scholarship, can we afford to relinquish it as central theoretical construct, without having to watch a decade of research effort go up in smoke?

It is precisely the intimate connection between gender and the constitution of relations between the sexes as a problem, a relevant topic for research, which makes it painful for any feminist scholar to question the adequacy of gender as theoretical construct. It accounts for my own hesitation as well. Nevertheless, my research indicates that some modification in the way gender is used theoretically is in order. Thus, it is with a somewhat heavy heart that I shall now return, once again, to my own initial theoretical assumptions about gender and power and indicate how I think the relationship between them should be conceptualized. It is my hope that this will contribute to a re-channelling of the feminist debate along a more profitable course - a course, which will enable a better understanding of asymmetrical power relations between the sexes.

Grounding gender in power relations

This inquiry began with the *explicit* assumption that power would be involved in any encounter between the sexes. Behind this assumption that gender involved power, lurked, however, the *implicit* assumption, based on the feminist critique of the health care system, that power was also gendered. Whereas I did not set out to prove this in any empirical sense, it informed both the way the inquiry was set up to begin with and was confirmed by my results as well. By showing how power works in my data, empirical support could be provided for the claim that power, indeed, pervades even the most routine instance of interaction between a male physician and a female patient.

Although my data substantiated the feminist critique of the health care system, providing some empirical grounding for earlier analyses of some of the problems facing women as patients, it did not give me a foothold for coming to terms with the specifically gendered character of power in medical encounters or, for that matter, in any other encounter. This is not simply because I did not compare various mixed-sex interactive combinations. For reasons set out at the beginning of the inquiry as well as in earlier on in this chapter, this was rejected as inadequate for coming to terms with the problem of power in gender relations in medical encounters. More importantly, the power practices delineated in the analysis, as well as how they work to constitute, sustain, or undermine asymmetrical relations between the participants, do not suggest that they are limited to female patients and male GPs or, for that matter, to females and males in general. In other words, there is no inherent theoretical reason based on the way power was exercised in the consultations to assume that the power practices I discovered were specific to relations between the sexes. The same conversational resources could obviously be used in other settings in order to exercise interaction control. Similar 'friendly' power practices might easily be found in other encounters between dominant and subordinate groups, where relations of domination need to be softened or resistance nipped in the bud. There is every reason to suspect that other social actors might employ similar strategies for 'nagging at the bases' of authority to the strategies employed by the female patients in my inquiry. In fact, the power practices seemed quite suited to a host of other encounters between dominant and subordinate groups, both within institutional contexts, as well as in everyday settings.

This is not to say that these practices might not occur more *frequently* between male practitioners and female patients and it certainly does not discount the particular relevance to women as patients. On the contrary, the simple fact that woman are frequently patients and do often engage in interaction with men makes any description of the workings of power in medical contexts or in interaction with men relevant to them.

In conclusion, my research findings do not provide fodder for the notion that power processes are themselves specific for gender relations, nor do they substantiate the need for a specific theory on power in gender relations.<25>

If power does not take on specific forms in gender relations, enabling us to differentiate gendered power from other structures of domination, what does this mean for a theory of power? And, more importantly, how does the relationship between power and gender need to be conceptualized within this theory? It is to these issues that I shall now direct my attention.

Grounding theories of power in empirical contexts

Theories of power require empirical grounding. No matter how comprehensive or sophisticated a formal theory of power is, and Giddens' structuration theory is certainly a case in point, it can never explain power as long as it remains devoid of any connection to concrete social contexts, where specific social actors drawn upon specific rules and resources organized in specifically structured ways. Power is always and everywhere embedded in actual contexts of social life; there can be no adequate analysis of power, which does not take the particulars of these contexts into account. Without contextual information, we only have the scaffolding. The building must still be erected.

In the present inquiry, an attempt has made to provide a theoretical account of power, which was grounded in the concrete context of medical encounters between male physicians and female patients. I began my investigation in the spirit of Glaser and Strauss (1967) by entering the field and listening to what my data had to say about power in medical interaction.<26> Whereas I used the feminist critique of the health care system as a kind of sensitizer to help me get started on my investigation, substantive issues and concepts relevant for conceptualizing power in the present inquiry emerged as I become more and more familiar with the consultations. When it became clear that I would need some additional theoretical help in order to understand what I saw happening in my data, I turned to a more formal theory on power, using it as a heuristic device, a kind of (grand) sensitizer for the further delineation of how asymmetrical power relations were being constructed, sustained, and undermined in and through in interaction.

Ultimately, a theoretical account of power could be provided, which was empirically grounded in a double sense. On the one hand, by describing the power practices involved in the process of defining, diagnosing, and treating women's problems in medical consultations, it supplied the feminist critique of power and gender in medical contexts with empirical grounding, providing empirical support, but also some needed elaboration concerning how power works in the interaction between male physicians and female patients. On the other hand, by using a formal theory of power; namely, Giddens' structuration theory, in the process of delineating power practices in medical consultations, a step could be made toward filling in some of the missing pieces of that theory as well, enabling it to become a grounded theory of power for one specific setting at the level of face-to-face interaction.

Thus, empirical grounding can solve both the problem of generating a theory of power, as well as the problem of disembodied or empty formal theories of power, enabling the investigator to uncover the specific features of domination and subordination in a specific social context.

If empirical grounding is a requirement for an adequate account of power, the next question is obviously: what is to be done about the problem of gender? In other words, what conclusions can be drawn from the present inquiry about the theoretical status of gender and how it needs to be theorized in relationship to power?

Grounding gender in theories of power

Gender relations (like relations between classes, different ethnic groups, nations or whatever) are invariably power relations. Since gender inequality belongs to the very fabric of social life as we know it, we would expect any 'grand theory'- and grand theory is what the theory of structuration claims to be - to provide some conceptual space for understanding this kind of social phenomenon. In particular, any theory concerned with how power needs to be conceptualized will have to provide some guidelines for how power and its inevitable accoutrements of gender (or class or ethnicity) should be theorized as well. Whereas empirical grounding is essential for delineating how structures of domination between various groups are constituted, sustained, and transformed, a satisfactory structural account will entail integrating the relationship between gender and power within the theoretical framework itself.

As previously mentioned, it was not only impossible to sort out power practices resulting from professional/layperson asymmetries and power practices resulting from gender asymmetries at the level of interaction, but the conflation could be found at a structural level as well. The emergence of the medical profession, its present organization, and the structural and cultural arrangements in which it is embedded reflect and constitute gender inequality. Professionalization and asymmetrical gender relations go hand in hand, each presupposing the other. In view of the fact that my data were taken from a medical context, I focussed on structures of domination involving professional power and gender.<27> In another setting, another focus might have been taken. However, regardless of which structures of domination are under consideration, there is no reason to believe that the structures themselves would be any less conflated.

If power practices cannot be specifically and exclusively relegated to interaction between the sexes and if structured domination between the sexes is intertwined with other forms of domination at a structural level, then an adequate analysis of gender, like an adequate analysis of power, needs to take this conflation as its starting point. This has several immediate consequences for any analysis of gender relations.

First, if gender is essentially about sexual difference, then a theory of gender should be used when difference is what one is interested in explaining. If, however, asymmetrical relations of power between the sexes are the issue, gender is not the appropriate theoretical starting point. In that case, gender relations are the location for investigating the research object (asymmetries) rather than the means for explaining it.

Second, if structures of domination based on gender are hopelessly conflated with other structures of domination, both at the level of interaction as well as at an institutional level, then any investigation of asymmetrical relations between the sexes needs to be conducted within the concrete context in which they occur. If we want to understand asymmetries in relations between the sexes, we cannot afford to divest gender of its particular context within a specific social arrangement. On the contrary, understanding gender as part of an asymmetrical power relationship requires that it be embedded in the concrete, the specific, the idiosyncratic and the momentary. In short, gender like power needs to be grounded in the concrete social practices of men and women in a specific social context. It is this grounding, which will enable the reflexive reformulation of theories of power and gender in the various areas of social life.

In conclusion, the use of gender as theoretical construct is in need of some revision. At a theoretical level, gender is suited to identifying layers of difference - between individual men and women, between social structures, between symbolic representations. If, however, asymmetrical relations involving domination and subordination are the issue, gender alone can never provide an adequate explanation for how asymmetrical relations between the sexes are constituted, maintained, or transformed. What it can provide is a locus for any study of power. Gender is part of the contextual grounding of relations involving domination and subordination-where to find them, but not how to account for them.

Taken together, this provides the direction in which theories on power and gender need to move. It is a direction, which not only runs counter to the tendency within social theory to produce 'grand theory,' but it runs counter to the historical feminist project devoted to developing or refining theories of gender as well (Harding 1986). It is my contention, however, that neither a grand theory on power, nor a feminist perspective on gender are the best approach for understanding gender and power. Both approaches presuppose that such a theory exists or might be developed - an assumption, which is, in and of itself, in need of revision. The relationship between power and gender is a complicated one; that much is clear. We need theories which can help us analyze asymmetrical relations involving power and gender in all areas and at all levels of social life. Considering the dazzling variety and complexity of social life in general, and gender relations in particular, it seems highly improbable that any one theory - whether it takes power or gender as its starting point - can ever hope to explain it all. Rather than putting our energies into the pursuit of the Perfect Theory; i.e. the one theory which will encompass *all* relations of power or *all* gender relations, our task should be quite another one.

Based on the present inquiry, it is my contention that we need to devote our efforts to developing theories on power and gender which are grounded on three levels.

First, they will have to be anchored in the concrete experiences of women and men, enabling us to account for their social practices in the contexts being investigated.

Second, theories on power require grounding in terms of a sophisticated and reflected feminist critique of how asymmetrical relations gender relations are constructed, maintained, but also undermined in these contexts. This necessitates, at times, letting our data help us revise our common-sense notions about gender and power.

Third, we need to use our experiences as scholars, grappling with specific research questions about power and gender, applied to specific areas and levels of social life, and analyzed with specific research methods, to help us generate social theory, which is both conceptually coherent and grounded in empirical data.

What is needed is not one theoretical perspective on gender and power - a kind of feminist 'grand theory'- but rather feminist theory, which is grounded. The present inquiry has been one such attempt.

NOTES

1. As set out in Chapter 1, there are *substantive* reasons for assuming that 'friendly' power practices are not only characteristic for my sample, but typical for young, urban, well-educated, male GPs in general. A sample of older GPs working in rural settings without the benefit of vocational training in therapeutic skills or integral approaches to diagnosis might well have provided instances of more authoritarian or straightforward forms of domination. See, for example, Holtrop (1983) for a good portrait of the rural Dutch GP, which supports this claim.

2. See, for example, Bal (1988), who discovered similar 'friendly' forms of power in his study of interaction in legal settings.

3. This is, by no means, limited to academics interested in power, but applies to the everyday theorist as well, who may also be curious about how power works to produce asymmetrical relations involving domination and subordination in her daily life-

situations and, more to the point, is concerned about how these may be transformed into more egalitarian forms of social interaction.

4. This is overly optimistic, of course. In view of the fact that validity has, and will always be, the undisputed linchpin of scientific inquiry, any researcher employing a small sample of cases to the end of being able to analyze them in-depth or 'qualitatively,' will find herself in a position of having to justify this enterprise against the critique of limited relevance. It goes beyond the scope of the present inquiry to do this at any length, and, in fact, this course of action is unnecessary as many before me have devoted considerable attention to the problem. See, especially, Glaser and Strauss (1967); Bogdan and Taylor (1975); Schwartz and Jacobs (1979); and Silverman (1985).

5. And, in fact, I have had plenty of time to think about them, since they are precisely the questions, which have been raised every time, *without exception*, that I have discussed the present research findings. This has been the case, whether the context was a lecture for students, practitioners in the health care field, or feminists. The same questions came up in research meetings; at conferences; and even in everyday conversations with friends. It seems to me that issues, which are raised with such regularity and persistence, deserve to be taken seriously and this is what I shall be doing in the remainder of this chapter.

6. Giddens (1984:377) distinguishes between *structure* as the skills and resources, recursively implicated in the reproduction of social systems (for example, a medical consultation), as the basis of the participants' knowledgeability, which is instantiated in their social practices, and *structures*, as sets of resources implicated in the institutional articulation of social systems, mediating and transforming relations which influences social and system integration. I am using 'structural account' here in the former sense.

7. Others are certainly conceivable: for example, structures of domination based on social class, ethnic background, age and so on. Since I had less access to this kind of information, these structures played a smaller role in the inquiry as a whole. Nevertheless, the point I am making here would definitely apply to power based on other structures aside from professionalization or gender.

8. A comprehensive structural account of power imbalances in medical encounters would need to include, at the very least, an analysis of the political economy of health care and its effect on the organization of medicine,allocation of resources, and expenditure on health care as well as a comparative international study of health policy. Examples of this kind of analysis are Navarro (1976), Doyal and Pennell (1979), Strong (1979), and, in The Netherlands, Huysmans *et al* (1973). Most of the authors dealt with in the following section also link their analyses of structural asymmetries in medical encounters to the social organization of commodity production and reproductive labor, cultural representations and ideology, class divisions and the like. I have, however, limited myself to those aspects of their structural accounts, which attempt to explain

power as based on *either* professional power or gendered power *or* as based on both.

9. Professionalization theories are broad in scope, dealing with professional hierarchies within the profession of medicine, as well as its effects on the interaction between practitioners and patients. I have concentrated here on the latter. The interested reader is referred to Larson (1977) for a good analysis of the various facets of the emergence of the modern professions.

10. Here are a few examples: Under the assumption that women needed to 'conserve' their energy for reproduction, they were put to bed at the onset of menstruation where they stayed, off and on, in order to 'regulate their periods' until menopause arrived and bed rest was, once again, the order of the day. Higher education was regarded as physically dangerous for women, causing the 'maternal organs' to atrophy, and, therefore, to be avoided at all costs. In an age of 'heroic' medical treatments like bleeding, amenorrhea was thought to be curable through the application of leeches to the labia before the period was expected and, if that failed to produce the desired results, to the cervix despite the danger of their occasional loss inside the uterus. Women's psychological problems were often taken to hand by surgical means; ovariotomies were performed by the score for indications like over-eating, attempted suicide, erotic tendencies, persecution mania, and just plain 'cussedness' (Ehrenreich and English 1973a,b; 1979).

11. Feminist scholars rarely regard gender as *sufficient* for understanding relations involving domination and subordination between the sexes, nor is it presented as monolithic explanation for asymmetries between men and women. The exceptions are, at a theoretical level, Daly (1978) and in terms of empirical research, investigators employing a sex-role-perspective (see Chapter 1). As was the case with the social roles perspective on professional power, explanations based on sex roles have proved hopelessly inadequate in accounting for structured power asymmetries between the sexes (Connell 1987).

    Feminist scholars have, however, employed gender as a focus, making it central to their structural accounts of power. I shall be returning to the some of the uses and abuses of gender as theoretical construct at the end of this chapter.

12. In fact, Hearn uses the terms 'patriarchy' rather than gender, defining it as

> shorthand for a complex set of social relations, within and by which men tend to dominate women (1987:187).

Whereas patriarchy has been a familiar stand-by in feminist literature and used in a similarly 'shorthand' fashion, it has since come under some criticism for its lack of precision. It does not refer, technically speaking, to any social system where men dominate women, but to a specific form where the father's authority is supreme (Rubin 1975). For this reason, I have avoided Hearn's use of patriarchy, returning to the more general category of gender relations.

13. Whereas studies on gender relations in professional encounters often provoke concern that other, presumably more relevant, 'factors' like class, professional or educational differences or even custom and what is 'natural' are being ignored, studies on professions rarely encounter similar objections that they have ignored gender. See a particularly thoughtful account of this by Morgan (1981), who did research on one specific profession, Anglican bishops. He wondered whether the 'massive maleness' of the clergy did not have equal influence to that of economic and social status on their professional attitudes, speculating that the often-heard remark from bishops that they did not 'suffer fools gladly' could easily be heard as a 'masculine trait in episcopal robes' (Morgan 1981:88).

14. I have not set Hearn's theory in any detail here. He provides a structural account of patriarchy, based on marxist understandings of women's position with regard to reproduction (sexual, biological, generative, physical). Its advantages do not lie in it being a good general theory of women's oppression and, in fact, there are other theories of this ilk, which are far more adequate (Rubin 1975; Hartsock 1983). His theory is of interest to me here because of his novel attempt to connect gender and the professions at a theoretical level as well as its application to the profession of medicine. As I shall be arguing shortly, this is a first, but essential step, toward developing a grounded theory of power and gender

15. This would be difficult to do in view of the mortality statistics, which indicate that men are, in fact, the more unhealthy sex, with 80% more men dying per year than their female counterparts. For an interesting plea that more attention be paid to the male patient, see van Oosten and Knegtering (1987).

16. This has been done much more comprehensively than Hearn by numerous feminist scholars, most notably Hartsock (1983), Keller (1985), Bleier (1986). Despite the merits of this work, it tends to run into some of the problems inherent in using gender as explanatory concept. In particular, a problematic polarization of women as emotional and men as rational is set up, as well as a de-contextualization of the concepts of rationality and emotionality. See Grimshaw (1986) for an excellent critique. I shall be returning to these issues in the remainder of the chapter.

17. This is, parenthetically, one of the unintended consequences of the feminist demand for more female physicians and specialists. Whereas this demand is certainly justified at the equal opportunity level, the suggestion that the substitution of male practitioners for female ones would automatically result in medical practice more attuned to the needs of female patients is erroneous. It places an overly-heavy burden on the shoulders of female physicians, who not only have had to fight their way up through the male-dominated ranks of medical school and specialist training, but are also expected to produce an alternative to medicine as well. Not only does this demand too much from the individual female physician, but, more importantly, it skirts the more relevant issue, which is what a *feminist* alternative to medical practice would look like. Presumably this would entail more than a change in the sex of the practitioner.

18. Male patients obviously have more power to mitigate this lowly status, particularly when they can drawn upon resources available to them by virtue of their class or ethnic background. See, West (1984), Fisher and Groce (1985, 1986). The more general point here is that power imbalances between various social groups tend to reinforce and intensify one another, whether at the level of structurally available resources, interpretative schemes (signification) or normative sanctions (legitimation).

19. The well-worn 'battle of the sexes' is a familiar phenomenon in all areas of cultural life - the subject of literature, art, music, and popular culture. Recently, it has extended beyond the boundaries of ordinary everyday life into the halls of academe, where not only practitioners of science do battle, but the subject itself has become a topic of scientific interest. See, also, Davis and Oldersma (1987).

20. As it turns out, very few social theories on power are, in this respect, 'worth their salt.' It is not simply that most social theories do not deal with gender or relations between the sexes specifically, but they do not provide any guidelines as to how their particular conception of power is to be used to understand relations of power. Often this is due to the fact that they do not regard gender relations as relations involving power, but rather something else, like, for example, 'structural determinism' (Lukes 1977) or 'affective individualism' (Giddens 1985). See, also, Davis and Oldersma (1987).

21. The notable exception is Hartsock (1983), who has developed a specifically feminist perspective on power in gender relations. See, also, the collected papers from the symposium Gender of Power, held in Leiden in september 1987, which explicitly address the issue of the relationship between power and gender (Leijenaar *et al* forthcoming).

22. For this particular line in feminist thinking on sex and gender, I am indebted to Annemieke v. Drenth. See, also, Van Drenth (1988) for a somewhat different, but, nevertheless, provocative rendition of recent developments within feminist scholarship on the subject of gender.

23. An excellent example of this can be found in Grimshaw (1986:260), who cites the experiences of two early feminist scholars attempting to write dissertations on 'women's issues.' Ann Oakley's proposal to do a study on women's housework was met with surprise and amusement on the part of her male sociologist-colleagues, who felt that this was not a 'proper topic' for academic research (Oakley 1974). Several years later, Dale Spender described similar reactions to her plan for a dissertation on language and sex (Spender 1980). It was rejected on the grounds that it was 'not a topic.'

24. It might be argued that the recent concern for 'shifts' in gender symbolism encompass the dynamic dimension of gender. I do not see, however, how the discovery *that* a shift in how we think about femininity or masculinity has taken place increases our knowledge concerning *how* the shift came into being or, more important, how it serves to construct asymmetrical power relations between the sexes. The most 'dynamic' account of gender remains the, in my opinion, sadly underrated ethnometho-

dological study on gender by Kessler and McKenna (1978) in which the authors show how gender is constructed as social category. Their research object is the transsexual, faced with the arduous task of having to learn how to 'do gender.'

25. In her discussion of the theoretical characteristics of power relations, Komter (forthcoming) maintains that power relations based on gender, ethnicity or class have more in common than that they are different. The characteristics of power itself do not provide a basis for distinguishing one kind of power relationship from another.

    > Although the conditions and circumstances of power inequality in gender relations are different from those in which power develops in other social relations, the mechanisms of power, its socio-cultural manifestations and its social consequences,are likely to be similar. (Komter forthcoming:13)

26. 'In spirit' as I did not use their comparative *method* for generating grounded theory. Their general orientation, as well as the presuppositions underlying their grounded-theory approach, did, however, provide a source of inspiration for the present inquiry. See, Glaser and Strauss (1967, Chapters 1-4).

27. Presumably, in my own material other structures of domination were involved as well. For example, the 'double uterus' tape provided ample evidence that inequalities based on age may well have been implicated in the paternalistic bend the power practices took. Social class will invariably be involved, albeit, once again, in conflated form. Medicine, as Ehrenreich notes, is a

    > system which throws women, blacks, working-class people into intimate and complete dependency on white, male, upper-middle and upper-class doctors... the relationship...is a powerful mechanism producing acquiescence in the overall social structure and its values. (1978:15)

REFERENCES

Aalderen, H.J. van (1982) Anders of steeds meer van hetzelfde.In J.Hollenbeek Brouwer and J. A.Knottnerus (eds) *Vertrekpunten in de gezondheidszorg*, Amsterdam: VU Uitgeverij .

Aalderen, H.J. van (1986) Geen verzetspotentieel in de patiëntenbeweging? *Gezondheid & Politiek* 3: 51-53.

Aalderen, H.J. van and Hollenbeek Brouwer, J. (1982) *Syllabus Methodisch werken*, HIVU uitgave, Amsterdam: VU Uitgeverij .

Aerts, M. (1986) Bijzondere openbaarheid: de macht van het feminisme. Een terreinverkenning. *Tijdschrift voor Vrouwenstudies* 28, 7 (4): 545-558.

Aghina, M.J. (1982) Patiëntenparticipatie in perspektief.In J. Hollenbeek Brouwer and J.A. Knottnerus (eds) *Vertrekpunten in de gezondheidszorg*, Amsterdam: VU Uitgeverij .

Arditti,R., Duelli Klein, R., and Minden, S. (1984) *Test-Tube Women*, London: Pandora Press.

Arms, S. (1975) *Immaculate Deception*, New York: Bantam Books.

Armstrong, D. (1983) *Political anatomy of the body: medical knowledge in Britain in the twentieth century*, Cambridge: Cambridge University Press.

Atkinson, J.M. & Drew, P. (1979) *Order in Court: the Organisation of Verbal Interaction in Judicial Settings*, London: MacMillan.

Atkinson, J.M. and Heath, C. (eds) (1981) *Medical Work. Realities and Routines*, Westmead, Farnborough, Hants:Gower.

Atkinson, J.M. & Heritage, J. (eds) (1984) *Structures of Social Action*, Cambridge: Cambridge Univ. Press.

Baart, I. and Derks, M. (1981) En de cirkel is weer rond...een ander licht op het huisartsenbezoek van vrouwen, *Tijdschrift voor Vrouwenstudies* 2 (4):532-548.

Bachrach, P. and Baratz, M.S. (1970) *Power and Poverty. Theory and Practice*, New York: Oxford Univ. Press.

Bal, P. (1988) *Dwangkommunikatie in de rechtszaal*, Arnhem: Gouda Quint.

Balint, M. (ed) (1970) *Treatment or Diagnosis. A Study of Repeat Prescriptions in General Practice*, London: Tavistock.

Barker, D.L. and Allen, S. (eds) (1976) *Sexual Divisions and Society: Process and Change*, London: Tavistock.

Barker-Benfield, B. (1972) The spermatic economy: a 19th century

view of sexuality, *Feminist Studies* 1 (1): 45-74.
Barrett, M. (1987) The Concept of 'Difference,' *Feminist Rev.* 26:29-41.
Barrett, M. and Roberts, H. (1978) Doctors and their patients. The Social Control of women in general practice. In C. Smart and B. Smart (eds), *Women, Sexuality, and Social Control*, London: Routledge and Kegan Paul.
Becker,H.S. (1961) *Boys in White: Student Culture in Medical School*, Chicago: University of Chicago Press.
Bendix, R. (1960) *Max Weber. An Intellectual Portrait*, New York: Doubleday & Company, Inc.
Bennett, W.L. and Feldman, M.S. (1981) *Reconstructing Reality in the Courtroom*, London and New York: Tavistock Publications.
Berger, P.L.and Luckmann, T. (1966) *The Social Construction of Reality*. Middlesex: Penguin Books.
Berliner, H.S. and Salmon, J.W. (1980) The Holistic Alternative to Scientific Medicine: History and Analysis. *Int. J. of Health Services* 10 (1):133-147.
Bernhard, J. (1975) *Women, Mothers, Values and Options*. Chicago: Aldine Publishing Co.
Bleier, R. (ed) (1986) *Feminist Approaches to Science*, Oxford: Pergamon Press.
Bliesener, T. (1980) Erzählen unerwünscht. Erzählversuche von Patienten in der Visite. In K. Ehlich (ed) *Erzählen im Alltag*, Frankfurt a.M.: Suhrkamp.
Bliesener, T. and Siegrist, J. (1981) Greasing the Wheels: Conflicts on the Rounds and How They Are Managed. *J. of Prag.* 5:181-204.
Bloom, S.W. and Wilson, R.N. (1979) Patient-practitioner relationships. In Freeman, H.E., Levine, S., and Reedek, L.G. (eds) *Handbook of Medical Sociology*, 3 ed., Englewood Cliffs, N.J.: Prentice Hall.
Boden, D. (1975) *The Business of Talk*, Cambridge: Polity Press.
Bogdan, R. and Taylor, S.J. (1975) *Introduction to Qualitative Research Methods*, New York: John Wiley & Sons.
Boston Women's Health Book Collective (1971) *Our Bodies, Ourselves*, New York: Simon and Schuster.
Bremer, G.J. (1980a) Over het gebruik van de huisartsen-anamneselijst bij vage, onduidelijke klachten. *Huisarts en Wetenschap* 23: 53-58.
Bremer, G.J. (1980b) Over klagen, in het bijzonder het klagen bij de dokter.Inaugural speech. Alphen aan de Rijn: Stafleu.
Bremer, G.J., Es, J.C. van and Hofmans, A. (eds) (1974) *Inleiding tot de huisartsgeneeskunde*, deel I. 2e druk, Leiden: Stenfert Kroese

b.v.
Brinkgreve, C., Onland, J.H., and de Swaan, A. (1979) *De opkomst van het psychotherapeutisch bedrijf*, Utrecht: Uitgeverij Het Spectrum.
Brouwer, W. and Dijkhuis J.H. (1967) *Praten met patiënten*, Utrecht: Erven J. Bijleveld.
Brouwer, W. and Touw-Otten, F. (1974) Van klacht tot klagen; een analyse van de premedische periode, *Huisarts & Wetenschap* 17 (9).
Broverman, I.K., Broverman, D.M., Clarkson, F.E., Rosenkrantz, P.S. and Vogel, S.R. (1970) Sex role stereotypes and clinical judgements of mental health, *J. of Consulting and Clinical Psychol.*4 (3):1-7.
Brown, P.and Levinson, S. (1978) Universals in language usage: Politeness phenomenon. In E.N.Goody (ed) *Questions and Politeness*, Strategies in Social Interaction, Cambridge: Cambridge Univ. Press.
Buma, J.T. (1950) *De huisarts en zijn patient*, Amsterdam: A.de Lange Universiteitspers.
Bury, M.R. (1986) Social Constructionism and the Development of Medical Sociology, *Soc. of Health & Illness* 8 (2):137-169.
Button, G. and Casey, N. (1984) Generating topic: the use of topic initial elicitors. In J.M. Atkinson and J. Heritage (eds) *Structures of Social Action*, Cambridge: Cambridge Univ. Press.
Byrne, P.S. and Long, B.E. (1976) *Doctors Talking to Patients* London: Her Majesty's Stationery Office.

Callinicos, A. (1985) Anthony Giddens. A Contemporary Critique. *Theory and Society* 14 (2):133-166.
Cameron, D. (1985) *Feminism and Linquistic Theory*, Houndmills, Basingstoke, Hampshire: MacMillan.
Cartwright, A., Lucas, S. and O'Brien, M. (1974) *Exploring Communications in General Practice: A Feasibility Study*, London: Report to the SSRC, Inst. for Social Studies in Medicine.
Centraal Bureau voor de Statistiek (1985) *Statistisch zakboek*, Den Haag: Staatsuitgeverij, CBS publicaties.
Centraal Bureau voor de Statistiek (1986) *Statistisch zakboek*, Den Haag: Staatsuitgeerij, CBS publicaties.
Chesler, P. (1972) *Women and Madness*, New York: Avon Books.
Chodorow, N. (1971) Being and Doing: A Cross-Cultural Examination of the Socialization of Males and Females. In V. Gornick and B.K. Moran (eds) *Women in Sexist Society*, New York: Basic Books, Inc.
Chodorow, N. (1978) *The Reproduction of Mothering*, Berkeley: Univ. of Calif. Press.

Chomsky, N. (1965) *Aspects of the Theory of Syntax*, Cambridge: MIT-Press.

Cicourel, A.V. (1972) *Cognitive Sociology*, Middlesex: Penguin Books.

Cicourel, A.V. (1980) Three Models of Discourse Analysis: The Role of Social Structure, *Discourse Processes* 3 (2):101-131.

Cicourel, A.V. (1982) Theoretical and methodological suggestions for using discourse to recreate aspects of social structure. Paper presented at the MAP-conference, Univ. of Penn., Nov. 1982.

Cicourel, A.V. (1983) Language in the structure of belief in medical communication. In Fisher, S. and Dundas Todd (eds) *The Social Organization of Doctor-Patient Communication*, Washington D.C.: Center of Applied Linguistics Press.

Cloward, R.A. and Piven, F.F. (1979) Hidden Protest: the Chanelling of Female Innovation and Resistance, *Signs* 4 (4): 651-669.

Commission on Chronic Illness (1957) *Chronic Illness in a Large City*, Cambridge: Harvard Univ. Press.

Connell, R.W. (1987) *Gender & Power*, Cambridge: Polity Press.

Conrad, P.and Kern, R. (eds) (1981) *The Sociology of Health and Illness: Critical Perspectives*, New York: St. Martins Press.

Conrad, P. and Schneider, J.W. (1981) *Deviance and Medicalization*, St. Louis: The C.V. Mosby Co.

Conte, R. (1981) Aspects of Interaction in a Medical Interview, *J. of Prag.* 5:113-144.

Cooperstock, R. (1978) Sex differences in psychotropic drug use, *Soc. Sci. Med.* 12B:179-186.

Cott, N. (1986) Feminist theory and feminist movements: The past before us. In: J. Mitchell and A. Oakley (eds) *What is feminism?*, New York: Pantheon.

Coulthard, M. and Ashby, M. (1976) A linguistic description of doctor-patient interviews. In M. Wadsworth and D. Robinson (eds) *Studies in Everyday Medical Life*, London: Martin Robertson.

Culpepper, E. (1978) Exploring menstrual attitudes. In M.S.Hennifin (ed) *Women Looking at Biology Looking at Women*, Cambridge, Mass.: Schenckman Publishing Co.

Daly, M. (1978) *Gyn/Ecology*, Boston: Beacon Press.

Davis, A. (ed) (1978) *Relationships between doctors and patients*, Westmead, Farnborough, Hants., England: Teakfield Ltd.

Davis, K. (1983) Feministische subjektiviteit en onderzoek in de Gezondheidszorg,*Congresbundel WUV*, Nijmegen.

Davis, K. (1984a) Women as patients: a problem for sex differences research, *Women's Studies Int. Forum* 7 (4):211-217.

Davis, K. (1984b) *Problem (Re)formulation in Psychotherapy. Some conversational aspects of the individualization of women's*

problems, Rotterdam: Konteksten.
Davis, K. (1985) Storytelling and the exercise of power. In T. Hak, J. Haafkens and G. Nijhof (eds) *Working Papers on Discourse and Conversational Analysis*, Rotterdam: Konteksten.
Davis, K. (1986) The Process of Problem (Re)formulation in Psychotherapy, *Sociology of Health and Illness* 8 (1):44-74.
Davis, K. and Oldersma, J. (1987) The Gender of Power: An Introduction. In M. Leijenaar, K. Davis, C. Helleman, J. Oldersma and D. Vos. *The Gender of Power. A Symposium,* Leiden, Vakgroep Vrouwenstudies and Vena.
Davis, K. (1988) Paternalism Under the Microscope. In A.D. Todd and S. Fisher (eds) *Gender and Discourse*, Norwood, N.J.: Ablex Publishing Co.
Davis, K. (forthcoming) The Janus-Face of Power. Some Theoretical Considerations involved in the study of gender and power. In M. Leijenaar, K. Davis, J. Oldersma and D. Vos (eds) *The Gender of Power: Selected Papers*, Leiden: DSWO Press.
Davis, M.S. (1968) Variations in patients' compliance with doctors' advice: An empirical analysis of patterns of communication, *Amer.J.of Public Health* 58:274-288.
Dohrenwend, B.and Dohrenwend, B. (1969) *Social Status and Psychological Disorder: A Causal Inquiry*, New York: Wiley.
Donnison, J. (1977) *Midwives and Medical Men*, London: Heinemann.
Donzelot, J. (1979) *The Policing of Families*, New York: Pantheon Books.
Dowrick, S. and Grundberg, S. (1980) *Why Children?* London: The Women's Press.
Doyal,L. and Pennell (1979) *The Political Economy of Health*, London: Pluto Press.
Drass, K.A. (1982) Negotiation and the structure of discourse in medical consultation, *Sociology of Health and Illness* 4 (3):320-341.
Dreifus, C. (ed) (1977) *Seizing Our Bodies. The Politics of Women's Health*, New York: Vintage Books.
Drenth, A.v. (1988) Vrouwenstudies en theorie. Over feministische ambivalenties en theorievorming binnen vrouwenstudies, paper for the Sociologische en Antropologische Conferentie, Antwerpen.
Drew, P. (1984) Speakers' Reportings in Invitation Sequences, In: J.M. Atkinson and J.C. Heritage (eds) *Structures of Social Action*, Cambridge: Cambridge Univ. Press.

Ehlich, K. (ed) (1980) *Erzählen im Alltag*, Frankfurt a.M.: Suhrkamp.
Ehlich, K. (ed) (1984) *Erzählen in der Schule*, Tübingen: Gunter Narr Verlag.

Ehrenreich, B. and Ehrenreich, J. (1978) Medicine and Social Control, In J.Ehrenreich (ed) *The Cultural Crisis of Modern Medicine*, New York: Monthy Review Press.

Ehrenreich, B. and English, D. (1973) *Witches, Midwives, and Nurses. A History of Women Healers*, Old Westbury, N.Y.: The Feminist Press.

Ehrenreich, B. and English, D. (1973) *Complaints and Disorders. The Sexual Politics of Sickness*, Old Westbury, N.Y.: The Feminist Press.

Ehrenreich, B. and English, D. (1979) *For Her Own Good*, London: Pluto Press.

Ehrenreich, J. (ed) (1978) *The Cultural Crisis of Modern Medicine*, New York: Monthy Review Press.

Es, J.C. van (1980) *Patiënt en huisarts. Een leerboek huisartsgeneeskunde*, Utrecht: Bohn, Scheltema & Holkema.

Fahrenfort, M. (1984) Transformaties in medische consulten, *Tijdschrift voor taal- en textwetenschap* 4 (1):25-39.

Fahrenfort, M. (1985) *Een doktersroman*, Amsterdam:VU Uitgeverij.

Fahrenfort, M. and Klinkert, J.J. (1978) Medicalisering, naar een omschrijving, *Huisarts en Wetenschap* 21:412-416.

Fee, E. (1983) Women and health care: A comparison of theories. In E.Fee (ed) *Women and Health: The Politics of Sex in Medicine*, Farmingdale, N.Y.: Baywood Publishing Co.

Fee, E. (1986) Critiques of Modern Science: The Relationship of Feminism to Other Radical Epistemologies. In R. Bleier (ed) *Feminist Approaches to Science*, New York: Pergamon Press.

Fidell, L.S. (1980) Sex Role Stereotypes and the American Physician, *Psychology of Women Quarterly* 4 (3):313-330.

Finch, J. and Groves, D. (eds) (1983) *A Labour of Love. Women, Work and Caring*, London: Routledge & Kegan Paul.

Fisher, S. (1983) Doctor Talk/Patient Talk: How Treatment Decisions are Negotiated in Doctor-Patient Communication. In S. Fisher and A.D.Todd (eds) *The Social Organization of Doctor-Patient Communication*, Washington D.C.: Center for Applied Linguistics Press.

Fisher, S. (1986) *In the Patient's Best Interest. Women and The Politics of Medical Decisions*, Rutgers University Press: New Brunswick, N.J.

Fisher, S. and Groce, S.B. (1985) Doctor-patient negotiation as cultural assumptions, *Sociology of Health and Illness* 7 (3):342-374.

Fisher, S. and Groce, S.B. (1986) Accounting Practices: Information Processing in Medical Interviews. Paper presented at the 11th World Conference of Sociology, New Delhi, India.

Fisher, S. and Dundas Todd, A. (eds) (1983) *The Social Organization of Doctor-Patient Communication*, Washington D.C.: Center of Applied Linguistics Press.

Fisher, S. and Dundas Todd, A. (1986) Friendly Persuasion: The Negotiation of Decisions to Use Oral Contraceptives. In S.Fisher and A. Dundas Todd (eds) *Discourse and Institutional Authority: Medicine, Education and Law*, Norwood, N.J.: Ablex Publishing Co.

Fishman, P. (1978) Interaction: the work women do, *Soc. Problems* 26:397-406.

Flader, D. and Giesecke, M. (1980) Erzählen im psychoanalytischen Erstinterview. In K. Ehlich (ed.) *Erzählen im Alltag*, Frankfurt a.M.: Suhrkamp.

Foucault, M. (1980) *Power/Knowledge: Selected Interviews and other Writings, 1972-1977*, New York, Pantheon.

Frankel,R.M. (1983) The laying on of hands:Aspects of the organization of gaze, touch, and talk in a medical encounter. In S. Fisher and A.D. Todd (eds) *The Social Organization of Doctor-Patient Communication*, Washington, D.C.:Center for Applied Linguistics.

Frankel, R.M. (forthcoming) Microanalysis and the Medical Encounter. In D. Helm, W.T. Anderson, and A.J. Mehan (eds) *New Directions in Ethnomethodology*, New York: Irvington.

Freidson, E. (1970) *The Profession of Medicine*, New York: Harper & Row Publishers, Inc.

Freidson, E. (1986) *Professional Power. A Study of the Institutionalization of Formal Knowledge*, Chicago: Chicago University Press.

Gabe, J. and Lipshitz-Phillips, S. (1984) Tranquillizers as social control? *Soc. Rev*. 32 (3):524-546.

Gallie, W.B. (1956) Essentially Contested Concepts, *Proceedings of the Aristotelian Society*, 167-98.

Garfinkel, H. (1967) *Studies in Ethnomethodology*, Englewood, Cliffs, N.J.: Prentice-Hall.

Garfinkel, H. and Sacks, H. (1970) On formal structures of practical actions. In J.C. McKinney and E.A. Tiryarian (eds) *Theoretical Sociology*, New York: Appleton Century Crofts.

Gerson, J.M. and Peiss, K. (1985) Boundaries, Negotiation, Consciousness: Reconceptualizing Gender Relations, *Social Problems* 32 (4):317-331.

Giddens, A. (1976) *New Rules of Sociological Method: A Positive Critique of Interpretative Sociologies*, London: Hutchinson.

Giddens, A. (1977) *Studies in Social and Political Theory*, London: Hutchinson.

Giddens, A. (1979) *Central Problems in Social Theory*, London: The MacMillan Press, Ltd.

Giddens, A. (1981) *A Contemporary Critique of Historical Materialism*, London: The MacMillan Press, Ltd.

Giddens, A. (1982) *Profiles and Critiques in Social Theory*, London: The MacMillan Press, Ltd.

Giddens, A. (1984) *The Constitution of Society*, Cambridge: Polity Press.

Giddens, A. (1985) Structuratietheorie en empirisch onderzoek (lecture), In O.J. Munters, E. Meijer, H. Mommaas, H. v.d. Poel, R. Rosendal, and G. Spaargaren, *Anthomy Giddens. Een kennismaking met de structuratietheorie*, Landbouwhogeschool Wageningen.

Giddens, A. (1987) *Social Theory and Modern Sociology*, Cambridge: Polity Press.

Glaser, B.G. and Strauss, A.L. (1967) *The Discovery of Grounded Theory: Strategies for Qualitative Research*, New York: Aldine Publ. Co.

Goffman, E. (1955) On face-work: an analysis of ritual element in sociale interaction, *Psychiatry* 18:213-31.

Goffman, E. (1961) *Asylums,* Middlesex: Penguins.

Goffman, E. (1967) *Interaction Ritual*, New York: Pantheon Books.

Goffman, E. (1971) *Relations in Public,* Middlesex: Penguins.

Goffman, E. (1974) *Frame Analysis*, New York: Harper Colophon Books.

Gordon, D.P. (1983) Hospital slang for patients: Crocks, Gomers, Gorks, and others, *Language in Society* 12 (2):173- 185.

Gordon, L. (1974) *Woman's Body, Woman's Right*, New York: Grossman.

Gove, W. (1978) Sex differences in mental illness among adult men and women: An evaluation of four questions raised regarding the evidence on the higher rates of women, *Soc. Sci.Med.* 12b:187-198.

Gove, W. and Hughes, M. (1979) Possible causes of the apperent sex differences in physical health: An empirical investigation, *Am. Soc. Review* 44:126-46.

Grimshaw, J. (1986) *Philosophy and Feminist Thinking*, Minneapolis: University of Minnesota Press.

Grunell, M. (1984) *Thuis in de jaren vijftig. Vrouwen over de hoeksteenvan die samenleving*, Amsterdam: Publicatiereeks Sociologisch Instituut.

Grunell, M. (1986) Vanzelfsprekendheid: scharnier tussen zelfbepaling en lot, *Lover* 3:162-69.

Haan, M. de and Bleys, F.C. (1980) *Moeheid,* Utrecht: Wetenschappelijke Uitgeverij Bunge.

Haan, M. de (1982) Vage klachten: wat verstaan patienten en huisartsen eronder? *Huisarts en Wetenschap* 25:190-95.

Hagemann-White, C. (1987) Gendered Modes of Behavior- a Sociological Strategy for Empirical Research. Paper presented at III International Interdisciplinary Congress on Women, Dublin, Ireland.

Haley, J. (1963) *Strategies of Psychotherapy*, New York: Grune & Stratton, Inc.

Harding, S. (1986) *The Science Question in Feminism*, Ithaca and London: Cornell University Press.

Harding, S. and Hintikka, M. (eds) (1983) *Reality Discovering*, Dordrecht: D.Reidel.

Hartman, M. and Banner, L.W. (eds) *Clio's Consciousness Raised. New Perspectives on the History of Women*, New York: Harper & Row.

Hartsock, N. (1983) *Money, Sex and Power. Toward a Feminist Historical Materialism*, New York: Longman.

Have, P. ten (1987) *Sequenties en Formuleringen. Aspecten van de interactionele organisatie van huisarts-spreekuur gesprekken*, dissertation, Amsterdam, UvA.

Hearn, J. (1982) Notes on Patriarchy, Professionalization and the Semi-Professions, *Sociology* 16 (2):184-202.

Hearn, J. (1987) *The Gender of Oppression. Men, Masculinity, and the Critique of Marxism*, Brighton: Wheatsheaf Books.

Heath, C. (1981) The opening sequence in doctor-patient interaction. In J.M. Atkinson and C.Heath (eds) *Medical Work. Realities and Routines*, Westmead, Farnborough, Hants: Gower.

Heath, C. (1982) Preserving the Consultation: Medical Record Cards and Professional Conduct, *Soc. of Health & Illness* 4:56-74.

Heath,C. (forthcoming) *The Partnership: Essays in the Social Organization in Speech and Body Movement in the Medical Consultation*, Cambridge: Cambridge Univ. Press.

Henley, N.M. (1977) *Body Politics. Power, sex and non-verbal communication*, Englewood Cliffs, N.J.: Prentice-Hall Inc.

Heritage, J.C. (1984) *Garfinkel and Ethnomethodology*, Cambridge: Polity Press.

Heritage, J.C. (1985) Recent Developments in Conversation Analysis, *Sociolinguistics* 15 (1):1-18.

Heritage, J.C. and Watson, D.R. (1979) Formulations as Everyday Objects. In G Psathas (ed) *Everyday Language: Studies in Ethnomethodology*, New York: Irvington Pub.,Inc.

Heydelberg,E. (1984) De ontdekking van de mondige patiënt, Tijdschrift voor *gezondheid en politiek* 5:14-17.

Hibbard, J.H. and Pope, C.R. (1986) Another Look at Sex Differences in the Youth of Medical Care: Illness Orientation and the Types of Morbidities for which Services are used, *Women & Health* 11 (2):21-36.

Hirst, P. (1982) The Social Theory of Anthony Giddens: A New Syncretism? *Theory, Culture & Society* 1 (2):68-82.

Hollenbeek Brouwer, J. (1982) Methodisch werken door huisartsen, en doorbraak naar een nieuwe ontwikkeling? In J.Hollenbeek Brouwer and J.A. Knottnerus (eds) *Vertrekpunten in de gezondheidszorg*, Amsterdam: VU Uitgeverij.

Hollenbeek Brouwer, J. and Knottnerus, J.A. (eds) (1982) *Vertrekpunten in de gezondheidszorg*, Amsterdam: VU Uitgeverij.

Holtrop, A. (1983) Bij de dokter, *Vrij Nederland* 10:3-35.

Homans, H. (ed) (1985) *The Sexual Politics of Reproduction*, Hampshire, England: Gower.

Honegger, C. and Heintz, B. (eds) (1981) *Listen der Ohnmacht*, Frankfurta.M.: Suhrkamp.

Hughes, D. (1982) Control in the Medical Consultation: Organizing Talk in a Situation where Co-Participants have Different Competence, *Sociology* 16 (3):359-376.

Huysmans, F., Juffersmans, P., Lagro, B., Niekerk, B.v., Smit, F., and Vlaar, H. (eds) (1973) *Gezondheidszorg in Nederland*, Nijmegen: SUN.

Hutter, B. and Williams, G. (eds) (1981) *Controlling Women. The Normal and the Deviant*, London: Croom Helm.

Illich, I. (1976) *Limits to Medicine*, London: Marion Bryars.

Jacobs, H.N., Kuyvenhoven, M.M. and Nunen, M.M. (1980) *Hulpverlening door de huisarts bij vage klachten*, Utrecht: UHI-Informatie, Utrechts Huisartseninstituut.

Janeway, E. (1981) *Powers of the Weak*, New York: Morrow.

Jefferson, G. (1972) Side Sequences. In D. Sudnow (ed) *Studies in Social Interaction*, New York: Free Press.

Jefferson, G. (1978a) At first I thought ....., lecture, University of California, Los Angeles.

Jefferson, G. (1978b) Sequential Aspects of Storytelling in Conversation. In J. Schenkhein (ed) *Studies in the Organization of Conversational Interaction*, New York: Academic Press, Inc.

Jefferson, G. (1980) On 'Trouble-Premonitory' Response to Inquiry, *Soc.Inquiry* 50 (3/4):153-85.

Jefferson, G. (1984a) On the Organization of Laughter in Talk about Troubles. In J.M. Atkinson and J.C. Heritage (eds) *Structures of Social Actions*, Cambridge: Cambridge Univ. Press.

Jefferson, G. (1984b) On step-wise transition from talk about a trouble to inappropriately next-positioned matters. In J.M.

Atkinson and J.C. Heritage (eds) *Structures of Social Actions*, Cambridge: Cambridge Univ. Press.
Jefferson, G. and Lee, J.R.E. (1981) The Rejection of Advice: Managing the Problematic Convergence of a 'Troubles-Telling' and a 'Service-Encounter.' *J. of Prag.* 5:399-422.

Kaag, M., Vreeburg, T. and Koffieberg, M. (1985) *Huisarts & Patiënt in Gesprek,* paper for Medical Sociology, Amsterdam, UvA.
Keller, E.F. (1985) *Reflections of Gender and Science*, New Haven: Yale Univ. Press.
Kessler, S. and McKenna, W. (1978) *Gender: An Ethnomethodological Approach*, New York: Wiley.
Klinkert, J.J. (1982) Medicalisering, *Medisch Contact* 51/52:1657-1662.
Kloosterman, G.J. (ed) (1983) *De voortplanting van de mens. Leerboek voor Obstetrie en Gynaecologie*, 6th edition Bussum: Uitgeverij Centen.
Knorr-Cetina, K.D. and Cicourel, A.V. (eds) (1982) *Advances in Social Theory and Methodology. Toward an Integration of Micro- and Macro-Sociologies*, Boston: Routledge & Kegan Paul.
Knottnerus, J.A. (1982) Krisis van de huisartsgeneeskunde. In J. Hollenbeek Brouwer and J.A. Knottnerus (eds) *Vertrekpunten in de gezondheidszorg*, Amsterdam: VU Uitgeverij.
Kocken, E.Th. and Til, K.v. (1982) Verantwoordelijkheid, een kwestie van leren. In J.Hollenbeek Brouwer and J.A. Knottnerus (eds) *Vertrekpunten in de gezondheidszorg*, Amsterdam: VU Uitgeverij.
Komter, A. (1985). *De macht van de zelfsprekendheid*, Leiden: VUGA.
Komter, A. (1986) Bespiegeling van een ongelovige. Reactie op Mieke Aerts 'Bijzondere openbaarheid: de macht van het feminisme.' *Tijdschrift voor Vrouwenstudies* 28, (7/4):559-563.
Komter, A. (forthcoming) Gender, Power and Feminist Theory. In M. Leijenaar, K. Davis, J. Oldersma, and D. Vos (eds) *The Gender of Power: Selected Papers*, Leiden: DSWO press.
Korsch, B.M. and Negrete, V.F. (1972) Doctor-Patient Communication, *Scientific American* 227:66-74.
Korsmeyer, C. (1978) The Hidden Joke: Generic Uses of Masculine Terminology. In M. Vetterling-Bragging, F.A. Elliston, and J. English (eds) *Feminism and Philosophy*, Totowa, N.J.: Littlefield, Adams & Co.
Kuiper, P.C. (1978) *Hoofdsom der Psychiatrie,* Utrecht: Bijleveld.

Labov, W. and Waletzky, J. (1967) Narrative Analyses: Oral Versions of Personal Experience. In J. Helm (ed) *Essays on the Verbal and Visual Arts*, Seattle: Univ. of Washington Press.
Labov, W. (1972) *Language in the Inner City: Studies in the Black English Vernacular*, Philadelphia: Univ. of Penn. Press.

Labov, W. and Fanshel, D. (1977) *Therapeutic Discourse. Psychotherapy as Conversation,* New York: Academic Press.
Lacoste, M. (1981) The Old Woman And The Doctor: A Contribution To The Analysis Of Unequal Linguistic Exchanges, *J. of Prag.5:169-180.*
Larson, M.S. (1977) *The Rise of Professionalism. A Sociological Analysis,* Berkeley: Univ. of Calif. Press.
Latour, B. and Woolgar, S. (1979) *Laboratory Life: The Social Construction of Scientific Facts,* Beverley Hills: Sage.
Laws, S. (1983) The Sexual Politics of Pre-Menstrual Tension, *Women's Studies International Forum* 6 (1):19-31.
Leeson, J. and Gray, J. (1978) *Women and Medicine,* London: Tavistock.
Leijenaar, M., Davis,K., Helleman, C., Oldersma, J., Vos, D. (eds) (1987) *The Gender of Power: A Symposium*, Leiden, Vakgroep Vrouwenstudies and VENA.
Leijenaar, M., Davis, K., Oldersma,J. and Vos,D. (eds) (forthcoming) *The Gender of Power: Selected Papers*, Leiden: DSWO press.
Lenanne, K.J. and Lenanne, J. (1973) Alleged Psychogenic Disorders in Women - A Possible Manifestation of Sexual Prejudice, *New England J. of Medicine* 288 (6):288-292.
Levinson, S.C. (1983) *Pragmatics*, Cambridge: Cambridge Univ. Press.
Lewin, E. and Olesen, V. (eds) (1985) *Women, Health and Healing,* New York: Tavistock.
Leydesdorff, S. (1984) Spinnewebben en spinrag, *Tijdschrift voor Vrouwenstudies 19:372-384.*
Lipset, D.R. (1982) The Painful Woman: Complaints, Symptoms and Illness. In M.T. Notman and C.C. Nadelson (eds) *The Woman Patient*, Vol. 3, New York: Plenum Press.
Lofland, J. (1971) *Analyzing Social Settings*, Belmont: Wadsworth Publ. Co.
Lorber, J. (1976) Women and Medical Sociology: Invisible Professional and Ubiquitous Patients. In M. Millman and R. Moss Kanter (eds) *In Another Voice. Feminist Perspectives on Social Life and Social Science*, New York: Octagon Books.
Lowe, M. and Hubbard, R. (eds) (1983) *Woman's Nature*, New York: Pergamon Press.
Ludwig, O. (1984) Berichten und erzählen. Variationen eines Musters. In K. Ehlich (ed) *Erzählen in der Schule*, Frankfurt a.M.: Suhrkamp.
Lukes, S. (1974) *Power, a radical view,* London: MacMillan Press.
Lukes, S. (1977) *Essays in Social Theory,* London: MacMillan Press.
Lukes, S. (1979a) On the Relativity of Power. In S.C. Brown (ed) *Philosophical Disputes in the Social Sciences,* Sussex: Harvester

Press, Ltd.
Lukes, S. (1979b) Power and Authority. In T. Bottelmore and R. Nisbet (eds) *A History of Sociological Analysis,* London: Heinemann Educational Books, Ltd.
Lukes, S. (1982) Panoptikon. Macht und Herrschaft bei Weber, Marx, Foucault. In K.M. Michel and T. Spengler (eds) *Kursbuch* 70,
Berlin: Rotbuch Verlag.

McConnel-Ginet, S. (1983) Intonation in a Man's World. In B. Thorne, C. Kramarae, and N. Henley (eds) *Language, Gender and Society,* Rowley, Mass.: Newbury House Publishers, Inc.
McKinlay, J.B. (1975) 'Who is really ignorant? Physician or Patient.' *J. of Health & Soc. Behav.* 16:3-11.
McLennan, G. (1984) Critical or Positive Theory? A Comment on the Status of Anthony Giddens' Social Theory, *Theory, Culture and Society* 2 (2):123-129.
McPherson, A. (ed) (1988) *Women's Problems in General Practice*, Oxford: Oxford University Press.
Maseide, P. (1981) Sincerity may frighten the patient: Medical dilemmas in patient care, *J.of Prag.* 5:145-167.
Mechanic, D. (1976) Sex,illness, illness behavior, and the use of health services, *J. of Human Stress* 2:29-40.
Mechanic, D. (1979) *Future Issues in Health Care,* New York: The Free Press.
Mechanic, D. (1980) Comment on Gove and Hughes, *Amer. Soc. Rev.* 45:513-514.
Mehan, H.B. and Wood, H.L. (1975) *The Reality of Ethnomethodology,* New York: Wiley.
Meigs, J.W. (1961) Occupational Medicine, *New England J. of Med.* 264:861-867.
Mendelsohn, R.S. (1981) *Male Practice: How Doctors Manipulate Women,* Chicago: Contemporary Books.
Menges, L.J. (1986) *Gezondheid en ziekte. Gedragswetenschappelijke verkenningen,* Alphen a.d. Rijn: Samsom Stafleu.
Merritt, M. (1976) On questions following questions in service encounters, *Language in Society* 5:315-357
Meyer, J.L. (1983) *Sekse als organisatie-principe: Veranderende normen in de asymmetrische machtsrelatie tussen vrouwen en mannen,* dissertations, Amsterdam, UvA.
Mills, C.Wright. (1959) *The Sociological Imagination,* New York: Oxford University Press.
Mishler, E.G. (1984) *The Discourse of Medicine*, Norwood,N.J.: Ablex Publishing Co.
Mitchell, J. (1971) *Women's Estate*, Middlesex: Penguin.
Molotch, H.L. and Boden, D. (1985) Talking Social Structure:

Discourse, Domination and the Watergate Hearings, *Amer.Soc. Rev.* 50:273-288.

Morgan, D. (1981) Men, masculinity and the process of sociological enquiry. In H. Roberts (ed) *Doing Feminist Research*, London: Routledge & Kegan Paul.

Munters, Q.J., Meijer, E., Mommaas, H., Poel,H.v.d.,Rosendal, R., Spaargaren, G. (1985) *Anthony Giddens. Een kennismaking met de structuratie-theorie,* Wageningen, Landbouwhogeschool.

Murcott, A. (1981) On the typification of 'bad patients.' In J.M. Atkinson and C.Heath (eds) *Medical Work. Realities and Routines,* Westmead, Farnborough, Hants: Gower.

Nathanson, C.A. (1975) Illness and the feminine role: a theoretical overview, *Soc. Sci. Med.*9:57-62.

Nathanson, C.A. (1977) Sex, illness, and medical care. A review of data,theory, and method, *Soc. Sci. Med.*11:13-25.

Navarro, V. (1976) *Medicine Under Capitalism,* New York: Prodist.

Nellis, M. (1980) *The Female Fix*, Middlesex: Penguin.

Northdurft, W. (1981) 'Ich komme nicht zu wort.'Austausch-Eigenschaften als Ausschluss-Mechanismen des Patienten in Krankenhaus-Visiten. In W. Frier (ed) *Pragmatik, Theorie und Praxis*, Amsterdam: Rodopi.

Oakley, A. (1972) *Sex, Gender & Society*, New York: Harper Colophon Books.

Oakley, A. (1974) *The Sociology of Housework*, Oxford: Martin Robertson.

Oakley, A. (1979) A case of maternity: paradigms of women as maternity cases, *Signs* 4 (4):607-631.

Oakley, A. and Mitchell, J. (eds) (1976) *Rights and Wrongs of Women*, Middlesex: Penguin.

Oosten, N. van and Knegtering, H. (1987) Een gezonde kerel...Gezondheidsverschillen tussen mannen en vrouwen, *Tijdschrift voor Gezondheid & Politiek* 5 (1):44-48.

Parsons, T. (1951) *The Social System*, New York: The Free Press.

Parsons, T. (1963) On the concept of political power, *Proceedings of the American Philosophical Society* 107:232-262.

Pease, I.H. and Crocker, L.H. (1954) *The Peckham Experiment*, London: Allen & Unwin.

Pendleton, D. (1983) Doctor-patient communication: A review. In D. Pendleton and J. Hasler (eds) *Doctor-Patient Communication*, London: Academic Press.

Petchesky. R.P. (1986) *Abortion and Woman's Choice*, London: Verso.

Phillips, D.L. and Segal, B.E. (1969) Sexual status and psychiatric symptoms,*Amer. Soc. Rev.* 34:58-72.

Polanyi, L. (1979) So What's the Point? *Semiotica* 25 (3/4):207-241.
Polanyi, L. (1983) *The American Story*, Norwood, N.J.: Ablex Publishing Co.
Pollner, M. (1975) The Very Coinage of Your Brain: The Anatomy of Reality Disjunctures, *Philosophy of Social Science* 5:411-430.
Pomerantz, A. (1975) Second Assessments. A Study of Some Features of Agreements and Disagreements, *dissertation*, Irvine, University of California in Irvine.
Pomerantz, A. (1984) Agreeing and disagreeing with assessments: Some features of preferred/dispreferred turn shapes. In J.M. Atkinson and J.C. Heritage (eds) *Structures of Social Actions*, Cambridge: Cambridge Univ. Press.
Prather, J. and Fidell, L. (1975) Sex differences in the content and style of medical advertisements, *Soc. Sci. Med.* 9:23-26.
Psathas, G. (ed) *Everyday Language: Studies in Ethnomethodology*, New York: Irvington Pub., Inc.

Quasthoff, U. (1979) Eine interaktive Funktion von Erzählungen. In H-G.Soeffner (ed) *Interpretative Verfahren in den Sozial- und Textwissenschaften*, Stuttgart: Metzler.
Quasthoff, U. (1980) *Erzählen in Gesprächen. Linguistische Untersuchungen zur Strukturen und Funktionen am Beispiel einer Kommunikationsform des Alltags*, Tübingen: Gunter Narr Verlag.

Rehbein, J. (1980) Sequentielles Erzählen - Erzählstrukturen von Immigranten bei Sozialberatungen in England. In K. Ehlich (ed) *Erzählen im Alltag*, Frankfurt a.M.: Suhrkamp.
Rehbein, J. (1984) Beschreiben, Berichten und Erzählen. In K. Ehlich (ed) *Erzählen in der Schule*, Tübingen: Gunter Narr Verlag.
Riessman, C.K. (1983) Women and Medicalization. A New Perspective, *Social Policy* 14 (1):3-19.
Riessman, C.K. (1988) Words of Difference: Contrasting Experience in Marriage and Narrative Style. In A.D. Todd and S. Fisher (eds) *Gender and Discourse*, Norwood, N.J.: Ablex Publishing Co.
Roberts, H. (ed) (1981a) *Doing Feminist Research*, London: Routledge and Kegan Paul.
Roberts, H. (1981b) Women and their doctors: power and powerlessness in the research process. In H. Roberts (ed) *Doing Feminist Research*, London: Routledge and Kegan Paul.
Roberts, H. (ed) (198lc) *Women, Health and Reproduction*, London: Routledge and Kegan Paul.
Rogers, C.R. and Kinget, M.C. (1960) *Psychotherapie en menselijke verhoudingen*, Utrecht: Uitgeverij Het Spectrum.
Rosaldo, M.Z. (1980) The Use and Abuse of Anthropology, *Signs* 5

(3):389-417.
Rosaldo, M.Z. (1983) Moral/Analytic Dilemmas posed by the intersection of Feminism and Social Science. In N. Haan, R.N. Bellah, R. Rabinow and W. Sullivan (eds) *Social Science as Moral Inquiry*, New York: Columbia University Press.
Rosaldo, M.Z. and Lamphere, L. (eds) (1974) *Women, Culture and Society*, Stanford: Stanford University Press.
Rothman, B.K. (1982) *In Labor: Women and Power in The Birth Place*, New York: W.W. Norton Co.
Royal College of General Practitioners (1972) *The Future General Practitioners*, London: British Medical Journal.
Rubin, G. (1975) The Traffic in Women: Notes on the 'Political Economy' of Sex. In R.R. Raiter (ed) *Towards an Anthropology of Women, New York: Monthly Review Press.*
Rummelhart, D.E. (1975) Notes on a Schema for Stories. In D.G. Bobrow and A. Collins (eds) *Representation and Understanding. Studies in Cognitive Science*, New York: Academic Press.
Ruzek, S. (1978) *The Women's Health Movement*, New York: Pranger.
Ryave,A.L. (1978) On the Achievement of Series of Stories. In J. Schenkhein (ed) *Studies in the Organization of Conversational Interaction*, New York: Academic Press, Inc.

Sacks, H. (1970) Facets of the Organization of Storytelling in Conversation. Paper presented at the VIIth World Congress of Sociology, Varna, Bulgaria.
Sacks, H. (1973) On some puns with some intimations.In R.W. Schuy (ed) *Report of the 32nd Annual Round Table Meeting on Linguistics and Language Studies*, Washington, D.C.: Georgetown Univ. Press.
Sacks, H. (1974a) An Analysis of the Course of a Joke's Telling in Conversation. In A. Bauman and J. Scherzer (eds) *Explorations in the Ethnography of Speaking*, Cambridge: Cambridge Univ. Press.
Sacks, H. (1974b) On the Analysability of Stories by Children. In R. Turner (ed) *Ethnomethodology*, Middlesex: Penguin.
Sacks, H. (1978) Some Technical Considerations of a Dirty Joke. In J.Schenkhein (ed) *Studies in the Organization of Conversational Interaction*, New York: Academic Press, Inc.
Sacks, H. (1984a) Notes on Methodology. In J.M. Atkinson and J.C. Heritage (eds) *Structures of Social Actions*, Cambridge: Cambridge Univ. Press.
Sacks, H. (1984b) On doing 'being ordinary.' In J.M. Atkinson and J.C. Heritage (eds) *Structures of Social Actions*, Cambridge: Cambridge Univ. Press.
Sacks, H., Schegloff, E., and Jefferson, G. (1974) A simplest systematics for the analysis of turn-taking in conversations,

Language 50:696-735.
Schatzman, L. and Strauss, A.L. (1973) *Field Research*, Englewood Cliffs, N.J.: Prentice-Hall, Inc.
Scheff, T.J. (1968) Negotiating Reality: Notes on Power in the Assessment of Responsibility, *Social Problems* 16 (1):3-17.
Schegloff, E.A. (1972) Sequencing in conversational openings. In J.J. Gumperz and D. Hymes (eds) *Directions in Sociolinguistics: The Ethnography of Communication*, New York: Holt, Rinehart and Winston.
Schegloff, E.A. (1979) Identification and recognition in telephone conversation openings. In G.Psathas (ed) *Everyday Language: Studies in Ethnomethodology*, New York: Irvington Pub.,Inc.
Schegloff, E.A. (1984) On some questions and ambiguities in conversation. In J.M. Atkinson and J.C. Heritage (eds) *Structures of Social Action*, Cambridge: Cambridge Univ. Press.
Schegloff, E.A., Jefferson, G. and Sacks, H. (1977) The Preference for Self-Correction in the Organization of Repair in Conversation, *Language* 53:361-382.
Schegloff, E.A. and Sacks, H. (1974) Opening Up Closings. In R. Turner (ed) *Ethnomethodology*, Middlesex: Penguin.
Scheman, N. (1983) Individualism and the objects of psychology. In Harding, S. and Hintikka, M. (eds) *Reality Discovering*, Dordrecht: D.Reidel.
Schenkhein, J. (ed) (1978) *Studies in the Organization of Conversational Interaction*, New York: Academic Press, Inc.
Scherzer, J. (1985) Puns and Jokes. In T.v. Dijk (ed) *Handbook of Discourse Analysis*, London: Academic Press.
Schutz, A. (1962) *Collected Papers*, vol.1, The Hague: Nijhoff.
Schwartz, H. (1979) On Recognizing Mistakes: A Case of Practical Reasoning in Psychotherapy. In H. Schwartz and J. Jacobs (eds) *Qualitative Sociology*, New York: The Free Press.
Schwartz, H. and Jacobs, J. (eds) (1979) *Qualitative Sociology*, New York: The Free Press.
Scully, D. (1980) *Men Who Control Women's Health*, Boston: Houghton Mifflin.
Scully, D. and Bart, P. (1973) A funny thing happened on the way to the orifice: Women in gynecological textbooks, *Amer. J. of Soc.* 78:1045-1050.
Seaman, B. and Seaman, G. (1977) *Women and the Crisis in Sex Hormones*, New York: Bantam Books, Inc.
Sharrock, W.W. and Turner, R. (1978) On a Conversational Environment for Equivocality. In J. Schenkhein (ed) *Studies in the Organization of Conversational Interaction*, New York: Academic Press, Inc.

Sherif, C. W. (1979) Bias in Psychology. In J.A. Sherman and E.T. Beck (eds). *The Prism of Sex*, Madison, Wisconsin: The University of Wisconsin Press.

Siegel, G.S. (1963) *Periodic Health Examinations*, Washington, D.C.: Public Health Service Publication.

Sigling, H.O. (1982) De persoon van de huisarts. Attitude-aspecten in het medisch onderwijs. In J. Hollenbeek Brouwer and J.A. Knottnerus (eds) *Vertrekpunten in de gezondheidszorg*, Amsterdam: VU Uitgeverij.

Silverman, D. (1985) *Qualitative Methodology and Sociology*, Hants: Gower.

Silverman, D. and Torode, B. (1980) *The Material Word*, London: Routledge and Kegan Paul.

Skopek, L. (1979) Doctor-Patient Conversation: A Way of Analyzing its Linguistics Problems, *Semiotica* 28 (3/4):301-311.

Smart, B. (1982) Foucault, Sociology, and the Problem of Human Agency, *Theory and Society* 11 (2):121-141.

Smith, D. (1978) K. is mentally ill *Sociology* 12 (1):23-54.

Smith, D. and David, S. (eds) (1975) *Women Look at Psychiatry*, Vancouver: Press Gang Publishing Co.

Smith, J. (1983) Feminist Analysis of Gender: A Critique. In M. Lowe and R. Hubbard (eds) *Woman's Nature*, New York: Pergamon Press.

Smith-Rosenberg, C. (1972) The hysterical woman: sex roles and role conflict in 19th century America, *Social Research* 39:652-678.

Smith-Rosenberg, C. (1973) Puberty to menopause, *Feminist Studies* 1:59-72.

Smith-Rosenberg, C. and Rosenberg, C. (1973) The female animal: Medical and biological views of woman and her role in 19th century America, *J. of Amer. History* 60 (2):332-356.

Smits, A., Beek, M., Laseur, M., Hendries, A. and Mokkink, H. (1983) Sekse-specifieke gedrag van de huisarts, *Gezondheid and Samenleving* 4 (3):154-160.

Sontag, S. (1978) *Illness as Metaphor*, New York: Farrar, Straus and Giroux.

Spender, D. (1980) *Man-made Language*, London: Routledge and Kegan Paul.

Spreeuwenberg, C. (1983) Met het oog op de huisarts, *Metamedica* 62:238-252.

Standing, H. (1980) Sickness is a woman's business? Reflections on the attribution of illness. In The Brighton Women and Science Group (eds) *Alice Through the Microscope*, London: Virago.

Stanley, L. and Wise, S. (1983) *Breaking out: Feminist Consciousness and feminist research*, London: Routledge and Kegan Paul.

Staveren, G. van (1984) *Taakopvatting en werkwijze bij huisartsen*, Amsterdam: VU Uitgeverij.
Stephen, W.J. (1979) *An Analysis of Primary Medical Care: An International Study*, Cambridge: Cambridge Univ. Press.
Stimson, G.V. and Webb, B. (1975) *Going to see the Doctor: The consultation process in general practice*, London: Routledge and Kegan Paul.
Stolk, J.G. (1975) Clinical implications of congenital uterine malformations. In T. Eskes (ed) *Aspects of Obstetrics Today*, Amsterdam: Excerpta Medica.
Stone, G.C. (1979) Patient compliance and the role of the expert, *J. of Soc.Issues* 35:34-59.
Strauss, A. (1963) The hospital and its 'negotiated order.' In E. Freidson (ed) *The Hospital in Modern Society*, New York: Free Press.
Stromberg, A. (ed) (1984) *Women, Health, and Medicine*, Palo Alto: Mayfield.
Strong,P.M. (1979) *The Ceremonial Order of the Clinic*, London: Routledge & Kegan Paul.
Summey, P.S. and Hurst, M. (1986) Ob/Gyn on the Rise: The Evolution of Professional Ideology in the Twentieth Century, I, *Women & Health* 11 (1):133-144.
Summey, P.S. and Hurst, M. (1986) Ob/Gyn on the Rise: The Evolution of Professional Ideology in the Twentieth Century, II, *Women & Health* 11 (2):103-122.
Swaan, A. de (1979) The Initial Interview as a Task. Studies in the Sociology of Psychotherapy, paper for XIth International Congress on Psychotherapy, Amsterdam.

Tannen, D. (1984) *Conversational Style: Analyzing Talk Among Friends,* Norwood, N.J.: Ablex Publishing Co.
Thomassen, J.F.M. and Woltring, F.J.M. (1982) Het TRIPT-project: een data-collectie uit zeventig huisartsenpraktijken, *Huisarts en Wetenschap* 25:394-397.
Thomassen, J.F.M. (1984) Spreekuurbezoek en klachtpresentatie. Een vergelijkende beschrijving van enkele kenmerken van spreekuur contacten van mannen en vrouwen met hun huisarts, working paper, Huisartsen Instituut, VU, Amsterdam.
Thomassen, J.F.M. (1986) *Consult Analyse*, Amsterdam: VU Uitgeverij.
Thompson, J.B. (1984) *Studies in the Theory of Ideology*, Cambridge: Polity Press.
Tielens, S.v., Beek, M., Koopmans, R., Van Trier, Y. and Mokkink, H. (1986) Behandelt de huisarts vrouwen anders dan mannen? *Huisarts en Wetenschap* 29:204-208.
Todd, A.D. and Fisher, S. (1988) *Gender and Discourse*, Norwood,

N.J.: Ablex Publishing Co.

Turner, R. (1972) Some Formal Properties of Therapy Talk. In D. Sudnow (ed) *Studies in Interaction*, New York: Free Press.

Turner, R. (1976) Utterance Positioning as an Interactional Resource, *Semiotica* 17:233-254.

Urry, J. (1982) Duality of Structure: Some Critical Issues, *Theory, Culture and Society* 1 (2):93-99.

Verbrugge, L. (1976) Females and illness: Recent trends in sex differences in the United States, *J. Health Soc. Behav.*17:387-403.

Verbrugge, L. (1979) Female illness rates and illness behavior. Testing hypotheses about sex differences in health, *Women and Health* 4 (1):51-78.

Verbrugge, L. (1984) How Physicians Treat Mentally Distressed Men and Women, *Soc.Sci.Med.*18 (1):1-9.

Verbruggen, M.H. (1976) Women and Medicine in 19th Century America, *Signs* 1:957-972.

Vries, P. de (1987) 'Het persoonlijke is politiek' en het ontstaan van de tweede golf in Nederland 1968-1973, *Socialisties-Feministies Teksten* 10:15-35.

Vogel, A. (1979) *De kleine dokter: een bonte reeks nuttige adviesen uit de zwitserse volksgeneeskunde*, 't Harde: Gezondheidsnieuws.

Wadsworth, M. and Robinson, D. (eds) (1976) *Studies in Everyday MedicalLife*, London: Martin Robertson.

Waitzkin, H. (1983) *The Second Sickness*, New York: The Free Press.

Watzlawick, P., Beavin, J.H. and Jackson, D.D. (1967) *Pragmatics of Human Communication*, New York: W.W. Norton and Co., Inc.

Webb, B. and Stimson, G. (1976) People's accounts of medical encounters. In M. Wadsworth and D. Robinson (eds) *Studies in Everyday Medical Life*, London: Martin Robertson.

Weber, M. (1956) *Wirtschaft und Gesellschaft*, Tübingen:Möhr.

West, C. (1984) *Routine Complications: Troubles with Talk Between Doctors and Patients*, Bloomington: Indiana Univ. Press.

Weterman-Hezemans, Th.C.P. (1982) Leren voor huisarts, een heroriëntatie. In J. Hollenbeek Brouwer and J.A. Knottnerus (eds) *Vertrekpunten in de gezondheidszorg*, Amsterdam: VU Uitgeverij.

Willems, L. (1983) *Gebruikersparticipatie in de hulpverlening*, Den Haag: Nimawo.

Willis, P. (1979) *Learning to Labour*, Farnborough: Saxon House.

Wilson, T.P. (1970) Conceptions of Interaction and Forms of Sociological Explanation, *Amer. Soc. Rev.* 35 (4):697-710.

Wise, S. and Stanley, L. (1987) *Georgie Porgie. Sexual Harassment in Everyday Life*, London: Pandora.

Wodak, R. (1980) Problemdarstellungen in gruppentherapeutischen Situationen. In K. Ehlich (ed) *Erzählen im Alltag*, Frankfurt a.M.: Suhrkamp.

Wodak, R. (1981) Women relate, men report: Sex differences in language behaviour in a therapeutic group, *J. of Prag.*5:261-285.

Wolffensberger, J. (forthcoming) Engendered Structure. A conceptualization of the alliance between gender and social system. In M. Leijenaar, K. Davis, J. Oldersma, and D. Vos (eds) *The Gender of Power: Selected Papers*, Leiden: DSWO press.

Wolinsky, F. and Wolinsky, S.R. (1981) Expecting Sick-Role Legitimation and Getting it, *J. Health Soc. Behav.* 22:229-242.

Wood, A.D. (1974) The fashionable diseases: women's complaints and their treatment in 19th century America. In M. Hartman and L.W. Banner (eds) *Clio's Consciousness Raised. New Perspectives on the History of Women*, New York: Harper and Row.

Wrong, D. (1967) The over-socialised conception of man. In N.J. Demerath and R.A. Peterson (eds) *System, Change and Conflict*, New York: Free Press.

Wrong, D. (1979) *Power: Its forms, bases, and uses*, Oxford: Basil Blackwell.

Wulff, H.R. (1976) *Rational Diagnosis and Treatment*, Oxford: Blackwell Scientific Publications.

Zimmerman, D.H. and West, C. (1975) Sex Roles, Interruptions and Silences in Conversation. In B. Thorne and N. Henley (eds) *Language and Sex: Difference and Dominance*, Rowley, Mass.: Newbury House.

Zola, I.K. (1972a) Medicine as an institution of social control, *Soc. Rev.* 20:487-503.

Zola, I.K. (1972b) Pathways to the Doctor - From Person to Patient, *Soc. Sci. Med.*7:677-689.

APPENDIX

TRANSCRIPT NOTATION

The transcription notation used in the present inquiry is an abridged version of the system developed by Gail Jefferson (Atkinson and Heritage 1984). Since the analysis was based on audio-tapes, symbols are not included, which represent non-vocal activities like gaze, gestures or applause. Furthermore, no attempt was made to provide a 'listener's transcript;' i.e. one that sounds the way it was spoken, since the transcripts were to be translated later for the text proper.

1. *Simultaneous or overlapping utterances*

(indicates simultaneous utterances as in:

P: (and because I
D: (what are you thinking of then...

where utterances start together, or as in:

D: (we <u>don't</u> do that because there'd be no end to it
P: (No she had it

where the overlap begins within a previous utterance.

\- indicates utterances, which are immediately adjacent without actually overlapping, as in:

P: Yeah, I had that same feeling too-
D: -Yeah, I know almost for sure that it's not supposed to be like this.

2. *Intervals within utterances*

((0.0)) indicates intervals in the stream of talk, timed in tenths of a second, either within an utterance as in:

P: The last time I took a bath in the middle of the night and ((1.0)) then I stick my finger down my throat

or between utterances as in:

D: Well, you can tell your husband that it really isn't physical.
((2.0))
P: Ohhhh.

3. *Characteristics of speech delivery*

: indicates an extension of the sound it follows
. indicates a stopping fall in tone
, indicates a pausing fall in tone
- indicates a halting, stammering cut off tone when at the end of a sound, or a connecting flow from the last sound when at the start of a sound
? indicates a rising inflection
! indicates an animated tone

Emphasis is indicated by underlining, capitalization, or both, as in:

P: BUT I DON'T WANT TO FEEL LIKE THAT AT ALL so I don't go down the street, do you understand?

(h) indicates audible aspiration, as in:

D: Yeah(hh)

((laughs)) indicates the occurrence of the phenomenon described within the double brackets or various characterizations of the talk, as in:

((complaining))

4. *Transcriptionist doubt*

(( ))	indicates the items enclosed within brackets are in doubt, as in: D: Just ((use)) this or: ((unclear))

5. *Other transcript symbols*

Lines in the transcript where phenomenon of interest occur are indicated by arrows in the left-hand margin, as in:

→	91	P	((complaining)) Those were real stomach cramps
	92		Then I was sick the whole day, just one time and
	93		that's when it goes away again and
→	94		((groans)) Ooh-h-h-h-h-h, then you double over.